SEA CHANGE

The Sustainability and the Environment series provides a comprehensive, independent, and critical evaluation of environmental and sustainability issues affecting Canada and the world today.

SUSTAINABILITY
AND THE
ENVIRONMENT

SEA CHANGE

Charting
a Sustainable Future
for Oceans in Canada

Edited by Ussif Rashid Sumaila, Derek Armitage,
Megan Bailey, and William W.L. Cheung

UBCPress · Vancouver

Printed in Canada on FSC-certified ancient-forest-free paper
(100% post-consumer recycled) that is processed chlorine- and acid-free.

UBC Press is a Benetech Global Certified Accessible™ publisher. The epub version of this book meets stringent accessibility standards, ensuring it is available to people with diverse needs.

Library and Archives Canada Cataloguing in Publication

Title: Sea change : charting a sustainable future for oceans in Canada /
    edited by U. Rashid Sumaila, Derek Armitage, Megan Bailey, and William Cheung.
Other titles: Sea change (2024)
Names: Sumaila, Ussif Rashid, editor. | Armitage, Derek R. (Derek Russel), editor. |
    Bailey, Megan (Fisheries economist), editor. | Cheung, William W.L., editor.
Description: Includes bibliographical references and index.
Identifiers: Canadiana (print) 20230560369 | Canadiana (ebook) 20230562574 |
    ISBN 9780774869041 (softcover) | ISBN 9780774869058 (PDF) | ISBN 9780774869065 (EPUB)
Subjects: LCSH: Marine resources conservation – Canada. | LCSH: Marine ecosystem
    management – Canada. | LCSH: Marine ecology – Canada. | LCSH: Ocean – Canada.
Classification: LCC GC1023.15 .S43 2024 | DDC 333.91/64160971—dc23

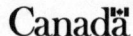

UBC Press gratefully acknowledges the financial support for our publishing program of the Government of Canada, the Canada Council for the Arts, and the British Columbia Arts Council.

This book has been published with the help of a grant from the Canadian Federation for the Humanities and Social Sciences, through the Awards to Scholarly Publications Program, using funds provided by the Social Sciences and Humanities Research Council of Canada, and with the help of the University of British Columbia through the K.D. Srivastava Fund.

UBC Press is situated on the traditional, ancestral, and unceded territory of the xwməθkwəy̓əm (Musqueam) people. This land has always been a place of learning for the xwməθkwəy̓əm, who have passed on their culture, history, and traditions for millennia, from one generation to the next.

UBC Press
The University of British Columbia
www.ubcpress.ca

To the ocean,

the marine life it supports, and the human lives

and livelihoods it sustains and nourishes

Contents

Tables and Figures

Preface

The journey to the publication of this book started back in 2013, right after Sumaila finished his term as director of the then Fisheries Centre (now Institute for the Oceans and Fisheries) at the University of British Columbia, when he initiated a conversation with members of his lab (the Fisheries Economics Research Unit) on how to develop a winning project proposal to address a significant challenge. This conversation was later expanded to leaders in the field of interdisciplinary social and natural sciences pertaining to oceans and fisheries. These conversations culminated in the formation of the OceanCanada Partnership (OCP), which went on to successfully win one of the coveted Social Sciences and Humanities Research Council (SSHRC) Partnership Grants. The main goals of the OCP are to: (1) develop an enhanced understanding of the uncertain changes occurring in Canada's coastal and ocean ecosystems and fisheries; (2) establish methodological frameworks and foundations for future research that supports the development of policies and regulations that can help Canada improve its overall performance in ocean science, management, and sustainability; and (3) integrate interdisciplinary research and approaches cohesively to produce results that can inform policy-makers.

The OCP provided us with the platform to develop the ideas, concepts, insights, and policy recommendations reported in this book. It also afforded us the opportunity to publish several contributions while training a sizable number of students and postdoctoral fellows who have gone on to create their own research programs or join government, nongovernmental organizations, and the private sector – contributing to sustaining fish and fisheries throughout Canada's three coasts and beyond (see www.oceancanada.org).

This volume aims to help Canadian society understand, prepare, and plan for the challenges that lie ahead for our oceans and our coasts. Its chapters are a core outcome of the OCP, composed of 22 formal research partners, including universities from coast to coast, community organizations, and Fisheries and Oceans Canada. They seek to identify future policy options and to develop shared understandings that are crucial for the sustainability of Canada's coastal and ocean ecosystems. While the insights and recommendations emerging from these chapters are situated in the present, they reflect decades of experience among the authors and their collaborators seeking positive social, economic, and ecological outcomes. The contributions thus offer insights that reflect a rich history of scholarship and action across our three oceans and coasts, as well as a direct recognition that the future of our oceans is fundamentally linked to reconciliation with Indigenous Peoples, partners, and nations.

Acknowledgments

Let us begin by stating that the work reported in this book was done across Indigenous lands and territories all over Canada. We acknowledge historical occupancy, ownership, and dispossession of these lands, and we acknowledge the historical treaties, ongoing negotiations, and modern land claims in support of Indigenous rights.

The OceanCanada Partnership (OCP) and this edited volume reflect the contributions of many individuals and organizations. The OCP was made possible by a Social Sciences and Humanities Research Council of Canada (SSHRC) Partnership Grant as well as supplementary contributions from our academic and nonacademic research partners. We are grateful to the many communities and organizations across Canada that have contributed to and participated in OCP research activities. In particular, we want to acknowledge the strong support of the University of British Columbia as the partnership's host institution.

Each of the chapters in this volume has been peer-reviewed by two individuals. We appreciate the constructive feedback and insights provided by anonymous reviewers and the following individuals in this process: Elena Bennett, Laurie Chan, Tony Charles, Kate Davies, Graham Epstein, Jeff Hutchings, Russ Jones, Ian Perry, Rachel Kelly, Inka Milewski, Stephan Schot, Adam Soliman, Wilf Swartz, Trevor Swerdfager, Gerald Singh, Tyler Eddy, and Jessica Blythe. In addition, Rosemary Ommer reviewed each chapter. We are vey grateful to all of you!

Several other individuals have made essential contributions to this book project and the OCP. First, the OCP is indebted to the leadership, direction, perseverance, and dedication of its Advisory Board members: Dr. Rosemary Ommer (chair), Dr. Fikret Berkes, Dr. Christopher Harvey (RIP), Haida Hereditary Chief Nang Jingwas (Russ Jones), Minister Herb Dhaliwal, and Dr. Gordon Munro. All these amazing people supported us in many different important ways. They have been with us since the inception of the OCP, guiding and pushing us forward. The leadership of Rosemary is worth special mention. Her contributions were simply invaluable, and we benefited especially from her deep knowledge of the SSHRC and how it works. We cannot thank you and your board members enough, Dr. Ommer!

Second, the OCP itself, and certainly this book as a product, would not have been possible without the ongoing support and contributions of Duncan Burnside (OCP Tech Manager) and Anne Marie Goodfellow (OCP Project Manager). Earlier project contributors are colleagues from Support Programs to Advance Research Capacity (SPARC), Sandra Ignagni and Ngao Hotte. Finally, we want to thank Melissa Pitts, the director of UBC Press, and James MacNevin, our UBC Press editor, for their enthusiasm and support for this book right from the first day we informed them about our plan – we very much appreciate your guidance throughout the publication process during the COVID-19 crisis.

Abbreviations

ACAR	Atlantic Coalition for Aquaculture Reform
ACCASP	Aquatic Climate Change Adaptation Services Program
AFPR	Atlantic Fishery Policy Review
AICFI	Atlantic Integrated Commercial Fisheries Initiative
AMAP	Arctic Monitoring and Assessment Programme
AMB	Archipelago Management Board (Gwaii Haanas)
ANPF	Arctic and Northern Policy Framework
APCFNC	Atlantic Policy Congress of First Nations Chiefs
ATK	Aboriginal Traditional Knowledge
ATP	Allocation Transfer Program
BSIMPI	Beaufort Sea Integrated Management Planning Initiative
CA	Controlling Agreement
CBPR	community-based participatory research
CCPFH	Canadian Council of Professional Fish Harvesters
CCRN	Community Conservation Research Network
CDQ	Community Development Quota
CFRN	Canadian Fisheries Research Network
CHN	Council of the Haida Nation
CIFHF	Canadian Independent Fish Harvesters' Federation
COPE	Committee for Original Peoples' Entitlement
COSEWIC	Committee on the Status of Endangered Wildlife in Canada
CPUE	catch per unit effort
CWN	Canadian Water Network
CWRC	Canadian Watershed Research Consortium
DBEM	dynamic bioclimatic envelope model
DIO	Designated Inuit Organization
EA	Enterprise Allocation

EBFM	ecosystem-based fisheries management
EBM	ecosystem-based management
EEZ	exclusive economic zone
EFF	Eastern Fishermen's Federation
EFJ	extended fisheries jurisdiction
EIA	environmental impact assessment
ESM	earth system model
ESSIMP	Eastern Scotian Shelf Integrated Management Plan
FFAW-Unifor	Fish, Food and Allied Workers Union
Fish-MIP	Fisheries and Marine Ecosystem Model Intercomparison Project
FJMC	Fisheries Joint Management Committee
FORCE	Fundy Ocean Research Centre for Energy
FPIC	free, prior, and informed consent
FPMB	Friends of Port Mouton Bay
FSC	food, social, and ceremonial
GH	Greenland halibut
GHG	greenhouse gas
HIRMD	Heiltsuk Integrated Resource Management Department
hPa	hectopascal
HTL	higher trophic level
HTO	hunters' and trappers' organization
ICC	Inuit Circumpolar Council
IFA	*Inuvialuit Final Agreement*
IFMP	integrated fisheries management plan
IM	integrated management
IOM	integrated ocean management
IOMPBS	Integrated Ocean Management Plan for the Beaufort Sea
IPA	Indigenous Protected Area
IPBES	Intergovernmental Science-Policy Platform on Biodiversity and Ecosystem Services
IPCA	Indigenous Protected and Conserved Area
IPCC	Intergovernmental Panel on Climate Change

IPO	Indigenous Peoples' Organizations	NEB	National Energy Board
IQ	individual quota	NFA	Nunavut Fisheries Association
IQ	Inuit Qaujimajatuqangit (Inuit traditional knowledge)	NG	Nunatsiavut Government
		NILCA	*Nunavik Inuit Land Claims Agreement*
ISO	International Organization for Standardization	NIRB	Nunavut Impact Review Board
		NMCA	National Marine Conservation Area
ISR	Inuvialuit Settlement Region	NMFS	National Marine Fisheries Service
ITK	Inuit Traditional Knowledge	NMR	Nunavik Marine Region
ITQ	individual transferable quota	NMRWB	Nunavik Marine Region Wildlife Board
KMWG	knowledge mobilization working group	NNFC	Northern Native Fishing Corporation
LCA	land claims agreement	NOAA	National Oceanic and Atmospheric Agency
LEO	Local Environmental Observer [Network]		
		NPV	net present value
LFA	lobster fishing area	NSA	Nunavut Settlement Area
LILCA	*Labrador Inuit Land Claims Agreement*	NSDFA	Nova Scotia Department of Fisheries and Aquaculture
LISA	Labrador Inuit Settlement Area		
LMMA	Locally Managed Marine Area	NWMB	Nunavut Wildlife Management Board
LNUK	Local Nunavimmi Umajulirijiit Katujjiqatigiinningit	OA	ocean acidification
		OCP	OceanCanada Partnership
LOMA	Large Ocean Management Area	OPP	Oceans Protection Plan
MaPP	Marine Plan Partnership	OWF	offshore wind farm
MDO	Makivik Designated Organization	PBIMP	Placentia Bay Integrated Management Plan
MEMPA	Musquash Estuary Marine Protected Area		
		PICFI	Pacific Integrated Commercial Fisheries Initiative
MFADSS	Marine Finfish Aquaculture Decision Support System		
		PIIFCAF	Policy for Preserving the Independence of the Inshore Fleet in Canada's Atlantic Fisheries
MFU	Maritime Fishermen's Union		
MPA	Marine Protected Area		
MRE	marine renewable energy	PNCIMA	Pacific North Coast Integrated Management Area
MREA	*Marine Renewable-energy Act* (Nova Scotia)		
		PP	primary production
MREEM	Marine Renewable Energy Enabling Measures	PV	participatory video
		RCAP	Royal Commission on Aboriginal Peoples
MSE	management strategy evaluation		
MSP	Marine Spatial Planning	RCP	Representative Concentration Pathway
MSY	maximum sustainable yield	RFA	Reconciliation Framework Agreement
NA	*Nunavut Agreement*	RNUK	Regional Nunavimmi Umajulirijiit Katujjiqatigiinninga
NAFO	Northwest Atlantic Fisheries Organization		
		RPPSG	Regroupement des pêcheurs professionnels du sud de la Gaspésie
NARW	North Atlantic right whale		

RWO	regional wildlife organization	SST	sea surface temperature
SARA	*Species at Risk Act*	TAC	Total Allowable Catch
SC	snow crab	TAH	Total Allowable Harvest
SDG	Sustainable Development Goal	TAT	Total Allowable Take
SEA	strategic environmental assessment	TBTI	Too Big To Ignore
SES	social-ecological system	Tg	teragram
SFA	shrimp fishing area	TJFB	Torngat Joint Fisheries Board
SJH-EMP	Saint John Harbour–Environmental Monitoring Partnership	TK	traditional knowledge
		TNMPA	Tarium Niryutait Marine Protected Area
SLO	social licence to operate	UNCLOS	*United Nations Convention on the Law of the Sea*
SOK	spawn on kelp		
SSHRC	Social Sciences and Humanities Research Council of Canada	UNDRIP	*United Nations Declaration on the Rights of Indigenous Peoples*
SSP	Shared Socio-Economic Pathway	WCA	West Coast Aquatic

PART 1

Setting the Stage

A Partnership Approach to the Study of Canada's Oceans and Coasts

U. Rashid Sumaila, Derek Armitage, Megan Bailey, and William W.L. Cheung

Canada is a maritime nation, bordered by the Arctic, Atlantic, and Pacific Oceans. With the world's longest coastline, the surface area of Canada's exclusive economic zones covers approximately 5.75 million km². Canada's oceans and the marine living resources within them are inextricably linked to the socio-cultural and economic well-being of Canadians across the country. Oceans help to regulate the climate, support diverse cultural practices and recreational activities, and are a source of food and nutritional security for tens of millions of people worldwide, including millions of Canadians (Srinivasan et al. 2010; Teh and Sumaila 2013; Hicks et al. 2019).

The Canadian economy remains closely tied to our oceans and coasts: industries working in, on, and around the oceans directly employ about 315,000 Canadians and contribute over $26 billion a year to the nation's wealth (DFO 2009). Specifically, gross revenues from Canadian ocean fisheries are estimated at about US$3.7 billion in 2018 (DFO 2018), generating economic and household income impacts throughout the Canadian economy of about US$9.1 billion and US$2.9 billion per year, respectively (Dyck and Sumaila 2010). Many coastal communities, and especially Indigenous communities, rely heavily on fish for food and employment as well as cultural and ceremonial uses (Berkes et al. 2005; Turner and Berkes 2006; Ommer 2007; Cisneros-Montemayor et al. 2016; Gibson and Sumaila 2017). Canada, therefore, has a huge responsibility to manage its oceans and coasts

sustainably for the benefit of all generations of Canadians (DFO 2009; Sumaila 2021).

Despite the diverse and significant benefits that the ocean brings, humans continue subjecting the ocean and the life it holds to multiple threats, including from overfishing (Pauly et al. 2002), pollution such as greenhouse gas (GHG) emissions (Sumaila and Tai 2020), oil spills, ocean plastic, and coastal development (Tilman et al. 1994; AMAP 2002; Halpern et al. 2008; IPBES 2019; Bindoff et al. 2019; Sumaila et al. 2012; Lau et al. 2020). Specifically, climate-induced stressors, such as ocean warming, ocean acidification, hypoxia, and sea-level rise, are impacting Canada's marine life and its ocean-coastal social-ecological systems (SESs) (Parry et al. 2007; Cheung et al. 2010; Denman et al. 2011; Bryndum-Buchholz et al. 2020). For example, ocean temperatures have been increasing in the last four decades and are expected to continue rising in the coming decades. In the Arctic Ocean, summer sea ice has declined to the lowest level on record. Mean sea levels along Canadian coasts are projected to rise by as much as 0.59, 0.75, and 0.96 m relative to 2010 in some parts of Pacific, Arctic, and Atlantic Canada, respectively, by the end of the 21st century under the "no mitigation" scenario (Han, Ma, and Slangen 2020).

In addition, concerns about the ecological and socio-economic consequences of ocean acidification through fisheries are growing rapidly (Denman et al. 2011; Steiner et al. 2018). These changes will exacerbate many current climate risks and present new risks and opportunities for fisheries (Lam et al. 2020; IPCC 2019; Sumaila et al. 2011), along with additional coastal erosion and retreat (Forbes et al. 2004), resulting in significant implications for communities, infrastructure, and ecosystems. Together, these threats and stressors compromise the health of ocean ecosystems, leading to economic and social impacts, including the loss of jobs, cultural and social identity, and economic benefits (Sumaila et al. 2011, 2019; Doney et al. 2012). Indeed, the Royal Society of Canada concluded that the nation has made little substantive progress in fulfilling national and international commitments to sustain marine biodiversity, such as the Aichi Biodiversity Target (Hutchings et al. 2012, 2020; Cisneros-Montemayor et al. 2018). There are also gaps in ocean governance and access to ocean resources, particularly for coastal and Indigenous communities (Bennett et al. 2018). Canada needs to do more.

Improving existing ocean management and governance, and integrating climate change into existing planning processes, using risk management, adaptive management, and novel governance strategies (Armitage et al. 2009), are necessary to secure the many benefits that our oceans are providing to Canadian economies, societies, and cultures. Existing inadequacy and inequity in ocean governance have meant that Canada's oceans may not be delivering on their potential to deliver food, health, environmental, and economic outcomes to Canadians. In some cases, mismanagement (Bavington 2010) or sustained inadequate management (Hutchings et al. 2020), in other cases inequitable management (Kourantidou et al. 2021) and misaligned policies (Kourantidou, Hoagland, and Bailey 2021), mean Canada should, and can, do better in helping to realize the potential of its ocean endowment (Sumaila 2021).

The OceanCanada Partnership (OCP) was established in 2014 and has been dedicated to building resilient and sustainable oceans on all three Canadian coasts, and supporting coastal communities as they respond to rapid and uncertain environmental and social changes. The OCP is a seven-year Social Sciences and Humanities Research Council of Canada (SSHRC) Partnership Grant–funded project composed of 22 formal research partners, including universities from coast to coast, community organizations, and Fisheries and Oceans Canada. The central goal of the OCP has been to understand and address threats facing Canada's Arctic, Atlantic, and Pacific Oceans and coastal regions and seek opportunities to develop a shared vision for the future of Canada's oceans – one that promotes the health and well-being of people living on coasts as well as the marine environment. Our highly interdisciplinary research consortium has brought together a wide range of expertise from many fields of study, including economics, law, geography, ethics, fisheries science, and oceanography, with the aim of integrating insights from across these broad fields with local and Indigenous knowledge, in order to help inform policies at the regional and national levels that are responsive to community needs. Our research synthesizes social, cultural, economic, and environmental knowledge about oceans and coasts nationally. Over the life of the project and beyond, we are taking stock of what we know about Canada's three oceans, building scenarios for the possible futures that await our ocean-coastal regions, and creating a national dialogue and shared vision for Canada's oceans. We are ultimately concerned with the health and well-being of communities that rely on the Pacific, Atlantic, and Arctic Oceans, and the livelihoods of those who gain sustenance from them, in both economic and cultural realms. Three major cross-cutting themes related to fisheries and oceans have emerged from our collective research: Changing Oceans, Access to Ocean Resources, and Ocean Governance.

This book is one key product of the OceanCanada Partnership that provides a "capstone" synthesis of diverse research by the OCP that addresses the current issues and challenges related to the future of Canada's oceans and coastal communities. Overall, the partnership has been amazingly productive, leading a top US scholar to state that our "list of outputs is dizzying, to say the least!" Outputs from the OCP include more than 440 publications, including a Special Feature of *Ecology and Society* titled "Canada and Transboundary Fisheries Management in Changing Oceans"; more than 540 presentations, meetings, and workshops; more than 50 films, documentaries, and videos; and at least 63 graduate students and postdoctoral fellows trained.

OceanCanada and this book build on the long history of ocean-related research in Canada, including a number of earlier high-profile projects such as Coasts Under Stress (see Ommer 2007), as well as the Community Conservation Research Network (CCRN, www.communityconservation.net), the Canadian Fisheries Research Network (Thompson et al. 2019), and the global initiative Too Big To Ignore (TBTI, www.toobigtoignore.net). These projects, and many other ocean-related initiatives in Canada, achieved significant progress on issues related to social-ecological health and governance of Canada's oceans and coasts. However, there were no research initiatives seeking to synthesize knowledge about Canada's three oceans, enabling us to consider the options for policy, planning, and management that could operate nationally and also capture significant regional differences. OceanCanada has sought to fill this gap by developing an enhanced understanding of the forthcoming uncertain changes occurring in Canada's coastal and ocean systems; establishing a methodological framework and foundation for future research that supports the development of policies and regulations that can help Canada improve its overall performance in ocean management and sustainability; and integrating interdisciplinary research and approaches cohesively to produce results that can inform policy. Specific research questions guiding the OCP and the chapters in this volume include: (1) How do changing oceans affect access to resources and governance? (2) How do social, economic, and governance responses to changing oceans impact ocean sustainability and coastal well-being? In responding to these questions, this book will be useful to scholars and policy-makers, as well as students of ocean science, fisheries, economics, and management.

A SOCIAL-ECOLOGICAL FRAMEWORK TO UNDERSTAND AND GOVERN CANADA'S OCEANS AND COASTS

An integrated perspective on our oceans and coasts is necessary to develop policy that will reflect a shared understanding of emerging threats, challenges, and opportunities among researchers, industry, Indigenous Peoples, and the Canadian public. Canada must identify future impacts, such as those related to climate change, on Canada's living marine resources and the resulting effects on livelihoods, communities, and economic sectors that depend on them. As reflected above, Canada's oceans and coasts are complex social-ecological systems that pose

major research, management, and policy challenges. An SES view emphasizes the unpredictable, dynamic, and evolving nature of interdependent social and ecological systems (Berkes et al. 2003). In SESs, conservation actions are immediately embedded in a complex web of social and ecological processes and interactions. Ostrom (2009) developed the first global SES framework to understand the processes that lead to changes in the stocks of renewable natural resources, and this perspective is relevant for how we understand Canada's oceans and coasts as well.

The breadth of our analysis requires a robust framework that will enable us to integrate linkages between socio-economic, cultural well-being, and biological conditions and governance characteristics, as well as examine drivers and responses to changes in the ocean and coastal environment. Such a framework needs to be comprehensive and broad to accommodate the research objectives and perspectives of the partnership. Our approach here adapts the bicoastal Coasts Under Stress SES framework (Ommer 2007) to meet OceanCanada's tricoastal conceptual framework (Figure 1.1a), using integrated social-ecological values to examine the management of ocean resources in the face of change and uncertainty (Berkes et al. 2003; Ostrom 2009).

As reflected in Figure 1.1a, Canada's ocean-coastal SESs are arranged geographically within the national system, which is thus composed of the three oceans and coasts (Arctic, Atlantic, and Pacific) as subsystems. Each regional SES reflects its own cultural, historical, social, economic, institutional, and biophysical characteristics; these interact through national-level policies, regional implementation, and inter-transfer of knowledge. The dynamics of the SESs are determined by the social, economic, and biophysical drivers at global, national, and regional scales, historical pathways of changes, and the current status of the SESs. Main direct and indirect drivers include climate change, access to the oceans and their resources, and changes in governance from local and national to international levels. Simultaneously, the dynamics of SESs also affect some of these drivers. Thus, our framework includes three cross-cutting themes – Changing Oceans, Access to Ocean Resources, and Ocean Governance – to address the interconnections between these direct and indirect drivers and Canada's SESs (Figure 1.1b).

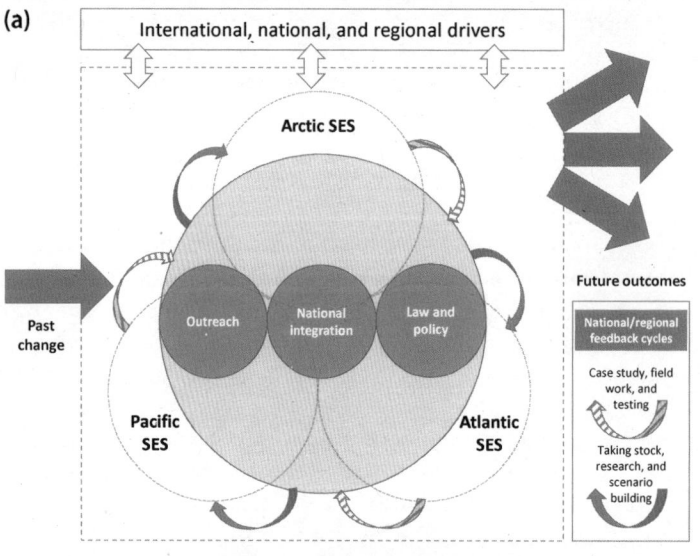

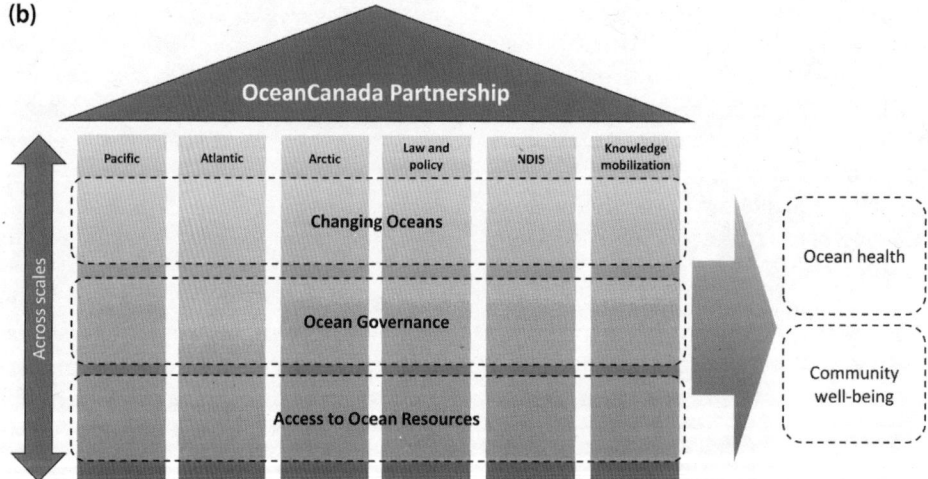

Figure 1.1
(a) Conceptual framework of OceanCanada, representing the ocean-coastal social-ecological system of Canada. (b) The OceanCanada Partnership Roof depicting cross-cutting themes of the partnership and how they relate to the OCP Working Groups. NDIS = National data and integrated scenarios.

The future of Canada's SESs is determined by the different responses of SES components to these drivers, moderated by related policies at the national and regional levels, and the inter-transfer of knowledge between the national and regional SESs. Building on our framework, each chapter reflects established methods in the social and natural sciences, including economics, community consultation and engagement, ecosystem and climate modelling, field-based interviews, and ocean governance in order to understand the past, present, and potential future of Canada's ocean-coastal SESs.

Accordingly, this book highlights the risks to Canadian society and marine ecosystems resulting from diverse drivers of change and the potential of different policies to reduce such risks. Ultimately, we hope to contribute to a shared vision among Canadians of different possible future states of human-coastal-ocean interactions. This is being accomplished by communicating our best understanding of the current and potential future states of Canada's oceans, and doing so with attention to and consideration of the complex socioeconomic, political, institutional, and cultural experiences that have shaped and continue to shape our uses of the oceans and coasts.

RESEARCH PATHWAYS AND POLICY PERSPECTIVES

Many aspects of this book make it unique: its inclusion of all three of Canada's coasts; its interdisciplinary nature (natural as well as social sciences and law); its clear focus on both the well-being of people and the health of the oceans; its central theme of changing oceans and how this change affects access to ocean resources by different coastal communities; and how changing oceans affect how Canada's oceans are governed. Furthermore, while other books take a regional approach, an important feature of this book is that it is national in scope and celebrates the similarities and differences in Canada's relationship with the ocean from coast to coast to coast. Finally, since its inception, OceanCanada has prioritized partnerships and encouraged diverse voices and contributions not only from academics but also from Indigenous and non-Indigenous communities, managers, and practitioners; consequently, this book draws on their expertise throughout. All these make this book a truly unique contribution to the study of changing oceans and their diverse impacts.

As climate change, resource overexploitation, and pollution continue to have immense effects on our oceans and coasts, an informed and engaged citizenry is becoming more aware of and concerned about the health of our planet in general. In terms of ocean change and citizen empowerment, our hope is that a volume of this nature will help broaden the discussion of the importance of our oceans, as well as contribute to dialogue around measures to be taken to ensure the sustainability of oceans and coastal communities.

In considering possible futures and pathways for Canada's oceans and coasts, four crucial insights have emerged and are reflected in this book.

First, the future of Canada's oceans and coasts is directly linked to our shared pathway toward reconciliation of Indigenous ocean issues, particularly in relation to climate change, governance, and economic access, which requires cross-scale consideration of the lasting effects of colonization on Indigenous Peoples. These effects include dispossession from land, ocean spaces, and marine resources, and continuing social, cultural, and economic impacts associated with a loss of access. Challenges to reconciliation include the mix of federal, provincial, and territorial jurisdictions, diversity of Indigenous populations, and political and structural resistance to power sharing. As reflected at various points in chapters in this volume, some positive examples of co-governance have emerged, with some limited progress to address the dispossession of Indigenous communities from fisheries and marine mammals. The field of reconciliation is rapidly changing, and Chapters 2 and 14 do not include developments beyond summer 2022, such as the Roman Catholic Church's repudiation of the fifteenth-century doctrine of discovery in April 2023. Commitments to recognize Indigenous law and governance are far from complete, however, and achieving ecologically sustainable and socially just coastal and ocean outcomes in Canada will necessitate equitable engagement of Indigenous communities in visioning and planning of ocean spaces. This may include, for example, better incorporation of and respect for Indigenous knowledge and world views, establishment of equitable and just co-management or co-governance arrangements that integrate knowledge systems and share power and responsibility, and transformation of relationships to support Indigenous self-determination.

The second important theme is the centrality of opportunities for new scholars and youth more generally. OceanCanada offered new opportunities for senior and junior scholars, students, and nonacademic members to undertake original, problem-driven, interdisciplinary research at multiple spatial scales using closely linked theoretical and methodological approaches. We have trained at least 63 students and postdoctoral fellows, at all times being mindful of preparing them with the knowledge, skills, collaborations, and partnerships they need to carry this research forward within Canada and beyond after the program ends. This volume reflects numerous contributions and a leadership role for a wide range of new and emerging scholars who are the next generation of Canada's science and policy community (see, for example, Chapters 3, 10, and 11).

A third crucial insight is the importance of balancing project execution as proposed and a respect for emergent properties. Funding for long-term science partnerships such as OceanCanada requires that objectives, deliverables, and hypotheses are clearly articulated upfront, with potential and likely sources of integration identified. It must be realized that these formulations will be based on the knowledge at the time of proposal writing, and thus, when the research is being conducted under a systems lens and across spatial and governance scales, should almost always change as a result of execution of the proposed research agenda. At the onset of OceanCanada, a regional compartmentalization made sense, with working groups delineated by Pacific, Arctic, and Atlantic geographies (Figure 1.1a). However, the reader will notice that the book is in fact not delineated as such, and that, rather, three cross-cutting themes emerged about halfway through the program (Figure 1.1b). These themes have become integration points across disciplines and geographies, and between academics and practitioners.

STRUCTURE OF THIS BOOK

The three major cross-cutting themes related to oceans and coasts that emerged from our collective research

Figure 1.2 Changing oceans and subsequent effects on access to ocean resources and ocean governance.

– Changing Oceans, Access to Ocean Resources, and Ocean Governance – provide organizational structure for this book (Figure 1.2). Drawing on these three themes, the book is organized into five parts: (1) Setting the Stage; (2) Changing Oceans; (3) Access to Ocean Resources; (4) Ocean Governance and (5) Into the Future. The starting point and central theme for the book is changing oceans, in both biophysical and social terms, which forms the basis for the parts on access and governance (Figure 1.2).

The central issues of each chapter and the connections between them highlight the book's integration, making it broader than the sum of its parts. Each section touches on all three coasts in order to provide a national scope.

In Part 1, "Setting the Stage," we draw attention to a central insight: that the future of Canada's oceans and coasts are tied to reconciliation and a new nation-to-nation perspective on their management and governance. In Chapter 2, Russ Jones and colleagues discuss how to achieve reconciliation between Indigenous Peoples and Canada, which is obviously crucial and central to Canada's ability to successfully manage and govern its ocean resource in an inclusive and equitable manner (Bennett et al. 2019). The authors identify reconciliation criteria based on the 2007 *United Nations Declaration on the Rights of Indigenous Peoples*. They also study how much progress has been made with regard to the well-being, economic conditions, and self-determination of Indigenous Peoples across Canada, and their results suggest that progress has varied widely.

In Part 2, "Changing Oceans," we present results, mainly but not exclusively, from the work of OceanCanada on our understanding of past and current changes taking place in the Arctic, Atlantic, and Pacific Oceans and how they are impacting Canada's ocean ecosystems, economies, and peoples. These topics are organized into three important themes describing different dimensions of Canada's changing oceans: rapid changes (Chapter 3), large changes (Chapter 4), and scenarios of changes (Chapter 5).

A characteristic of the changing oceans that challenges Canada's ocean-dependent marine life and human communities is the rapid rate of change. Chapter 3, by Travis Tai and colleagues, explores how the fast pace of environmental changes in Canada's three oceans impact and elevate risks on coastal communities from scientific, cultural, and societal perspectives. The chapter particularly highlights many hazards in relation to rapid and episodic changes such as marine heat waves (Cheung et al. 2022) that have fisheries repercussions, through harmful algal bloom events affecting biodiversity, to extreme storms threatening coastal structures, and also points to the hazards to these social-ecological systems posed by nonclimatic events such as oil spills and sediment runoffs. The chapter concludes that preparation for occurrences of these events, supported by improved knowledge generation, integration, and communication across the scientific, cultural, Indigenous, local, and societal perspectives, is needed to develop effective adaptations that will enable social-ecological systems to avoid the worst damage from these rapid changes.

In addition to the rapid pace of ocean changes that is challenging Canada's ocean-related SESs, the large magnitude of such changes is also important to consider. Chapter 4, by Nadja Steiner and colleagues, draws from existing knowledge (including Indigenous and local knowledge) on some of the observed and emerging large environmental changes in Canada's oceans and their impacts on dependent human communities. They highlight that Canada is experiencing dramatic changes in ocean conditions, from ocean warming and acidification to loss of sea ice. These are affecting marine life in the Pacific, Arctic, and Atlantic coasts of Canada differently, resulting in serious and diverse impacts on fisheries, cultures, and ecosystem services that are important to many coastal communities. Thus, changing oceans are impacting the ability of Canada to achieve sustainable development. The chapter explains the need for concerted effort in climate mitigation as well as adaptation by local communities, government institutions, law, and policies in order to jointly enable Canada to achieve a "healthy oceans, healthy people" vision.

Building on the earlier chapters, Chapter 5, by Louise Teh and colleagues, presents more comprehensive scenarios and projections of potential outlooks for Canadian oceans and coastal communities. In particular, the chapter uses scenarios and projections, available for different spatial and organization scales, that are generated from different perspectives (scientific, Indigenous, local, and societal) to articulate alternative visions about the future. Such multiscale scenario syntheses highlight specific local-scale challenges that coastal communities are facing under changing ocean conditions, as well as the potential match and mismatches with national-scale narratives and outlooks. This chapter provides several important new insights to help inform the development of sustainable pathways for Canada's coastal communities, including the potential synergies and trade-offs between local and national scales of development, and the need to reconcile competing goals of ocean resource management and adaptation to the changing oceans. Some of these insights pose specific governance challenges that are examined in subsequent chapters.

In Part 3, "Access to Ocean Resources," we discuss the implications of changing oceans for access to ocean resources, and the challenges and opportunities that such changes may bring to rights holders and stakeholders, both within coastal communities and throughout the country as a whole. In Chapter 6, Megan Bailey and Anthony Charles use an access lens to analyze ongoing conflict in the harvesting of American lobster. Under the Peace and Friendship Treaties signed by Britain with different Indigenous Peoples in the 1700s, codified in Section 35 of the Constitution,[1] and reaffirmed by the Supreme Court of Canada in the *Marshall* decision of 1999,[2] Mi'kmaq have a right to earn a moderate livelihood from fishing. Despite this, in the fall of 2020, conflict erupted over who has a right to fish, where, and when, capturing the attention of the media and the public across Canada and even internationally. The authors draw on an OCP access framework (see Bennett et al. 2018) to explain how and why benefiting from coastal fisheries remains a challenge for Mi'kmaq.

In Chapter 7, Evelyn Pinkerton and colleagues focus on a specific type of access program – quotas – popular in some of Canada's fisheries. The authors summarize the history of quota implementation across Canada, and highlight the impact that quotas have had on fisheries and fishers. As a novel contribution, a set of scenarios are developed for transitioning away from quota fisheries, with the authors leaning on the most recent *Fisheries Act* amendments as precedent setting in their support for owner-operator fisheries, which, unlike our history with quotas, have more effectively kept the benefits of the fisheries in coastal communities.

In Chapter 8, Carie Hoover and colleagues focus on access in the Arctic, where climate change, commercial fisheries, reconciliation, and allocative policies all come to a head. They focus on co-management, the mandated management framework for commercial fisheries across Inuit Nunangat (Inuit homelands) based on various land claims agreements. Co-management boards, some of whose members are co-authors of this chapter, make recommendations to the fisheries minister, including for access and allocation, with economic and social implications for land claims beneficiaries. The history of inequitable access is reviewed in this chapter, and options for a more equitable future, one that involves greater recognition of the role co-management boards play in governing resources for the benefit of Inuit, are discussed.

In Part 4, "Ocean Governance," we present our results on how changing oceans affect and challenge ocean governance, policies, and laws in Canada. In Chapter 9, Derek Armitage and colleagues identify some of the ingredients needed for coastal communities to transform how they interact with and govern their ocean resources and coasts in the context of change, and in ways that sustain social and ecological systems. Insights from this chapter point to the processes, relationships, and capacities required to support governance transformation from the ground up, as well as the interjurisdictional engagement, leadership, and knowledge (including notably Indigenous leadership) needed to move through phases of transformative change.

In Chapter 10, Evan Andrews and colleagues examine the links among coastal fisheries' rebuilding, knowledge, and "governance fit." Specifically, they consider how diverse knowledge and knowledge co-production processes can catalyze governance arrangements that better "fit" the challenges of fisheries rebuilding. Insights from this chapter connect across multiple chapters with an important message: efforts to rebuild fisheries and coastal communities, and to recover the abundance of marine life, require diverse knowledge to better fit governance to contexts of rapid change and uncertainty. Finally, in Chapter 11, Sondra Eger and colleagues outline some crucial opportunities and challenges associated with integrated management of Canada's oceans and coasts. Notably, however, they draw attention to "bright spots,"

or initiatives that have led to, or are anticipated to lead to, positive ecological, social, economic, and governance outcomes important for social and ecological coastal sustainability. In doing so, the chapter helps us to understand the conditions in which integrated ocean and coastal management might emerge and persist.

In Part 5, we synthesize the main points of the book, look forward, and conclude with policy implications of the work of the OceanCanada Partnership. In Chapter 12, Cecilia Engler and colleagues assess the capacity that legal and policy frameworks in selected ocean sectors in Canada have to integrate climate change considerations and respond to changing systems. They draw lessons from ocean-based renewable energies as a potential contributor to mitigation efforts, the protection of aquatic species at risk, and resource-oriented activities sustaining Canadian livelihoods: fisheries and marine aquaculture. They conclude with a clear message: Canada has made progress but it is time to pick up the pace and to ensure that our legal and policy frameworks are ready for the implications of a changing climate.

In Chapter 13, Vincent L'Hérault and colleagues draw attention to the importance of methodology. They outline the OCP approach and document some of the creative ways in which communities are engaged in ocean and coastal research (e.g., participatory video projects). As they show, participatory methods are more inclusive and able to bridge knowledge and epistemological gaps between local communities and research, and have demonstrated their ability to contribute to meaningful, trust-based relationships that lead to genuine collaboration. These methodologies are crucial to navigating a path forward that connects research with those most affected by ocean change, access, and governance challenges.

In Chapter 14, Russ Jones and colleagues conclude with a core message: reconciliation and Indigenous ocean management is the path forward in Canada. Specifically, they discuss the necessary changes underway in governance, resource access, and protecting culture and values that are having mixed success at transforming relationships. Policy recommendations that emerge in this chapter focus on changes needed to establish a just and equitable reconciliation framework and the measures to advance shared management, planning, and governance of ocean spaces. Reconciliation in Canada remains an unfinished business. Much more effort is needed to confront and address injustices from colonization, including political domination, loss of territory, and cultural imposition.

In the final chapter, Chapter 15, our aim is to synthesize insights and recommendations in ways that resonate with all Canadians concerned about the long-term sustainability of our oceans and coasts and the social, cultural, and economic activities that depend on them. Specifically, we ask how we can and should navigate pathways forward to foster viable and desirable ocean and coastal futures. In response, we summarize the main findings reported in each of the preceding chapters and draw attention to some of the core themes that have emerged from the collective efforts of the OceanCanada Partnership: reconciliation, changing oceans, changing access, changing governance, and the relationship among law, policy, and knowledge mobilization. As well, we provide practical pathways and recommendations to achieve a healthy ocean while supporting thriving coastal communities in Canada.

Finally, it is worth mentioning a few developments, both national and international, that took place since the submission of our manuscript to UBC Press that can have important implications for the sustainable management of Arctic, Atlantic, and Pacific social-ecological systems off the coasts of Canada. Nationally, Canada's National Climate Adaptation Strategy and Action Plan was published in November 2022. If this is implemented, it would help the country shake off its image as a country that is lagging in incorporating climate change consideration in fisheries management (e.g., Boyce et al. 2021; Pepin et al. 2022). Internationally, the COP27 agreement on climate change; the *World Trade Organization Fisheries Subsidies Agreement*; the *United Nations High Seas Treaty*; and the *Kunming-Montreal Global Biodiversity Framework* were all approved by the nations of the world. All governments in Canada, businesses, NGOs, civil society more generally, and scientists all should contribute to the effective implementation of these agreements and plans to help ensure that we achieve Infinity Fish, i.e., the notion that, if managed wisely, fish can continue to nourish humans forever, thereby generating infinity benefits (Sumaila 2021).

NOTES

1 *Constitution Act, 1982,* being Schedule B to the *Canada Act 1982* (UK), 1982, c 11.
2 *R v Marshall,* [1999] 3 SCR 456.

REFERENCES

AMAP (Arctic Monitoring and Assessment Programme). 2002. *Arctic Pollution 2002: Persistent Organic Pollutants, Heavy Metals, Radioactivity, Human Health, Changing Pathways.* Oslo: AMAP.

Armitage, D., R. Plummer, F. Berkes, R. Arthur, A. Charles, I. Davidson-Hunt, A. Diduck, et al. 2009. "Adaptive Co-Management for Social-Ecological Complexity." *Frontiers in Ecology and the Environment* 7 (2): 95–102.

Bavington, D. 2010. *Managed Annihilation: An Unnatural History of the Newfoundland Cod Collapse.* Vancouver: UBC Press.

Bennett, N.J., A.M. Cisneros-Montemayor, J. Blythe, J.J. Silver, G. Singh, N. Andrews, A. Calò, et al. 2019. "Towards a Sustainable and Equitable Blue Economy." *Nature Sustainability* 2 (11): 991–93.

Bennett, N.J., M. Kaplan-Hallam, G. Augustine, N. Ban, D. Belhabib, I. Brueckner-Irwin, A. Charles, et al. 2018. "Coastal and Indigenous Community Access to Marine Resources and the Ocean: A Policy Imperative for Canada." *Marine Policy* 87: 186–93.

Berkes, F., J. Colding, and C. Folke, eds. 2003. *Navigating Social-Ecological Systems: Building Resilience for Complexity and Change.* Cambridge: Cambridge University Press.

Berkes, F., R. Huebert, H. Fast, M. Manseau, and A. Diduck, eds. 2005. *Breaking Ice: Renewable Resource and Ocean Management in the Canadian North.* Calgary: University of Calgary Press.

Bindoff, N.L., W.W.L. Cheung, J.G. Kairo, J. Arístegui, V.A. Guinder, R. Hallberg, N. Hilmi et al. 2019. "Changing Ocean, Marine Ecosystems, and Dependent Communities." In *IPCC Special Report on the Ocean and Cryosphere in a Changing Climate,* edited by H.-O. Pörtner, D.C. Roberts, V. Masson-Delmotte, P. Zhai, M. Tignor, E. Poloczanska, K. Mintenbeck, et al., 447–587. Cambridge: Cambridge University Press.

Boyce, D.G., S. Fuller, C. Karbowski, K. Schleit, and B. Worm. 2021. "Leading or Lagging: How Well Are Climate Change Considerations Being Incorporated into Canadian Fisheries Management?" *Canadian Journal of Fisheries and Aquatic Sciences* 78 (8): 1120–29.

Bryndum-Buchholz, A., F. Prentice, D.P. Tittensor, J.L. Blanchard, W.W.L. Cheung, V. Christensen, E.D. Galbraith, O. Maury, and H.K. Lotze. 2020. "Differing Marine Animal Biomass Shifts under 21st Century Climate Change between Canada's Three Oceans." *FACETS* 5 (1): 105–22.

Cheung, W.W.L., T.L. Frölicher, V.W.Y. Lam, M.A. Oyinlola, G. Reygondeau, U.R. Sumaila, T.C. Tai, et al. 2021. "Marine High Temperature Extremes Amplify the Impacts of Climate Change on Fish and Fisheries." *Science Advances* 7, 40: eabh0895.

Cheung, W.W.L., V.W.Y. Lam, J.L. Sarmiento, K. Kearney, R. Watson, D. Zeller, and D. Pauly. 2010. "Large-Scale Redistribution of Maximum Fisheries Catch Potential in the Global Ocean under Climate Change." *Global Change Biology* 16: 24–35.

Cisneros-Montemayor, A.M., D. Pauly, L.V. Weatherdon, and Y. Ota. 2016. "A Global Estimate of Seafood Consumption by Coastal Indigenous Peoples." *PLOS One* 11 (12): e0166681.

Cisneros-Montemayor, A.M., G.G. Singh, and W.W. Cheung. 2018. "A Fuzzy Logic Expert System for Evaluating Policy Progress towards Sustainability Goals." *Ambio* 47 (5): 595–607.

Denman, K., J.R. Christian, N. Steiner, H.-O. Pörtner, and Y. Nojiri. 2011. "Potential Impacts of Future Ocean Acidification on Marine Ecosystems and Fisheries: Current Knowledge and Recommendations for Future Research." *ICES Journal of Marine Science* 68 (6): 1019–29. http://doi.org/10.1093/icesjms/fsr074.

DFO (Fisheries and Oceans Canada). 2009. *Our Ocean, Our Future: Federal Programs and Activities.* DFO/2009–1 581. Ottawa: Fisheries and Oceans Canada.

–. 2018. "Seafisheries Landed Value by Province." https://www.dfo-mpo.gc.ca/stats/commercial/land-debarq/sea-maritimes/s2018pv-eng.htm.

Doney, S.C., M. Ruckelshaus, D.J. Emmett, J.P. Barry, F. Chan, C.A. English, H.M. Galindo, et al. 2012. "Climate Change Impacts on Marine Ecosystems." *Annual Review of Marine Science* 4: 11–37.

Dyck, A.J., and U.R. Sumaila. 2010. "Economic Impact of Ocean Fish Populations in Global Fishery." *Journal of Bioeconomics* 12: 227–43.

Forbes, D.L., G.S. Parkes, G.K. Manson, and L.A. Ketch. 2004. "Storms and Shoreline Retreat in the Southern Gulf of St. Lawrence." *Marine Geology* 210 (1–4): 169–204.

Gibson, D., and U.R. Sumaila. 2017. "Determining the Degree of 'Small-Scaleness' Using Fisheries in British Columbia as an Example." *Marine Policy* 86: 121–26.

Halpern, B.S., S. Walbridge, K.A. Selkoe, C.V. Kappel, F. Micheli, C. d'Agrosa, et al. 2008. "A Global Map of Human Impact on Marine Ecosystems." *Science* 319 (5865): 948–52.

Han, G., Z. Ma, and A.B.A. Slangen. 2020. "Scenarios of Twenty-First Century Mean Sea Level Rise at Tide-Gauge Stations across Canada." *Atmosphere-Ocean* 58: 287–301.

Hicks, C.C., P.J. Cohen, N.A. Graham, K.L. Nash, E.H. Allison, C. D'Lima, D.J. Mills, et al. 2019. "Harnessing Global Fisheries to Tackle Micronutrient Deficiencies." *Nature* 574 (7776): 95–98. https://doi.org/10.1038/s41586-019-1592-6.

Hutchings, J.A., J.K. Baum, S.D. Fuller, J. Laughren, and D.L. VanderZwaag. 2020. "Sustaining Canadian Marine Biodiversity: Policy and Statutory Progress." *FACETS* 5 (1): 264–88. https://doi.org/10.1139/facets-2020-0006.

Hutchings, J.A., I.M. Côté, J.J. Dodson, I.A. Fleming, S. Jennings, N.J. Mantua, R.M. Peterman, et al. 2012. *Sustaining Canadian Marine Biodiversity: Responding to the Challenges Posed by Climate Change, Fisheries, and Aquaculture.* Ottawa: Royal Society of Canada.

IPBES (Intergovernmental Science-Policy Platform on Biodiversity and Ecosystem Services). 2019. *Summary for Policymakers of the Global Assessment Report on Biodiversity and Ecosystem Services of the Intergovernmental Science-Policy Platform on Biodiversity and Ecosystem Services.* Bonn: IPBES Secretariat.

IPCC (Intergovernmental Panel on Climate Change). 2019. "Summary for Policymakers." In *IPCC Special Report on the Ocean and Cryosphere in a Changing Climate.* Bremen, Germany: IPCC.

Kourantidou, M., P. Hoagland, and M. Bailey. 2021. "Inuit Food Insecurity as a Consequence of Fragmented Marine Resource Management Policies? Emerging Lessons from Nunatsiavut." *Arctic* 74: 40–55.

Kourantidou, M., P. Hoagland, A. Dale, and M. Bailey. 2021. "Equitable Allocations in Northern Fisheries: Bridging the Divide for Labrador Inuit." *Frontiers in Marine Science* 8 (93): 1–19.

Lam, V.W., E.H. Allison, J.D. Bell, J. Blythe, W.W.L. Cheung, T.L. Frölicher, M.A. Gasalla, and U.R. Sumaila. 2020. "Climate Change, Tropical Fisheries and Prospects for Sustainable Development." *Nature Reviews Earth and Environment* 1 (9): 440–54.

Lau, W.W.Y., Y. Shiran, R.M. Bailey, E. Cook, M. Stuchtey, J. Koskella, C.A. Velis, et al. 2020. "Evaluating Scenarios toward Zero Plastic Pollution." *Science* 369 (6510): 1455–61.

Ommer, R. 2007. *Coasts under Stress: Restructuring and Social-Ecological Health*. Montreal and Kingston: McGill-Queen's University Press.

Ostrom, E. 2009. "A General Framework for Analyzing Sustainability of Social-Ecological Systems." *Science* 24 (5939): 419–22.

Parry, M.L., O.F. Canziani, J.P. Palutikof, P.J. van der Linden, and C.E. Hanson, eds. 2007. *Impacts, Adaptation and Vulnerability*. Cambridge: Cambridge University Press.

Pauly, D., V. Christensen, S. Guénette, T.J. Pitcher, U.R. Sumaila, C.J. Walters, et al. 2002. 'Towards Sustainability in World Fisheries." *Nature* 418 (6898): 689-95.

Pepin, P., J. King, C. Holt, H. Gurney-Smith, N. Shackell, K. Hedges, and A. Bundy. 2022. "Incorporating Knowledge of Changes in Climatic, Oceanographic and Ecological Conditions in Canadian Stock Assessments." *Fish and Fisheries* 23 (6): 1332–46.

Srinivasan, U.T., W.W.L. Cheung, R. Watson, and U.R. Sumaila. 2010. "Food Security Implications of Global Marine Catch Losses due to Overfishing." *Journal of Bioeconomics* 12 (3): 183–200.

Steiner, N.S., W.W.L. Cheung, H. Drost, C. Hoover, J. Lam, L. Miller, A. Cisneros-Montemayor, et al. 2018. *AMAP Report on Arctic Ocean Acidification: Case 5: Climate Change Impacts on Subsistence Fisheries in the Western Canadian Arctic – A Framework Linking Climate Model Projections to Local Communities*. Tromsø, Norway: Arctic Council.

Sumaila, U.R. 2021. *Infinity Fish: Economics and the Future of Fish and Fisheries*. Amsterdam: Elsevier.

Sumaila, U.R., W.W.L. Cheung, V.W.Y. Lam, D. Pauly, and S. Herrick. 2011. "Climate Change Impacts on the Biophysics and Economics of World Fisheries." *Nature Climate Change* 1: 449–56.

Sumaila, U.R., A.M. Cisneros-Montemayor, A. Dyck, L. Huang, W.W.L. Cheung, J. Jacquet, K. Kleisner, et al. 2012. "Impact of the Deepwater Horizon Well Blowout on the Economics of US Gulf Fisheries." *Canadian Journal of Fisheries and Aquatic Sciences* 69 (3): 499–510.

Sumaila, U.R., and T.C. Tai. 2020. "End Overfishing and Increase the Resilience of the Ocean to Climate Change." *Frontiers in Marine Science* 7: 523.

Sumaila, U.R., T.C. Tai, V.W.Y. Lam, W.W.L Cheung, M. Bailey, A.M. Cisneros-Montemayor, O.L. Chen, and S.S. Gulati. 2019. "Benefits of the Paris Agreement to Ocean Life, Economies, and People." *Science Advances* 5 (2): eaau3855. https://doi.org/10.1126/sciadv.aau3855.

Teh, L.C., and U.R. Sumaila. 2013. "Contribution of Marine Fisheries to Worldwide Employment." *Fish and Fisheries* 14 (1): 77–88.

Thompson, S.A., R.L. Stephenson, G.A. Rose, and S.D. Paul. 2019. "Collaborative Fisheries Research: The Canadian Fisheries Research Network Experience." *Canadian Journal of Fisheries and Aquatic Sciences* 76 (5): 671–81.

Tilman, D., R.M. May, C.L. Lehman, and M.A. Nowak. 1994. "Habitat Destruction and the Extinction Debt." *Nature* 371 (6492): 65–66. https://doi.org/10.1038/371065a0.

Turner, N.J., and F. Berkes. 2006. "Coming to Understanding: Developing Conservation through Incremental Learning in the Pacific Northwest." *Human Ecology* 34: 495–513.

2

Status of Reconciliation and Indigenous Ocean Management in Canada

Russ Jones, Nancy Doubleday, Megan Bailey, Ken Paul, Fraser Taylor, and Peter Pulsifer

Reconciliation of Indigenous ocean issues, particularly in relation to climate change, governance, and economic access (in essence the three themes of the OceanCanada Partnership) requires cross-scale consideration of the lasting effects of colonization on Indigenous Peoples, including dispossession from land, ocean spaces, and marine resources, and continuing social, cultural, and economic impacts (Royal Commission on Aboriginal Peoples 1996; Truth and Reconciliation Commission of Canada 2015). In this chapter, we develop a framework for reconciliation and assess progress on ocean issues by analyzing the extent to which the injustices of colonization have been, or are being, overcome. Current relationships are guided by a mix of historical and modern treaties and are being redefined through new processes and agreements, as well as court decisions, court challenges, negotiations, and political actions. We propose reconciliation criteria based on the *United Nations Declaration on the Rights of Indigenous Peoples* (UNDRIP) (2007) and examine best practices, including progress in well-being, economic conditions, and self-determination of Indigenous Peoples across Canadian coastlines, which has varied widely.

Indigenous Peoples make up 4.9% of Canada's population, with many living in 683 communities, some of which are located along Canada's three coasts (Figure 2.1).[1] They are a fast-growing young population compared with Canada as a whole, and form a higher proportion of the population in remote areas such as the Arctic (67%) and north coastal British Columbia (45%).[2]

Reconciliation has become a major driver of changes in ocean management, including governance arrangements and marine resource access, and is beginning to ameliorate impacts of colonization on Indigenous relationships to ocean spaces and resources. The vignette by the lead author in the text box "Herring and protected area management in Gwaii Haanas" illustrates some long-standing issues with management and resource use in Haida Gwaii ("Islands of the People") and provides an example of steps toward redress.

This vignette illustrates the struggle to achieve reconciliation in one small part of Canada's coastline. Drivers for policy change are complex and have included political changes, Haida direct action, legal challenges, and negotiation. Conflicts have led to negotiated agreements and management plans that resulted in structural changes to management (i.e., creation of a consensus-based management board). Core issues remain to be resolved, such as the Haida jurisdiction and role in fisheries management, and just and fair Haida access to fisheries. Progress has been gradual, sometimes requiring years of progressive litigation, or to negotiate agreements or develop plans, with the result that after 30 years of working together in Gwaii Haanas, the Haida and Canada are still on their journey toward reconciliation.

The lack of agreement on the meaning of "reconciliation" has been identified as a problem for researchers (Rouhana 2011, 292). Rouhana (2011) defined it as follows: "a process that seeks a genuine, just, and enduring end to the conflict between the parties and transformation

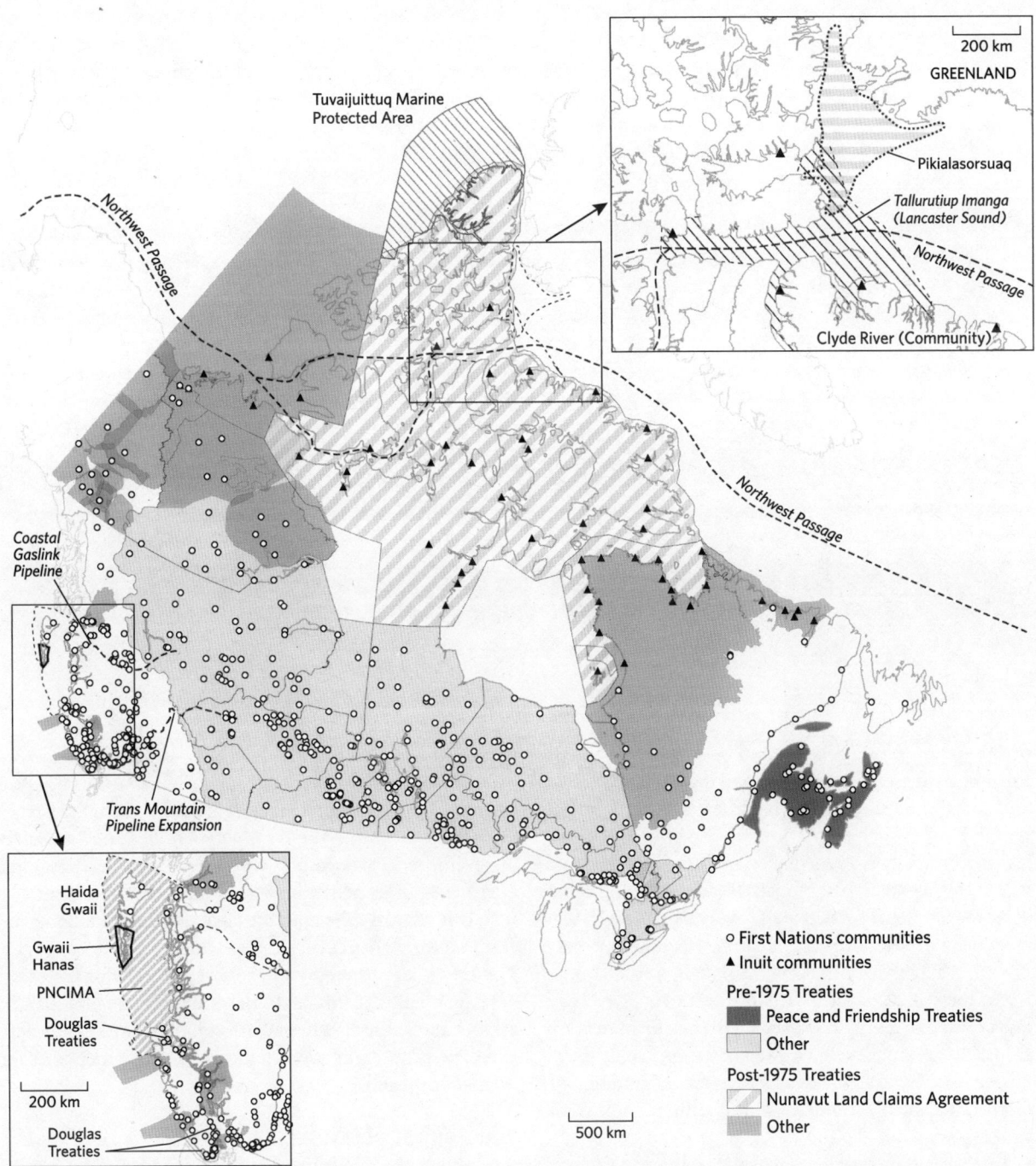

Figure 2.1 Indigenous communities and historical and modern-day treaties in Canada.

HERRING AND PROTECTED AREA MANAGEMENT IN GWAII HAANAS

Despite never signing a treaty, the Haida Nation and Canada are making progress toward co-governance of marine areas in advance of treaties.

In 1985, conflicts over clearcut logging led to a standoff between the Haida and loggers at Athlii Gwaay (Lyell Island) that resulted in protection and joint management of the Gwaii Haanas ("Islands of Beauty") area (Figure 2.1). Negotiations led to agreements and establishment of a consensus-based Archipelago Management Board (AMB) in 1993, whose mandate was expanded to include some aspects of fisheries in 2009.

Herring stocks in Gwaii Haanas have been depressed for close to two decades and have been closed to commercial fisheries since 2004. A disagreement over opening the commercial herring fishery in 2015 led to the Haida obtaining an injunction that stopped a commercial fishery opening by Fisheries and Oceans Canada (DFO). In granting the injunction, the judge placed weight on the long-term co-management relationship between Canada and the Haida Nation in Gwaii Haanas, concluding that "there is a heightened duty for DFO and the Minister to accommodate the Haida Nation in negotiating and determining the roe herring fishery in Haida Gwaii, given the existing Gwaii Haanas Agreement, the unique Haida Gwaii marine conservation area, the ecological concerns, and the duty to foster reconciliation with and protection of the constitutional rights of the Haida Nation," thereby recognizing the Gwaii Haanas agreements as interim steps toward reconciliation. In an earlier forestry case, the courts determined that the Haida had a strong prima facie case for Aboriginal title.

In 2018, a comprehensive Gwaii Haanas Gina 'Waadluxan KilGuhlGa Land-Sea-People Management Plan was completed that protects 40% of the Gwaii Haanas marine area from extraction but allows Haida traditional fisheries. Efforts to resolve fisheries conflicts are ongoing and include developing a joint herring rebuilding plan within an ecosystem-based management framework (expected for the 2024 fishing season), reconciliation negotiations to address issues such as the AMB's role in fisheries management, and a Haida title case that includes the marine area that began in 2002 and, despite delays, is nearing trial.[3]

---

of the nature of the relationship between the societies through a course of action involving intertwined political and social changes and which addresses both politically tangible issues such as distribution of power and historical responsibility." Reconciliation in Canada has been advanced by negotiation of modern treaties and land claims agreements. Court decisions and evolving Aboriginal law have been drivers for reconciliation. Negotiations have, however, largely occurred within symmetrical frameworks that disadvantage Indigenous parties who may have less power or capacity and may fail to transform relationships.[4] Lack of parity between negotiation tables means that incentives and tools available in one negotiation are often not available in others, and learning is not transferred among jurisdictions.

This review of reconciliation of ocean issues on Canada's Atlantic, Pacific, and Arctic coasts examines the extent to which the injustices of political domination, loss of territory, and cultural imposition arising from colonialism (Moore 2016) have been, or are being, overcome. We do this by drawing on the relevant articles from the *United Nations Declaration on the Rights of Indigenous Peoples* to develop criteria against which to discuss the ongoing process of reconciliation across Canada's three coasts.

This chapter describes the current status of reconciliation across Canada, including a scan of seven ocean issues and five case studies, and proposes criteria for assessing reconciliation. Chapter 14 describes how reconciliation can address the core impacts of colonization and presents policy recommendations for how a lasting reconciliation can be realized.

HISTORY OF RELATIONSHIP

Canada's dispossession of Indigenous Peoples has generally followed four stages – (1) separate worlds; (2) contact and cooperation; (3) displacement and assimilation; and

(4) negotiation and renewal – but vary regionally and temporally along Atlantic, Pacific, and Arctic shorelines (Royal Commission on Aboriginal Peoples 1996, 1: 40–44; Havemann 1999).

Indigenous Peoples on the Atlantic, Pacific, and Arctic coasts have distinct cultures, languages, and practices that are tied intimately to the land and resources. They had traditional territories based on occupancy and use, and boundaries were demarcated. These separate worlds changed after contact with European explorers and traders and led to cooperation as well as conflict. Indigenous groups played a key role in determining Canada's boundaries through alliances with the British in the Atlantic region that prevented incursions by the United States. Likewise, on the Pacific coast, British colonies sought peace with Indigenous groups during the period when borders were being defined with the United States. Some alliances were cemented through historical treaties (Figure 2.1). As well, the presence of Inuit helped to advance Canada's claims to Arctic sovereignty, waters, and resources. The policy of treaty making continued in inland portions of Canada until 1923, but was not followed in coastal areas of British Columbia or the Arctic from Confederation until the mid-1970s.[5]

Indigenous communities were contained through the establishment of Indian reserves in Atlantic and Pacific regions. A period of assimilation followed under the *Indian Act* of 1876, which treated Indians as wards of the state subject to paternalistic policies that included separating children from their families and sending them to live in residential schools (see, e.g., Pyne and Taylor 2019; GCRC 2020). This resulted in the disruption of families, loss of language, and intergenerational trauma (Truth and Reconciliation Commission of Canada 2015).

A major shift in policy and relationships with Indigenous Peoples took place with the passage of Canada's *Constitution Act, 1982*, which recognized and affirmed the existing rights of Canada's Aboriginal peoples, supported by many successful court challenges. The recent stage of negotiation and renewal (identified by Royal Commission on Aboriginal Peoples 1996) has been called an era of confrontation or pluralism.[6] But a parallel approach to treaties aimed at recognition of rights, including negotiation of reconciliation agreements for key issues, is currently in play.[7] There has been a recent sea change in political and societal attitudes toward Indigenous issues, and this provides a promising context for significant progress.

Significant milestones marking the four stages in the relationship for each coast are summarized below. A brief timeline of recent legal, policy, and political changes related to ocean reconciliation is shown in Figure 2.1 and further described in Table 2.1.

Atlantic Timeline

Separate Worlds

Pre-contact Indigenous Nations in the Atlantic region (not including Labrador) included Mi'kmaq, Maliseet,[8] Innu (also known as Montagnais-Naskapi), Dorset, and Beothuk. Mi'kmaq and Maliseet were fishing and hunting societies. Beothuk and Innu were largely cariboo-hunting cultures.

Contact/Cooperation

Norse settlement about 1000 AD in the Eastern Arctic and northern Atlantic was unsuccessful and was followed by European commercial fisheries and settlement in the 14th and 15th centuries. Mi'kmaq participated in wars and made alliances with the French and English from 1613 to 1761. This led to a series of Peace and Friendship Treaties between Mi'kmaq, Maliseet, and the British Crown from 1725 to 1794. Innu were active partners in the fur trade.

Displacement/Assimilation

Beothuk were purged from Newfoundland, which was declared a colony in 1824, with the last known woman dying in 1829. Immigration and settlement increased following the American Revolution, leading to displacement of Mi'kmaq and Maliseet from their land and resource base. At this time, they also suffered population decline due to disease. The colonies of Nova Scotia, New Brunswick, and Lower and Upper Canada (which divided into the provinces of Ontario and Quebec) joined Confederation in 1867. Prince Edward Island joined in 1873. A period of treaty denial followed with the *Indian Act* of 1876 and the *British North America Act* of 1867 (Knockwood 2003, 48). Reserves in New Brunswick, Nova Scotia, and Prince Edward Island were established

Table 2.1

Timeline of selected policies and court decisions related to ocean reconciliation

Date	Policies and court decisions
1725–94	Peace and Friendship Treaties in Atlantic (Figure 2.1)
1969–2011	Most Northern/Arctic land claims agreements (Figure 2.1)
1976–98	*Calder v British Columbia (AG)* decision (1973) leads to Nisga'a Final Agreement negotiations (Figure 2.1)
1984–93	Federal Conservative government (Brian Mulroney, Kim Campbell)
1990	*R v Sparrow* decision – Indigenous priority to fish for food, social, and ceremonial purposes
1990–present	Federal Aboriginal Fisheries Strategy – framework for implementing fishing rights
1992–present	BC Treaty Process established with goal of negotiating treaties within 10 years
1993–2006	Federal Liberal government (Jean Chrétien, Paul Martin)
1996	• *R v Gladstone* decision – recognition of Indigenous commercial fishing rights • *Oceans Act* commits to integrated ocean management (IOM) and development of a national network of Marine Protected Areas • Royal Commission on Aboriginal Peoples Report (>400 recommendations)
1999	*R v Marshall* decision – treaty commercial fishing rights
2001	Marshall Response Initiative
2002	*Species at Risk Act* requires consideration of Indigenous traditional knowledge
2004–present	Federal Aboriginal Aquatic Resource and Oceans Management Program
2006–15	Federal Conservative government (Stephen Harper)
2007	Pacific Integrated Commercial Fisheries Initiative; Atlantic Integrated Commercial Fisheries Initiative
2010	As signatory to *Convention on Biological Diversity,* Canada commits to Aichi target of protecting 10% of coastal and marine areas by 2020
2015–present	Federal Liberal government (Justin Trudeau) with platform to advance Indigenous reconciliation
2016	• Report of Canada's Truth and Reconciliation Commission (94 calls to action) • Canada becomes a party to *United Nations Declaration on the Rights of Indigenous Peoples* (UNDRIP)
2016–present	Oceans Protection Plan announced and implemented; renewed in 2022
2018	Canada adopts 10 principles for reconciliation (Department of Justice Canada 2018)
2019	BC *Declaration on the Rights of Indigenous People Act* for implementation of UNDRIP
2020	With 21 other countries, Canada commits to protecting 25% of coastal and marine areas by 2025 and 30% by 2030
2021	Federal *UNDRIP Act* for implementation of UNDRIP
2022	Kunming-Montreal Global Biodiversity Framework commits to protecting 30% of coastal and marine areas by 2030 (UNEP 2022)

by colonial authorities rather than according to treaties as in other parts of the country. Newfoundland did not become part of Canada until 1949.

Negotiation/Renewal

A series of hunting and fishing rights cases led to the Supreme Court of Canada's *Marshall* decision in 1999, which recognized the Mi'kmaq and Maliseet treaty right to fish.[9] Mi'kmaq and Maliseet are negotiating implementation of Peace and Friendship Treaties.[10] Innu in Quebec have been engaged in negotiation of a treaty for over 20 years. The Crown has failed to recognize the rights of Mi'kmaq who occupied Newfoundland concurrently with Beothuk.

Pacific Timeline

Separate Worlds

The Pacific coast is home to numerous Indigenous Nations with distinct cultures speaking a variety of languages[11] who controlled territories and access to marine and terrestrial resources.

Contact/Cooperation

First contact with the Spanish in 1774 was followed by further Spanish and British exploration. Both a land-based and maritime fur trade occurred, the latter from about 1790 to 1820. The US border was established at the 49th parallel by the *Oregon Treaty* in 1846, after which the Hudson's Bay Company relocated from the Columbia River to Victoria. From 1849 to 1856, the British Crown established colonies[12] that were combined in 1856 before joining the Dominion of Canada in 1871. A handful of Douglas Treaties were signed on Vancouver Island from 1850 to 1854 (Harris 2008, 21–23). The northern boundary and the Alaska panhandle dispute were not resolved until 1903.

Displacement/Assimilation

Indigenous populations were drastically reduced by epidemics, and reached a low point about 1929.[13] Canada imposed Indian reserves from about 1850 to the 1930s as a means of reducing conflict between settlers and Indigenous Nations over land and fisheries (Harris 2008, 92–105, 164–86). Reserves on the coast were generally smaller due to coastal Indigenous Peoples' reliance on fisheries (Harris 2008, 6). Imposed regulations gradually dispossessed Indigenous People from their fishing places and fisheries (Pearse 1982, 176–81; Newell 1993; Harris 2001, 196–208).

Negotiation/Renewal

This stage was gradual, beginning with court decisions such as the Supreme Court's 1973 *Calder* decision, which recognized the existence of Aboriginal title prior to colonization but split on whether it had been extinguished.[14] Efforts followed to negotiate a treaty with the Nisga'a, under the policy of negotiating one BC treaty at a time. A series of court decisions recognizing Aboriginal fishing and hunting rights, supported by the *Constitution Act,* *1982,* helped to change the dynamics (Table 2.1). In the 1997 *Delgamuukw* case, the Supreme Court of Canada held that Indigenous title had not been extinguished in British Columbia.[15] Subsequent cases found that Indigenous laws pre-existed and survived the assertion of Crown sovereignty.[16] The *Nisga'a Treaty* (also known as the *Nisga'a Final Agreement*) was signed in 1998 after 30 years of negotiation. A BC Treaty Process that began in 1991 with the goal of completing treaties throughout the province within 10 years has proven difficult. As of 2022, there were 7 groups in British Columbia implementing a treaty (Figure 2.1) and 31 in various stages of negotiation. About 44% of the *Indian Act* bands in the British Columbia are not currently involved in negotiations. Legal decisions continue to be important, such as the recognition of Tsilhqot'in title to lands in central British Columbia.[17]

Arctic Timeline

Separate Worlds

Inuit are a transnational people who occupy the circumpolar region and reside in homelands in what is now Canada, Greenland, the United States, and the Russian Federation.[18] Inuit traditionally rely primarily on seal, whale, walrus, fish, and caribou, and displaced earlier Dorset and Thule peoples.

Contact/Cooperation

First contact with Europeans in search of the Northwest Passage began with Martin Frobisher in 1576. The Hudson's Bay Company, established in 1670, opened trade with Indians of the western forests, which led to penetration of the Arctic and contact with the Inuit. Mapping of the Mackenzie Delta began in 1826. The Royal Proclamation of 1763 recognized Indigenous Peoples but also created commercial domain and governance. Traders and missionaries began arriving in the Arctic thereafter, and Inuit were encouraged to trap furs. Near the end of this stage, the Hudson's Bay Company opened 15 new trading posts from 1921 to 1931.

Displacement/Assimilation

From 1934 to 1959, a number of families were transferred from Cape Dorset, Pond Inlet, and Northern Quebec to

Devon, Ellesmere, and Cornwallis Islands in the High Arctic, as well as within Labrador, in part to advance Canada's Arctic claims and in part as a response to food scarcity. Initially, Arctic governance was modelled on the colonial practice of the British Empire, with territorial councils appointed by the federal government and reporting to the Minister of Indian Affairs and Northern Development (or equivalent, depending on the time period) in the Yukon and Northwest Territories.

Negotiation/Renewal

Negotiations of land claims agreements across the Canadian Arctic occurred over several decades beginning in 1969.[19] Inuvialuit negotiations were triggered by Justice Thomas Berger's inquiry into the Mackenzie Valley pipeline and launched by a petition presented in the House of Commons by Mary Carpenter, a young Inuvialuit woman.[20] The *Inuvialuit Final Agreement* was completed and ratified in 1984, and became law in Canada in 1985, with constitutional protection. It was followed by division of the Northwest Territories, the settlement of the Tungavik Federation of Nunavut land claim, and the creation of Nunavut. These agreements were negotiated pursuant to the federal Comprehensive Land Claims Policy of 1974. Negotiation of both agreements considered interest in offshore issues, but the federal position was that seabed resources belonged to the federal government.[21] The Quebec Inuit land claim, the *James Bay and Northern Quebec Agreement,* was settled in 1975 under a different legal regime. The Labrador Inuit (Nunatsiavut region) and Nunavik Inuit (Northern Quebec) completed land claims agreements in 2005 and 2006, respectively. In the southeastern part of Hudson Bay lies the area covered by the *Eeyou Marine Region Land Claims Agreement,* signed in 2010. The Innu of Labrador signed a treaty in 2011.[22] Southern Labrador Inuit are currently negotiating a land claim that would be separate from that of Nunatsiavut region (Bell 2020).

RECONCILIATION

Reconciliation and Canadian Policy

Colonization and the creation of what is now Canada has had multiple and intergenerational effects on Indigenous Peoples, creating a ripple effect in statistics on Indigenous health and human well-being, including life expectancy, unemployment, high school graduation, incarceration in prisons, and suicide rates (Royal Commission on Aboriginal Peoples 1996; Truth and Reconciliation Commission of Canada 2015; Cooke et al. 2007). The history of colonization and its effects are detailed by both the Royal Commission on Aboriginal Peoples (RCAP) and the Truth and Reconciliation Commission of Canada (TRC), including recommended steps toward reconciliation, such as self-government and apology.[23] Measures of Indigenous well-being have improved in recent years but continue to be significantly lower than those of the general population and subject to wide regional disparities (Cooke et al. 2007).

Canada recently made a political commitment to "a renewed nation-to-nation relationship with Indigenous Peoples based on recognition of rights, respect, cooperation and partnership" (Prime Minister of Canada 2016a, 2018), a commitment that was included in Mandate Letters to Ministers, including the December 2019 letter to the Minister of Fisheries, Oceans and the Canadian Coast Guard (Prime Minister of Canada 2019). As well, Canada recently adopted 10 principles for achieving reconciliation by renewing Indigenous-Crown relationships (Department of Justice Canada 2018). Canada became a signatory to UNDRIP in 2016 and incorporated it into Canadian legislation.[24] Although the direction is positive, the effects of these new policies of recognition of rights and the adoption of UNDRIP on federal, provincial, and territorial law and policy in Canada remain to be seen (see, e.g., Assembly of First Nations 2018).

Indigenous Peoples across Canada have diverse perspectives on reconciliation. By negotiating with groups of Indigenous Peoples, the Canadian approach allows for recognition and accommodation of these differences. There is no one Indigenous world view or value system, but there are commonalities, including the understanding of the place of humans in the natural world, and cultural and spiritual relationships to territories and living things. Indigenous history and teachings are passed on from their ancestors and importance is placed on future generations (e.g., Sterritt 2016; Kinnear 2007). For example, the Haida Nation identified six Haida ethics and values that define the Haida world view and guide marine planning, including respect, responsibility, interconnectedness, balance,

seeking wise counsel, and reciprocity, that were considered the foundation of the Haida Gwaii Marine Plan
(MPPI 2015). *Yahguudang* (respect) is defined as follows:
"Respect for each other and all living things is rooted in
our culture. We take only what we need, we give thanks,
and we acknowledge those who behave accordingly"
(MPPI 2015, 11). Similarly, the Mi'kmaq world view of
Netukulimk sees the world as a connected web, and recognizes that humans do not have dominion over nature
but are just a part of it (Prosper et al. 2011). When understood properly, *Netukulimk* recognizes that nature can
provide for the well-being of both the individual and the
community as a whole.

Canada's response to Indigenous demands for power
and resource sharing, including economic access and
compensation, varies, depending on the existence of
prior treaties or land claims agreements, and on the situation itself. Gaps in the existing treaties and land claims
agreements concerning ocean management make reconciliation in the context of oceans and marine issues
complicated.

As a result, Canadian courts are called upon to help
interpret historical and modern treaties. Canadian
courts have identified reconciliation as an objective of
negotiations between the Crown and Indigenous groups,
as a test for infringement of Aboriginal rights, and as a
standard to hold governments to account for wrong-
doing. However, the courts' interpretation of reconciliation may differ from political or social interpretation
(Walters 2008).

Framework for Assessing Reconciliation Progress in Ocean Management

Canada's relationship with Indigenous Peoples along the
country's three coasts is at differing stages of recovery
from colonization. Indigenous rights are influencing
several processes, including ocean management and
planning, establishment of Marine Protected Areas (with
increasing calls for Indigenous protected areas), development of integrated ocean management plans, recovery
plans for species at risk, and environmental and socio-
economic assessments of projects such as pipelines and
ports. Fundamental to reconciliation of ocean issues is
finding measures to overcome three key injustices associated with colonialism in the history of fisheries and ocean

management in Canada: political domination, loss of
territory, and cultural imposition (Moore 2016). Political
domination and denial of self-determination are central
to colonialism injustice (452), but taking of Indigenous
land or territory may be just as great a wrong of settler
colonization (455).

To achieve its goals, a reconciliation process must address four key issues: justice, truth, responsibility, and
restructuring of the social and political relationship between the parties (Rouhana 2011). Justice is the frame of
reference or guiding principle for each step in reconciliation. Progress in reconciliation is reviewed in this chapter
and further analyzed relative to the four issues in Chapter
14, where we propose a pathway toward a just and equitable reconciliation. Canada's approach to reconciliation
has been informed by its Ten Principles (Department of
Justice Canada 2018), but more specific guidance is provided in UNDRIP. In this chapter, we identify criteria that
should be evident on the path to reconciliation, as informed by best practices described later in this chapter
and by drawing on the relevant UNDRIP articles (Table
2.2). We use these criteria to assess current progress and
identify future options for successful reconciliation by
means of an overview of seven ocean issues and five case
studies from Canada's three coasts. The ocean issues
represent common sources of conflict or tension between
Indigenous Peoples and states. Note that Table 2.2 shows
the relevant UNDRIP articles by article number only. For
example, self-determination as well as land, territories,
and resources are all fundamental rights captured in
UNDRIP, and thus criteria for reconciliation could include things like compensation for loss, and processes for
securing consent.[25]

For each coast, Table 2.3 highlights key historical,
political, and legal elements in relation to reconciliation
of seven issues that capture common sources of conflict
over ocean activities: fisheries and marine mammals,
integrated ocean management, Marine Protected Areas,
species at risk, shipping, oil and gas, and aquaculture. The
current status of each issue is presented from historical,
legal, policy, and political perspectives, followed by a brief
summary of reconciliation actions by region that mostly
align with the above criteria. We then use case studies
from across the three coasts to highlight the presence and
absence of specific reconciliation criteria. In this way, a

Table 2.2

Proposed criteria for reconciliation based on UNDRIP

Type of injustice and reconciliation criteria	*UNDRIP articles*
Political domination	
Effective Indigenous organizations in place at appropriate scales	18
Self-government or management agreements in place	4
Mechanisms and resources to implement agreements and treaties	29, 37, 39
Development of joint policies and plans	5, 29
Processes and practices in place to secure Indigenous consent	10, 19, 28, 32
Meaningful engagement in development of relevant legislation, regulations, and/or designations/listings	19, 38
Incorporation of Indigenous laws into decision making	27
Incorporation of Indigenous priorities and strategies into decision making	32
Resorting to courts to resolve disputes	32, 37, 40
Indigenous capacity to govern or manage, including financial autonomy	39
Loss of territory (and benefits thereof)	
Consent for allocations, licences, tenures, or plans in a territory related to an activity	19
Agreements on share or proportion of a resource or activity; or jointly approved plans in place	17, 19
Allocation policies or plans or targets account for Indigenous title and rights to specific territories	19, 26
Compensation for loss	10, 20, 28, 32
Revenue sharing or management funding for new or existing activities or uses	26, 32
Joint assessments of activities to account for environmental, social, cultural, and economic impacts	23, 32
Sustainable use and/or species recovery over the long term as determined through assessments	25, 29, 32
Cultural imposition	
Ability to practise rights and culture	8, 15
Incorporation of traditional knowledge into policies and plans	31
Uses Indigenous language in negotiation and decision making	13
Contributes to an equal standard of living, e.g., income, benefits, traditional food	21, 24
Activity occurs consistent with community values	23, 25
Policies and plans incorporate Indigenous world view	25

narrative examination and application of the framework developed based on UNDRIP can be developed and used to explain the varying degrees of reconciliation in ocean governance evident within Canada.

Case 1 – Integrated Marine Planning on Canada's North Pacific Coast

Collaborative marine planning work in northern British Columbia has been driven by commitments on the government side to integrated management in Canada's *Oceans Act* of 1996,[26] the need to reconcile Aboriginal rights where no treaties exist, and the skillful self-organization of many Indigenous groups on scales conducive to planning (Jones, Rigg, and Lee 2010).[27] This case study illustrates many of the reconciliation criteria during the planning phase.

In 2005, Canada announced the Pacific North Coast Integrated Management Area (PNCIMA) on the northern British Columbia coast as one of five pilot areas for integrated marine use planning. Over three years, Canada, British Columbia, and 17 First Nations negotiated a collaborative letter of intent that established principles and a governance framework for planning.[28] Enabling conditions addressed during the pre-planning phase were:

Table 2.3

Reconciliation of ocean issues in Canada

Issue	Current status				Reconciliation action by region and measure
	Historical	*Legal*	*Policy*	*Political*	
FISHERIES AND MARINE MAMMALS	Fisheries and marine mammals integral to Indigenous societies for food and trade; settlers introduce new markets and fisheries, e.g., Atlantic cod, Pacific salmon, etc.; discriminatory regulations imposed, e.g., *Fisheries Act* (Harris 2008; Pearse 1982). Arctic whaling banned commercially in 1964, preventing harvest of large whales by Inuit. Small coastal species still taken for food by Inuit.	Indigenous rights and priority over other fishing largely defined through court decisions, e.g., food, social, and ceremonial fisheries, *R v Sparrow* (1990); commercial fishing rights, *R v Gladstone* (1996), *R v Marshall* (1999), *Ahousaht Indian Band and Nation v R* (2013). Regulated through food, social, and ceremonial licences and communal commercial licences (unique to First Nations) issued by Fisheries and Oceans Canada (DFO).	Most Atlantic and Pacific fisheries managed by limited entry licences and/or quota shares by the 1990s; several active federal programs enabling negotiation of fisheries agreements and transfer of access.[1] Since 2018 reconciliation agreements based on recognition of rights being explored as a way to accelerate transfer of benefits and progress toward treaties.[2]	Comprehensive land claims agreements (LCA) in Arctic in 1990s; BC Treaty Process starts in 1990 but held up in part by lack of progress on fisheries; policies support court-determined priorities subject to limits (Case 2); co-management agreements vary by region; Aboriginal commercial catch shares imposed by Canada in asymmetric power relationship (limited negotiation). Inuvialuit regain recognition of right to hunt bowhead whales in 1984 *Inuvialuit Final Agreement*.	**Atlantic** *Access* – Right to fish formerly recognized in treaties but ignored in practice; catch shares transferred through Marshall Response Initiative (Case 2).[3] *Management* – Limited Indigenous management control except some traditional fisheries such as eel. **Pacific** *Access* – About 13% of all BC licences and quotas transferred through treaty or initiatives (amount not negotiated).[4] A few treaties have been reached but some Indigenous groups have gone to court to further define the nature of their fishing rights.[5] *Management* – Some inland and a few coastal fisheries managed separately from commercial fisheries through negotiated agreements under federal Aboriginal Fisheries Strategy (see, e.g., Jones 2006). **Arctic** *Access* – Priority of Inuit hunting rights recognized at early stage; access to marine mammals codified through LCAs; new commercial fishing licences issued to Inuit organizations and non-Indigenous corporations but Inuit are seeking a more equitable share (see, e.g., Chapter 10).[6] *Management* – Wildlife co-management boards established through LCAs; variable participation by Inuit organizations in marine and environmental policy issues, negotiations, national and international affairs, legal reviews, hearings, and studies. For example, the Inuit Circumpolar Council is a permanent participant in the Arctic Council (1996), and current initiatives like the Pikialasorsuaq Commission (Case 3) are an important factor in recognition of Inuit agency in marine affairs domestically and internationally (Pikialasorsuaq Commission 2017).

▼ Table 2.3

Issue	Current status				Reconciliation action by region and measure
	Historical	Legal	Policy	Political	
SHIPPING	Coastal Indigenous groups make extensive use of marine resources and waterways, e.g., intertribal trade; waterway use increases for trade and development, e.g., Grand Banks and Atlantic cod fishery; maritime fur trade in Pacific; search for Northwest Passage opens Arctic (Figure 2.1); rail, road, and container traffic fuel port development.	Canada expands exclusive economic zone (EEZ) to 200 nautical miles, 1982; right to innocent passage in the EEZ in accordance with Law of the Sea; port-state control maintains international and domestic standards; conventions under International Maritime Organization, e.g., ballast water, sulfur dioxide (SO_2) emissions, places of refuge.	State has more control over territorial sea (to 12 nautical miles); shipping not typically addressed in LCAs or treaties; voluntary Tanker Exclusion Zone keeps tankers en route from Valdez, Alaska, to Washington away from BC coast (1985); lack of comprehensive oil spill response plans; concern for oil spills in Arctic; Liberal government establishes $1.5 billion Oceans Protection Plan to improve marine safety (2016; see also n52 at chapter end).	Port developments on west coast; loss of sea ice opening shipping traffic in Northwest Passage; concerns about environmental impacts; Indigenous groups block projects due to lack of environmental assessments for shipping, e.g., Northern Gateway (2017), Trans Mountain Pipeline (2019); challenges to Canadian sovereignty in Arctic; Canada files claim with United Nations to extend Arctic jurisdiction relating to resources, management, and transportation (George 2019).	**Atlantic** The Crown has been slow to recognize Indigenous rights and interests in shipping, and there are currently no agreements with Atlantic Indigenous groups. However, in July 2019, Transport Canada (2019a) announced funding and capacity for Indigenous organizations in Nova Scotia (2), Newfoundland (1), and Quebec (5), for engagement. **Pacific** Indigenous groups have delayed or stopped several major projects through court processes (Case 4). A *Reconciliation Framework Agreement for Northern Shelf Bioregion* (2018) supporting cooperative planning of shipping and marine response by Indigenous partners and Canada is in the implementation stage (Case 5); Pacific Places of Refuge Contingency Plan updated by Haida Nation and Transport Canada (2018); moratorium on oil tankers in northern British Columbia (2009) in response in part to Indigenous concerns. **Arctic** Shipping not explicitly addressed in LCAs; other forms of engagement at different scales, e.g., review of the *Canada Shipping Act* Regulations (multiple years), review of the 1986 Nanassivik Mine shipping accident affecting Inuit hunters, Northwest Passage shipping season and ship regulations, offshore oil and gas, and provision of mining services. Inuit in Nunavut receiving funding to develop capacity and participate in activities under Oceans Protection Plan (OPP) (Transport Canada 2019a). Limited capacity or control of shipping through Northwest Passage or resources for response to Arctic oil spills.

INTEGRATED OCEAN MANAGEMENT (IOM)			
Indigenous Peoples maintain connections to ocean spaces (cultural, spiritual, social, economic); graduated assertion of Crown authority through legislation and policy; relevant to international migratory fish and mammal species that have always been essential to cultural identity, e.g., bowhead whales, protection of coastal habitat.	*Fisheries Act* (1880s) regulates fisheries; habitat protection measures since 1970s; Canada's *Oceans Act* of 1996 commits to integrated planning and decision making together with "affected aboriginal organizations."	Oceans Strategy and Oceans Action Plan outline collaborative planning approach; mixed progress on five pilot areas that DFO identifies for integrated planning in 2003, e.g., northern British Columbia case study (Case 1); Canada is supporting international conventions through Marine Spatial Planning approach that has been initiated at a regional level (UNESCO 2021).	Integrated ocean management plans completed for four ocean pilots identified by DFO in 2003. Indigenous participation varied by ocean area. Canada's assertion of sovereignty over Arctic waters was a key factor in negotiation of LCAs.

Atlantic

Planning – Indigenous participation in two Atlantic IOM pilots (ESSIMP and PBIMP)[7] have been minimal and dominated by stakeholders. DFO has identified three Large Ocean Management Areas in the Atlantic Region as candidates for Marine Spatial Planning – Gulf of St. Lawrence, Scotian Shelf/Bay of Fundy, East Coast of Newfoundland – and has started to engage Indigenous groups and support technical capacity.

Decision making – Aboriginal parties not included in ESSIMP and PBIMP governance structures. Unequal Indigenous participation in planning tables. Limited funding for Indigenous capacity.

Pacific

Planning – Pacific North Coast Integrated Management Area (PNCIMA) Plan, IOM pilot for northern British Columbia, completed in 2017 (Case 1). Endorsed by DFO, Province of British Columbia, and a group of Indigenous organizations. First Nations and the province have approved and are implementing subregional marine plans (Haida Gwaii/North Coast/Central Coast/North Vancouver Island) that identify Marine Protected Area (MPA) network candidates. Although not an IOM pilot, Canada, British Columbia, and Nuu-chah-nulth Nation partner in Aquatic Management Board on West Coast of Vancouver Island.[8]

Decision making – PNCIMA Letter of Intent (2008) outlines collaborative governance structure (Canada/BC/First Nations) based on consensus.

Arctic

Planning – Pilot for Integrated Ocean Management Plan for the Beaufort Sea (IOMPBS) worked within framework of existing LCAs in Arctic; governance approach adapted for integrated planning.

Decision making – IOMPBS uses processes established through LCA with oil and gas sector involved in steering committee.

▼ Table 2.3

Issue	Historical	Legal	Policy	Political	Reconciliation action by region and measure
		Current status			
		Legal	Policy	Political	
AQUACULTURE	Indigenous cultivation of some species, e.g., Pacific "clam gardens," Mi'kmaq (Wolastoqey) "oyster gardens"[9]; salmon hatcheries since 1900s; Atlantic salmon farms on Atlantic and Pacific coasts since 1912; Pacific oysters since 1970s; rapid growth of mussel farming in Atlantic since 1980s; freshwater Arctic char mainly raised in southern aquaculture facilities; world capture fisheries reached peak in 1990s, with aquaculture representing about 50% of world production by 2016 (FAO 2018).	Regulatory regimes vary by province; DFO has lead role with regard to management practices; BC Supreme Court defined federal responsibility for aquaculture management in 2009. DFO manages aquaculture in British Columbia and Prince Edward Island. All other provinces have delegated jurisdiction for aquaculture monitoring and regulations.	Federal policies support research; included as economic development in some modern treaties. Cohen Inquiry (2012) looks at impact of salmon aquaculture as a vector for transfer of sea lice and disease affecting BC wild salmon; ocean acidification due to climate change poses a risk to shellfish survival and growth.	Some agreements with Indigenous groups in British Columbia, e.g., Kitasoo, Ahousaht; opposition to salmon farms in Broughton Archipelago. Most Atlantic First Nations either participate or are seeking funding to participate in the economic benefits of aquaculture. Only one finfish operation in Nova Scotia, and one kelp operation. The rest of the Atlantic First Nations operations are in mussels, oysters, and scallops. No large-scale closed containment operations as yet.[10] Farming of Arctic char in southern aquaculture facilities represents loss of Arctic genetic resources and local economic opportunities.	**Atlantic** *Access* – Rapid expansion of salmon and shellfish farming since 1980s; only a few operations occur with Indigenous consent. A few Indigenous-owned operations. Licensing regimes controlled by federal and provincial regulators. Nova Scotia and New Brunswick manage leases and do not consult or give preferred access to First Nations. Most First Nations seek access to diversify economic benefits. Atlantic Integrated Commercial Fisheries Initiative (AICFI) encourages Indigenous participation in Atlantic aquaculture. No remediation for biodiversity loss due to invasive species such as tunicates (sea squirts). *Decision making* – Federal and provincial governments do not consult when approving leases for non-Indigenous and foreign lease requests. First Nation operators are charged annual licence fees. **Pacific** Rapid expansion of salmon and shellfish farming since 1980s followed by consolidation; only a few facilities operate with Indigenous consent; only a few partnerships. New BC salmon aquaculture policy requires consent by First Nations for new and existing salmon farms (BC Ministry of Agriculture 2019). **Arctic** Limited potential for marine aquaculture due to ice cover and slow growth. No Indigenous involvement in freshwater Arctic char produced in Yukon or southern Canada, including export of eyed eggs (Ethier 2014). Marine mammal harvesting and live capture and aquaria display of marine mammals have stirred controversy and faced bans in various circumstances.

Inuit reliance on marine mammals; Atlantic and Pacific Indigenous reliance on fish; offshore O&G exploration; known reserves (West Coast, Arctic); East Coast development (Hibernia); revenue-sharing agreements with Newfoundland and Labrador for Hibernia drilling in 1980s.

EEZ expands to 200 nautical miles in 1976; provincial jurisdiction over seabed limited (within the jaws of the land); wrangles over federal and provincial jurisdiction, e.g., court rules for British Columbia in Strait of Georgia; conflict in Hecate Strait unresolved.

Canada has not recognized Indigenous rights to seabed resources; some offshore O&G wells not explicitly addressed in modern LCAs or treaties; influenced by Indigenous and environmental concerns; United States has O&G wells in Beaufort Sea and recently stopped approving new drilling leases in the Arctic National Wildlife Refuge (June 2021) (Gibbens 2021).

Atlantic

Conflicts and inquiries – "Old Harry" in the Gulf of St. Lawrence approved by Natural Resources Canada in 2010 without consultation with First Nations (Séguin 2010).

Decisions/status – No Indigenous members on Canada-Newfoundland/Labrador Offshore Petroleum Board although a few are on fisheries working group. No Indigenous resource revenue-sharing agreements or policies exist in the Atlantic.

Pacific

Conflicts and inquiries – Indigenous and environmental groups opposing offshore O&G development and pipelines (Case 4); independent Federal-Provincial Environmental Review Panel recommends O&G exploration subject to 92 recommendations (1984–86); BC Offshore Oil and Gas Task Force concludes science and technology adequate for offshore O&G extraction.

Decisions/status – 1972 federal policy decision not to explore for offshore oil on West Coast due to environmental and Indigenous concerns; *Oil Tanker Moratorium Act* (2019) bans tankers at northern BC ports.[11]

Arctic

Conflicts and inquiries – Mackenzie Valley Pipeline Inquiry results in decision not to build; Integrated Ocean Management Plan for the Beaufort Sea developed with O&G as a partner on steering committee.

Decisions/status – Exploration, development, and production of O&G resources subject to LCAs; discussions regarding offshore O&G exploration involved federal government, Northwest Territories, and Inuvialuit (Canadian Press 2018).

Issue	Current status				Reconciliation action by region and measure
	Historical	Legal	Policy	Political	
MARINE PROTECTED AREAS (MPAs)	Indigenous reliance on marine places for spiritual, cultural, socioeconomic purposes; lack of trust as terrestrial parks excluded Indigenous uses in the past; trend toward joint management of protected areas and Indigenous involvement in planning and management, e.g., UNDRIP Articles 27 and 29.	Consultation required on potential infringements of Aboriginal rights, e.g., right to fish, right to hunt; *Oceans Act of 1996* and *Canada National Marine Conservation Areas (NMCA) Act of 2002* may require engagement and require agreements with Indigenous organizations prior to MPA designation; can designate NMCA "reserves" subject to settlement of land claims or treaty negotiations.	Canada commits under *Convention on Biological Diversity* (CBD) to protect 10% of coastal and marine areas by 2020 and was advocating for 30% protection (DFO 2020); federal government achieves 13.8% protection by 2019, not including regional initiatives in progress;[12] CBD sets global target to protect 30% by 2030 (UNEP 2022). MPAs currently led by federal or provincial agencies based on reconciliation with Indigenous Peoples, including use of traditional knowledge (TK) (DFO 2023b); political commitments such as regional Marine Spatial Planning (MSP) expected to inform future MPA establishment. Policies for Indigenous Protected and Conserved Areas (IPCAs) in the marine environment under discussion but may require new legislation to meet expectations for Indigenous leadership.[13]	Co-governance of MPAs largely ad hoc without defining policy; commitment to collaborative planning with Indigenous groups identified in Ministerial Mandate Letters (2016); linkages to development (e.g., Arctic O&G) and regional MSP; governance regimes for MSP and IPCAs will require agreements between Indigenous groups and federal authorities.	**Atlantic** *Status* – Limited engagement and no co-governance agreements to date. Atlantic First Nations were not consulted at the onset of the 2017 Atlantic MPA program. Intervention by the Nova Scotia Mi'kmaq in 2017 forced DFO to consult on MPAs and planning (Withers 2018). Indigenous involvement in MPA establishment presently at the consultation stage. DFO proposing to include MPAs as part of MSP and seems open to some type of co-management arrangement. *Collaborative initiatives* – None. **Pacific** *Status* – Several Indigenous partnership agreements and collaborative management plans in place. Indigenous groups were engaged in but not signatories to Canada–British Columbia MPA Network Strategy (2014); collaborative MPA network planning underway for Northern Shelf Bioregion (Case 1; identified as priority in collaborative Pacific North Coast Integrated Management Area [PNCIMA] plan [2017]); designation of Tang.ɢwan – ḥačxʷiqak – Tsigis Marine Protected Area in progress (could contribute 2.43% to national MPA conservation target), and is subject to a co-management agreement with affected Indigenous groups (DFO 2023a). *Collaborative initiatives* – 2 MPAs (out of 5 established in British Columbia[14]) involve joint management plans, i.e., Gwaii Haanas (2019), SGaan Kinghlas–Bowie Seamount (2019); agreements and collaborative management plans developed in partnership with Indigenous groups, e.g., Haida: Gwaii Haanas (1999) and SGaan Kinghlas–Bowie Seamount (2006). **Arctic** *Status* – Only two Arctic MPAs by 2017 in Beaufort Sea. Significant steps toward protection of two large MPAs in Nunavut since: Tallurutiup Imanga (Lancaster Sound) contributes 1.9% and Tuvaijuittuq contributes 5.55% toward NMCA targets. Inuit interest in additional protections such as Pikialasorsuaq (Case 3).

Historical reliance on fish and marine mammals; international focus on species loss leads to international CBD.	Arm's-length assessments by Committee on Status of Endangered Wildlife in Canada (COSEWIC); listings under *Species at Risk Act* (SARA) of 2002; requirement to consult Indigenous organizations and use ATK or IQ[17] in development of management plans and recovery strategies.	CBD Goal A: by 2020, Canada to plan and manage using an ecosystem approach; resistance to listing commercial species, e.g., BC salmon, BC rockfish; Indigenous involvement in recovery plans low despite legal requirements (Hill et al. 2019). Comparatively little use of TK or IQ in development of recovery plans. Extirpated species and habitat rebuilding will require recognition of First Nations title and require federal and provincial governments to justify infringements that have led to the disappearance of species due to habitat destruction.	*Collaborative initiatives* –Tarium Niryutait MPA in Beaufort Sea protects beluga whales; Anguniaqvia niqiqyuam MPA protects ecological values and has conservation objective based on Indigenous traditional and local knowledge; NMCA boundary for Tallurutiup Imanga (Lancaster Sound) established in 2017 with completion of an *Inuit Impact and Benefit Agreement*.[15] Tuvaijuittuq in the High Arctic, partially within the Nunavut Settlement Region, was designated for interim protection in 2019, while the Qikiqtani Inuit Association, the Government of Nunavut, and the Government of Canada work with Inuit and northern partners to explore the feasibility of longer-term protection. In February 2022, the governments of Canada and Nunatsiavut committed to assessing the feasibility of establishing an Indigenous Protected Area along the northern Labrador coast under the *Canada NMCA Act* (Parks Canada 2022a).[16] **Atlantic** *Collaborative initiatives* – Atlantic salmon is endangered and has been extirpated in many areas. First Nations have worked on recovery plans and hatcheries with provinces, Parks Canada, and DFO with limited success. American eel has been listed as a species of concern and there are currently only food, social, ceremonial fisheries in some areas. First Nations in New Brunswick are proposing elver ranching (raising of small eels). **Pacific** *Collaborative initiatives* – Indigenous groups contributing to northern abalone recovery plans, e.g., Abalone Recovery Implementation Group, Haida Gwaii Community Action Plan (DFO 2007); COSEWIC assessment of Okanagan chinook.[18] **Arctic** *Collaborative initiatives* – Use of IQ in COSEWIC polar bear assessment (COSEWIC 2018) results in continued special concern status.[19]

Notes:

1 Federal programs related to Indigenous fisheries include: Aboriginal Fisheries Strategy (1990); Aboriginal Aquatic Resource and Ocean Management Program (1994); Marshall Response Initiative (2001); Aboriginal Aquatic Resource and Ocean Management Program (2004); Atlantic Integrated Commercial Fisheries Initiative (2007); Pacific Integrated Commercial Fisheries Initiative (2007).

2 More than 80 negotiating tables have been created since 2015, including some for Pacific fisheries and Atlantic fisheries (CIRNA 2020).

3 In 2016, Atlantic First Nations accounted for $122 million in commercial landings in Eastern Canada (6% of total landings), including $50 million from lobster (4% of all lobster) and $48 million from snow crab (15% of all snow crab) (Coates 2019, 20). However, a framework for a moderate livelihood as determined by *R v Marshall* (1999) has not yet been determined. See, e.g., note 38 at chapter end.

4 The Pacific Integrated Commercial Fisheries Initiative (PICFI) was established in 2005 following two policy reviews of Pacific fisheries (McRae and Pearse 2004; First Nation Panel on Fisheries 2004). Allocation Transfer Program (ATP) and PICFI licence and quota purchases totalled $154 million from 2008 to 2016 and increased commercial fishing access controlled by First Nations from 3% to 13% (DFO 2016). Total BC licence value as of 2017 was $2.3 billion (DFO 2022a).

5 The Nisg̱a'a, Tsawwassen, and Maa-nulth First Nations *Final Agreements* define specific Indigenous roles in fisheries management, along with defined shares of commercial fisheries by species (Figure 2.1). In 2009, several Nuu-chah-nulth Nations established a right to fish multiple species for the purpose of sale in a defined territory that extended nine miles from shore that applied to all species except geoduck clam (*Ahousaht Indian Band and Nation v Canada*, 2018 BCSC 633). The right to fish was interpreted to consist of a small-boat fishery with wide community participation, but was the subject of an appeal about the lack of progress in negotiations about harvest shares and fishing regimes that largely upheld the decision (*Ahousaht Indian Band and Nation v Canada (Attorney General)*, 2021 BCCA 155).

6 Arctic fishery allocations to Inuit vary by species and region, with less access for Inuit in the more southerly fishing areas as a result of late entry of Canadian Inuit into commercial marine fishing, incomplete resolution of fisheries issues in LCAs, and unilateral federal decisions.

7 The Eastern Scotian Shelf Integrated Management Plan (ESSIMP) was completed through DFO-led collaborative process in 2006 with the Province of Nova Scotia. It focuses on an offshore area that has numerous fisheries and fisheries management plans. However, ESSIMP was not approved or implemented by DFO. The Placentia Bay Integrated Management Plan (PBIMP) was completed by DFO and the Province of Newfoundland and Labrador in 2011. Planning area includes coastal waters of Newfoundland and the Grand Banks.

8 As described by Pinkerton (2007), the Aquatic Management Board led dialogue on local fisheries and marine issues through a multi-stakeholder process in the 1990s. This resulted in some data gathering and assessment but no joint plans.

9 Clam gardens are human-induced mud flats or terraces created by building a stone wall in the intertidal zone (e.g., Thomson 2015). "Traditional" aquaculture methods on oysters, mussels, and clams keep stocks abundant and purify waters (e.g., Denny et al. 2016).

10 A steelhead farm in Powell River, BC, was converted to a floating closed containment in 2019 (DFO 2019b).

11 Transport Canada 2019b. The tanker ban was supported by almost all coastal Indigenous groups.

12 Includes 14 MPAs under the *Oceans Act*, three National Marine Conservation Areas, one marine National Wildlife Area, and 59 marine refuges (DFO 2019a).

13 An Indigenous Circle of Elders identifies three defining elements for IPCAs in Canada: "they are Indigenous-led, they represent a long-term commitment to conservation; and they elevate Indigenous rights and responsibilities." See ICE 2018, 5; Zurba et al. 2019.

14 Others include Hecate Strait and Queen Charlotte Sound Glass Sponge Reefs MPA, Endeavour Hydrothermal Vents MPA, and Scott Islands Marine National Wildlife Area.

15 Inuit in Nunavut had lobbied for protection since the National Energy Board hearings in 1974; the *Inuit Impact and Benefit Agreement* established a consensus-based joint Inuit–Government of Canada cooperative management board and supports an Inuit Stewardship program. See Parks Canada 2022b, 2022c.

16 In the Beaufort Sea, Inuvialuit participated in hunts by commercial whalers in a symbiotic form of whaling: Inuvialuit got access to ships and gear, whalers benefited from the Inuit hunting and sewing skills, and both shared in the kills. This symbiosis ended in 1934, when the last whale boat left (Raddi and Weeks 1985). In the Pacific, Nuu-chah-nulth on the West Coast of Vancouver Island were active whalers until the late1920s, when whales began to be commercially depleted (Coté 2010).

17 ATK = Aboriginal Traditional Knowledge; IQ = Inuit Qaujimajatuqangit (Inuit traditional knowledge).

18 The COSEWIC assessment was supported by an ATK report (personal communication by Gloria Goulet, co-chair, COSEWIC ATK Subcommittee, December 30, 2019).

19 ATK and IQ have been critical for assessing population trends.

(1) development of Indigenous partnerships and governance structures; (2) completion of Indigenous marine traditional knowledge studies; and (3) independent funding of Indigenous capacity for marine planning.

Progress slowed in 2011 when the federal government withdrew from its initial commitments to develop detailed coastal plans in the first phase of PNCIMA planning. However, a Marine Plan Partnership (MaPP), made up of the Province of British Columbia and the initial 17 First Nations partners in PNCIMA, developed marine spatial plans for four subregions without federal involvement. These were endorsed in 2015 and candidate sites for Marine Protected Areas (MPAs) were identified; they are being implemented through modified regional and subregional governance structures that do not include the federal government.[29] In 2017, the governance partners, including most of the original First Nations, endorsed a high-level PNCIMA plan that outlines an ecosystem-based management framework, goals, objectives, and five planning priorities, including MPA network planning.[30] This did not address fishing, shipping, and oil and gas development issues. A similar governance structure is now being applied to shipping, including Canada's new Oceans Protection Plan, and is expected to be used for both MPA network planning and PNCIMA implementation (Case 5).[31] A tripartite MPA network planning process is underway, and the federal, provincial, and First Nations partners completed a network action plan in February 2023 that sets targets for establishment by 2025 and 2030.[32]

Elements of reconciliation in the PNCIMA and MaPP processes include governance structures and jointly agreed plans to address political domination and, to a lesser degree, cultural imposition. Structures for planning and implementation are based on consensus decision making.[33] The PNCIMA plan and MaPP plans are jointly endorsed by Canada and/or the Province of British Columbia and participating First Nations organizations, consistent with the principle of Indigenous consent. MaPP plans are being implemented and some progress has been made on MPA network planning, which is one of the PNCIMA priorities. Contentious issues such as fisheries and marine shipping have been set aside but continue to be part of a political and Aboriginal rights-based dialogue (see Table 2.3 and Case 5).

Case 2 – Reconciliation of Fishing Rights in Atlantic Canada

Indigenous rights in fisheries have led to numerous legal and policy conflicts across Canada relating to management and access (Table 2.3). Colonial policies are only gradually changing as a result of court decisions and reconciliation processes such as negotiation of treaties or land claims agreements. Loss of fisheries access has affected cultural well-being, particularly when rights are infringed over more than one generation.

A series of Peace and Friendship Treaties in the Atlantic from 1752 to 1794 recognized the importance of fisheries to Mi'kmaq and Maliseet Nations (Knockwood 2003), but despite the treaties, colonial systems dispossessed these nations from their territories and restricted their access to fisheries resources. In the 1980s, courts affirmed that the treaties had not extinguished Aboriginal rights to food, social, and ceremonial fisheries, and in 1999 the Supreme Court of Canada affirmed their treaty right to fish for commercial purposes to achieve a moderate livelihood.[34] A period of uncertainty and conflict followed, requiring clarification by the court that the fishery was subject to federal regulation. A federal program (the Marshall Response Initiative) was created to transfer commercial licences and commercial fishing quota to Indigenous groups through agreements (Table 2.3, footnote 3). While a variety of federal programs seek to negotiate fisheries agreements,[35] results vary across Nations and regions and generally fall short of reconciliation criteria related to resource access and governance identified in Table 2.2.[36] In fact, 2 out of 34 eligible Mi'kmaq and Maliseet Nations did not participate in the Atlantic Integrated Commercial Fisheries Initiative (AICFI), which delivered the program. However, the fisheries agreements generated significant benefits, including own-source revenue that supports community benefits and rebuilding (Coates 2019). Federal programs such as the 2004 DFO Aboriginal Aquatic Resource and Oceans Management Program supported development of technical and scientific capacity but fell short of a substantive approach to Indigenous fisheries management, with the result that First Nations still have limited capacity to manage natural resources within their traditional lands and waters.

Mi'kmaq identify *Netukulimk* ("take only what you need") as a guiding concept for stewardship of resources that encompasses principles of respect, reverence, responsibility, and reciprocity (Prosper et al. 2011; McMillan and Prosper 2016, 641). Atlantic salmon are classified as a species at risk, and agreements such as those under the federal Aboriginal Fisheries Strategy restrict how Atlantic salmon can be fished, how many can be taken, and how they can be removed (Shelley, Denny, and Fanning 2016). In 1993, the Listuguj community, now under the authority of Listuguj Mi'gmaq Government (LMG), abandoned the federal approach, and exercised their jurisdiction over Atlantic salmon, passing their own law governing salmon fishing for both food and sale on the Restigouche River, which is situated in both New Brunswick and Quebec, in order to better respect their traditional values and harvesting practices (Centre for First Nations Governance 2011). The law, consistent with *Netukulimk,* addresses conservation and management of fisheries, and has since been recognized as a model for co-management. In 2019, the LMG launched its own community-based treaty fishery, with its own lobster law and lobster management plan.[37] The fishery focused on food distribution, with some of the catch sold to offset the costs of fishing. In 2020, due to lack of progress in negotiations to implement the *Marshall* decision, including developing a framework for a moderate livelihood fishery, several other Mi'kmaq communities in Nova Scotia took the approach of developing and implementing their own management plans for the lobster fishery as a means of exercising their fishing rights.[38] The initial federal response was to avoid confrontation, but some non-Indigenous fishers cut traps or took Mi'kmaq fishing gear in attempts to intimidate Mi'kmaq fishers, emboldened by the lack of a clear federal statement (Maher 2020). During the 2021 fishing season, DFO reinitiated a policy to negotiate rights-based fisheries through agreements, and attempted to limit the fishery to the existing commercial fishing season, which has been met with resistance by First Nations.[39]

Case 3 – Wildlife Management and Indigenous Protected Areas in Nunavut

Although it has been 24 years since the formation of Nunavut, and more than 30 years since the signing of the 1992 *Nunavut Land Claims Agreement,* the process of reconciliation in the region continues, particularly with respect to fisheries, wildlife and ocean management, oil and gas development, and the establishment of protected areas.[40] A system for co-management that ensures active participation of Inuit in all decisions related to wildlife in Nunavut was established, but new issues continue to arise.[41]

In 2016, community members of Clyde River, Nunavut (Figure 2.1) brought a case to the Supreme Court of Canada contesting the decision of the National Energy Board (NEB) to allow seismic testing for offshore oil and gas exploration in Baffin Bay and Davis Strait. Evidence indicates that seismic blasting can harm marine life, affecting many areas of Inuit rights, including Inuit rights to harvest marine species.[42] The Supreme Court's decision quashed the exploration licence on grounds of inadequate consultation and failure on the part of the NEB to fulfill the Crown's duty to consult Inuit.

Inuit are leading Arctic efforts to secure alternatives using broad-scale models for marine management in the region, such as an international initiative that was the focus of an Inuit Circumpolar Council workshop in Nuuk, Greenland, in 2003. Pikialasorsuaq, or North Water Polynya, a region between the east coast of Baffin Island, Nunavut, and Greenland (Figure 2.1) has been a critical area of shared resources for millennia, and was a primary target for European and American whalers and sealers. Pikialasorsuaq is the largest Arctic polynya – an area of open water surrounded by sea ice – and the most biologically productive region north of the Arctic Circle.[43] The initiative resulted in formation of the Pikialasorsuaq Commission, an Inuit-led body formed to conduct consultations in Nunavut and Greenlandic communities closest to Pikialasorsuaq, to continue multilingual dialogue (Inuktut, Kalaallisut, and English), and to evaluate strategies for managing and safeguarding the region.

Recognizing that Pikialasorsuaq is "seriously threatened by rapid change in the region including climatic and environmental change, increased shipping activities, tourism, oil and gas exploration and development" (Pikialasorsuaq Commission 2017, ii), a foundational priority of the commission is to mitigate threats to the ecosystem using the best available knowledge and management practices.[44]

This consensus-based process under Inuit leadership brings together all key actors,[45] and places Inuit knowledge, perspectives, language, culture, and established land (including sea ice and ocean) use at the forefront of the dialogue. To this end, the commission currently makes three key recommendations for the establishment of: (1) an Inuit Management Authority; (2) a protected area managed by Inuit to support the Inuit vision of a working seascape, comprising the polynya itself and a larger management zone; and (3) a free travel zone for Inuit across the Pikialasorsuaq region (Pikialasorsuaq Commission 2017, xii).[46]

Inuit protection and management of Pikialasorsuaq may be seen as a process of reconciliation that enhances Inuit self-determination, through exercise of Inuit rights over Inuit territory, while recognizing Inuit priorities, languages, and traditional and contemporary practices.

Case 4 – Conflicts over Pipelines and Offshore Oil and Gas Development

Since the 1970s, several major oil and gas development projects in Canada have been cancelled or delayed as a result of disagreements with Indigenous Peoples. While decisions have been political, the issues have primarily related to Aboriginal rights and justice and requirements for reconciliation.

The Mackenzie Valley Pipeline Inquiry of 1974–77 highlighted the impacts of development on the sensitive Arctic environment as well as Indigenous culture and way of life (Berger 1977, xi–xix), and recommended that any pipeline construction be postponed until Native claims were settled. The pipeline, which would have stimulated oil and gas exploration in the Beaufort Sea, was largely opposed by northern Indigenous groups and did not proceed. Development issues resurfaced with completion of land claims agreements in the late 1980s and 1990s. In a 2016 policy decision, the United States and Canada acknowledged the vulnerability of Arctic ecosystems and designated large portions of Arctic waters off limits to future licensing.[47] Strategic environmental assessments of oil and gas development in the Arctic began in Labrador in 2008[48] and other areas in 2016–17 (CIRNA 2021). The latter assessments are supporting negotiations associated with land claims agreements that will contribute to a Canadian review of Arctic development in 2021.

On the Pacific coast, Indigenous groups supported a federal moratorium on offshore oil and gas drilling in 1972. The Province of British Columbia tried unsuccess-

fully to revive discussions following a 2002 review.[49] In the mid-2000s, the focus shifted to pipelines from the Alberta oil sands and northwestern BC gas fields for export of crude oil and gas from BC ports. Indigenous groups played a prominent role in hearings to review marine impacts of the Enbridge Northern Gateway Pipelines and Trans Mountain Pipeline projects through processes established by the National Energy Board. The 2016 federal approval of the Enbridge Northern Gateway pipeline from Alberta to Kitimat was overturned by the courts due to inadequate consultation about Aboriginal rights with two of the coastal First Nations appellants. The Liberal government elected in 2015 subsequently rejected the project and placed a moratorium on oil tanker traffic to northern BC ports.[50] Expansion of the Trans Mountain oil pipeline from Edmonton to Vancouver was approved by the Liberal government in 2017. The Province of British Columbia and several First Nations challenged the decision, and the court initially overturned the approval due to lack of consultation. The federal government purchased the pipeline in 2018, claiming the project was in the national interest, and construction is proceeding after having weathered several court challenges.[51]

Pipelines and oil and gas development have been controversial and little progress has been made towards reconciliation of Indigenous title and rights where projects have proceeded. New environmental assessment legislation was approved in 2019 that outlines requirements for early engagement with Indigenous Peoples on major projects and supports agreements or partnerships that could avoid future court challenges in this industry (Government of Canada 2019; McCarthy Tetrault 2018, 2019).

Case 5 – Shipping, the Reconciliation Framework Agreement for the Northern Shelf Bioregion, and Benefit Agreements

Ports and shipping on Canada's Pacific coast have been expanding since the mid-2000s, although several controversial development projects have been cancelled or delayed due to impacts on Aboriginal rights (Case 4). In 2017, the federal government announced a major new national Oceans Protection Plan (OPP) that would invest in improved infrastructure and cooperative planning, including with Indigenous groups.[52] This was followed in 2018 by a Reconciliation Framework Agreement (RFA) to advance collaborative governance and management on ocean topics, including shipping, negotiated between 14 First Nations in northern British Columbia and five federal agencies (including Transport Canada). The RFA builds on Canada's Ten Principles (Department of Justice Canada 2018), and offers an alternative approach to reconciliation of shipping issues through direct negotiation and enhanced consultation. However, engagement has been slow, with uncertainty about Indigenous roles in national issues, and few reconciliation criteria or results have been met or achieved as yet.

The RFA creates a new governance structure for engagement on shipping that addresses waterway management, marine response, and environmental issues. It builds on a multi-level consensus-based governance structure developed for marine planning by Indigenous Nations in Northern British Columbia (Case 1).

RFA initiatives include development of collaborative regional and local oil spill response plans and pilots for enhanced maritime awareness systems in Indigenous communities and vessel management. As well, two ocean rescue tugs were deployed on Canada's West Coast in 2018 on a three-year contract to address gaps in marine safety. The RFA facilitates dialogue and supports local Indigenous capacity.

The RFA has been implemented for several years now and the reconciliation process continues to be refined. The RFA approach provides a potential model for collaboration with Indigenous Nations (rather than litigation) that is applicable to other regions, including the Arctic and Atlantic.

Separately from the RFA, one liquefied natural gas project has navigated many of the hurdles for approval of both a pipeline and shipping terminal in northern British Columbia. Over the past decade, up to 16 projects have been proposed for the area, with many being cancelled due to market conditions (Northwest Institute 2019). The LNG Canada project for a pipeline from northeastern British Columbia to Kitimat received federal and provincial approval in 2015. The company negotiated approximately 25 benefit agreements with individual First Nations for the pipeline, and announced its investment decision in September 2018 (Government of British

Columbia n.d.b). One of the benefit agreements with coastal nations related to the port facilities and the marine passageway.[53] Examples of partnership agreements include a towing contract with the Haisla and equity agreements with various nations.[54] Another agreement between the Province of British Columbia and some of the nations along the tanker route provides annual fees to the nations, including base fees and additional fees based on investment decisions and LNG production (Government of British Columbia n.d.a). Pipeline construction began in 2018 and is due to be completed in the mid-2020s. Wet'suwet'en hereditary chiefs continue to oppose the construction and marine traffic, and their blockade of the Coastal GasLink Pipeline led to an agreement with Canada and the province in 2021 that it was hoped would provide a process for resolving future disputes, but in early 2022, despite equity-sharing agreements with some First Nations, Wet'suwet'en hereditary chiefs continued to oppose the project.[55] The agreement with Canada and the province included political recognition of the Wet'suwet'en hereditary governance system that was the basis for the *Delgamuukw* decision on Indigenous title in 1997.[56]

SUMMARY

Reconciliation in Canada is an ongoing process to confront and address injustices from colonization, including political domination, loss of territory, and cultural imposition. We have provided a brief history of reconciliation in Canada and examined reconciliation of ocean issues based on 23 criteria for reconciliation in ocean management identified from UNDRIP. Progress was assessed based on a scan for the presence of these criteria across seven ocean issues and five case studies. Based on our scan, we note that Canada has made mixed progress toward reconciliation of ocean issues. The case studies highlight best practices in reconciliation as well as some of the challenges from addressing reconciliation in a symmetrical framework. With the exception of fisheries in the Atlantic and Pacific, and marine mammals and marine conservation strategies in the Arctic, treaties and land claims settlements have focused on land and terrestrial resources rather than ocean spaces and marine resources, although this recently shows signs of changing. The situation is complicated by the mix of federal, provincial, and territorial jurisdictions, diversity of Indigenous populations, and political and structural resistance to power sharing. In Chapter 14, we analyze how reconciliation can address the core impacts of colonization and present policy recommendations for how a lasting reconciliation can be realized.

ACKNOWLEDGMENTS

Our appreciation goes to Louise Mandell, QC, who reviewed an earlier version of this chapter, and to Steve Diggon and Miles Richardson for their helpful review and comments.

NOTES

1 Made up of 630 First Nation and 53 Inuit communities. As Métis do not have reserve lands, their communities are not clearly defined (CIRNA 2022).

2 According to the 2016 Census, "Arctic" includes Nunavut (30,550 identify as Aboriginal out of 35,580 reporting), Northwest Territories (20,860 out of 41,135 reporting), and Skeena Queen Charlotte Regional District (8,035 out of 17,895 reporting) (Statistics Canada 2019).

3 Haida herring use and marine use in Gwaii Haanas are further described in a Haida Marine Traditional Knowledge Study (Winbourne et al. 2011). Gwaii Haanas was designated by the Haida as a Haida Heritage Site (1985) and by Canada as a National Park Reserve (1988) and National Marine Conservation Area Reserve (2013). Agreements include the *Gwaii Haanas Agreement* (1993) and *Gwaii Haanas Marine Agreement* (2009). Joint management plans include a terrestrial plan (2003), an interim marine plan (2010), and a Gwaii Haanas Land-Sea-People plan (Government of Canada 2021). Jones, Rigg, and Pinkerton (2017) describe the herring conflict and dispute resolution processes. See *Council of the Haida Nation and Others v DFO*, Federal Court, March 5, 2015, Docket T-73-15, 2015, para 10 for details about the injunction. More recently, Haida Gwaii herring was listed as a stock requiring a rebuilding plan under revisions to the *Fisheries Act* in 2019 that came into force in April 2022.

4 Rouhana (2011) differentiates between reconciliation and conflict settlement, which does not typically seek deeper transformation of relations between societies, and conflict resolution, which seeks coexistence and cooperation but avoids the core issues due to negotiation in a symmetrical framework (see Chapter 14, n. 3).

5 Quebec and the Atlantic provinces of New Brunswick and Nova Scotia were part of the original Confederation of Canada in 1867. The colony of British Columbia joined in 1871. The colony of Prince Edward Island became part of Canada in 1873. Rupert's Land and the North-West Territories (now comprising most of the Northwest Territories and Nunavut) were sold to Canada by the Hudson's Bay Company in 1870. The Arctic Islands were transferred from Britain to Canada in 1880. And Newfoundland and Labrador finally joined Canada in 1949.

6 Havemann (1999, 22–2 3) refers to classifications by J.R. Miller (1990) and Andrew Armitage (1995), respectively.

7 See CIRNA 2020.

8 "Maliseet" had been a common term to name the nation and will be used in this chapter. However, recent cultural and political leaders are reverting back to "Wolastoqey Nation," which is the traditional name of the nation and is translated as "the beautiful river."

9 *R v Marshall*, [1999] 3 SCR 456.

10 See CIRNA 2015.

11 The Pacific Northwest Coast (from the Gulf of Alaska to Oregon) is the second most diverse linguistic area of Indigenous North America after California (Thompson and Kincade 1990). Coastal languages or linguistic families include Haida, Tsimshian, Heiltsuk, Kwakwaka'wakw, Nuu-chah-nulth, and Salish.

12 Crown Colony of Vancouver Island, British Columbia, Queen Charlotte Islands, and Stikine.

13 According to Duff (1964, 39), the population fell from about 70,000 in 1835 to a low point of about 22,600 in 1929.

14 *Calder v British Columbia (AG)*, [1973] SCR 313, [1973] 4 WWR 1.

15 *Delgamuukw v British Columbia*, 1997 SCC 3.

16 According to Louise Mandell (2018, 58), "the court [has] held that Indigenous laws and rights to land are inherent collective rights that pre-existed and survived the assertion of Crown sovereignty, that have never been extinguished and that find expression today in the Constitution."

17 *Tsilhqot'in Nation v British Columbia*, 2014 SCC 44.

18 The "Arctic Timeline" section is drawn in part from Neatby 1984.

19 It is important to note that while Northern Quebec and Labrador appear as "Sub-Arctic" under administrative boundaries and geophysical regions, the Inuit of Northern Quebec and Labrador are polar maritime peoples, and connect with Inuit across the Arctic Region, sharing genealogy, culture, and political aspirations for self-determination.

20 See "Case 4: Conflicts Over Pipelines and Offshore Oil and Gas Development" in Chapter 3. The Inuvialuit, Loucheaux, and many of the long-term white settlers of the Mackenzie Delta formed the Committee for Original Peoples' Entitlement (COPE) to seek land claims and self-government from the Canadian government in response to increased oil and gas exploration and multiple pipeline proposals in the Beaufort Region. Mary Carpenter, a residential school survivor, launched a petition for COPE/Inuvialuit rights in 1969. See Carpenter 2017.

21 Some ocean issues, such as Indigenous ownership of the seabed and offshore oil and gas, have been off the table in negotiations. However, there are signals that this may change. In 2009, Canada's courts ruled that Nuu-chah-nulth territory for the purpose of the commercial fishery extended nine miles offshore, and several ongoing court cases in British Columbia may test Aboriginal title to the seabed and ocean resources (e.g., Haida title case).

22 This followed 15 years of lobbying against mining at Voisey's Bay and objections to a low-level jet training facility at Goose Bay.

23 The RCAP makes more than 400 recommendations; the TRC identifies 94 calls to action.

24 The Province of British Columbia and Canada passed separate legislation in 2019 and 2021, respectively. See *Declaration on the Rights of Indigenous Peoples Act,* SBC 2019, c 44; and *United Nations Declaration on the Rights of Indigenous Peoples Act,* SC 2021, c 14. The purpose of the BC legislation is to align provincial laws and policies with the aims of UNDRIP. Measures include development of an action plan in consultation and cooperation with Indigenous peoples and annual reporting on progress. A five-year action plan completed in March 2022 identifies initial actions to be taken by British Columbia in consultation and collaboration with Indigenous peoples (BC Ministry of Indigenous Relations and Reconciliation n.d.). The purpose of Canada's legislation is to affirm and provide a framework for implementation of the declaration. It requires, in consultation and cooperation with Indigenous peoples, taking all measures necessary to ensure that the laws of Canada are consistent with the declaration and implementing an action plan to achieve the objectives of the declaration.

25 According to Kymlicka (2007, 595), UNDRIP captured a broad international consensus that targeted norms are needed for Indigenous peoples and that these norms should address claims related to history and territory; Indigenous peoples are increasingly acknowledged to have a wide range of rights relating to their areas of traditional settlement, including rights to political self-government, control over natural resources, language rights, and autonomous legal, cultural, and educational institutions. UNDRIP was not seen as posing the same threat to the state as recognition of national minorities.

26 The *Oceans Act* (SC 1996, c 31) commits to developing integrated management plans in collaboration with "affected aboriginal organizations."

27 A total of 17 First Nations participated in PNCIMA or Marine Plan Partnership (MaPP) planning initiatives. Some nations near Prince Rupert and Northern Vancouver Island did not participate or withdrew before a plan was endorsed. A small portion of the planning area was subject to a treaty (*Nisga'a Final Agreement*) and was addressed through a separate process.

28 The initial memorandum of understanding was signed in 2008. The province signed on in 2010. The Nanwakolas Council signed in 2011 but did not sign off on the final PNCIMA plan.

29 Implementation agreements were signed in 2016.

30 Priorities include: updating the collaborative governance structure, MPA network planning, monitoring/adaptive management, cumulative effects assessment, and work planning for PNCIMA implementation.

31 MaPP plans are being brought forward to Canada by First Nations and the Province of British Columbia through both MPA and PNCIMA discussions. In 2021, the province began participating in the Reconciliation Framework Agreement (RFA) process (Case 5) as a governance partner (personal communication by Steve Diggon, Coastal First Nations, April 9, 2022).

32 https://mpanetwork.ca/nap/.

33 This includes MaPP and RFA governance structures. MPA network planning and future PNCIMA implementation will be under the RFA governance structure.

34 *R v Marshall,* [1999] 3 SCR 456.

35 For a list of federal programs, see Table 2.3, footnote 1.

36 Negotiations of Aboriginal and treaty rights and self-government between Canada and Mi'kmaq and Maliseet groups in Nova Scotia, New Brunswick, Prince Edward Island, and the Gaspé region of Quebec have been in progress over the past 20 years (since the *Marshall* decision), and several milestone agreements have been completed (CIRNA 2019).

37 LMG is one of the Peace and Friendship Treaties nations and a participant in the DFO communal licence fishery. The *Lestiguj Lobster Law* asserted Lestiguj authority over the fishery and did not threaten conservation. See https://

listuguj.ca/wp-content/uploads/2019/04/Migmaq-Fishing-Law.pdf.

38 Negotiations over more than 20 years have failed to make progress at defining a moderate livelihood fishery based on the *Marshall* decision. See, e.g., Cooke 2020; Googoo 2020.

39 DFO's plan resulted in some agreements but conflicts continue. Minister Jordan's "New Path" statement requires First Nations to participate in the existing commercial fishery, which only partially accommodates their access rights. This approach has been unacceptable to some First Nations and has been met with criticism by First Nations that DFO has not properly consulted and lacks a mandate to negotiate the implementation of Atlantic First Nations treaty rights for fishing. See, e.g., DFO 2021a, 2021b; Plowman 2021.

40 Nunavut was created on April 1, 1999, through the *Nunavut Act* and the *Nunavut Land Claims Agreement*. It is a vast territory of over 1.9 million square kilometres of land and 0.15 million square kilometres of water, and accounts for over a fifth of Canada's land mass. The coastline is extensive, with more than 100 major islands.

41 The Nunavut Wildlife Management Board is responsible for conservation and management of wildlife, including bowhead whale, seal, bears, walrus, beluga, and narwhal in the Nunavut Settlement Area, and has an advisory role in marine management of ocean areas. Three Regional Wildlife Boards oversee harvesting at the regional level. See *Nunavut Land Claims Agreement* and the mandate of the Nunavut Wildlife Management Board (n.d.).

42 *Clyde River (Hamlet) v Petroleum Geo-Services Inc*, 2017 SCC 40, [2017] 1 SCR 1069.

43 For more information see Pikialasorsuaq Commission n.d.

44 "Each community emphasized that Inuit who live in the region are best placed to monitor and manage the region. These communities want to set and lead the research agenda, study the indicators of change, and establish realistic hunting regulations that will continue to sustain their communities" (Pikialasorsuaq Commission 2017, x).

45 The Pikialasorsuaq Commission brings together local, regional, national, and international organizations, all levels of government bodies and local authorities, scientific institutions and universities, and international bodies such as the Arctic Council and the International Maritime Organization (IMO).

46 As well, DFO supports Indigenous leadership and collaboration on marine conservation initiatives and views the Pikialasorsuaq as a new integrated ocean management initiative (DFO 2022b).

47 In Canada, the decision would be reviewed every five years (Prime Minister of Canada 2016).

48 Conducted by the Canada–Newfoundland and Labrador Offshore Oil Petroleum Board (n.d.).

49 Numerous offshore exploration permits were issued prior to 1972 in Hecate Strait, Queen Charlotte Sound, and Dixon Entrance.

50 The federal *Oil Tanker Moratorium Act* (2019) bans shipping of oil products on large tankers through northern BC ports.

51 Forty-three First Nations have made agreements with Trans Mountain Pipeline (Trans Mountain 2018). In July 2020, the Supreme Court of Canada declined to hear an appeal of unsuccessful challenges of federal approvals by three First Nations (*CBC News* 2020).

52 OPP is a five-year, $1.5 billion program (Transport Canada 2022). The 2022 federal budget committed a further $2.0 billion over nine years to renew and expand the program.

53 See Government of British Columbia n.d.b. Two of the coastal nations that signed agreements, Haisla and Gitxaala, had previous court declarations disallowing federal approval of the Enbridge Northern Gateway Pipelines.

54 See Haisla Nation 2019. First Nations are reported to be seeking at least 22.5% ownership in the pipeline project, according to Greaves and Lackenbauer (2019, 5).

55 Hereditary leaders of the Wet'suwet'en oppose the pipeline although elected leaders signed an agreement (Greaves and Lackenbauer 2019), and blockaded construction despite court injunctions. The dispute sparked widespread protest and railway blockades throughout Canada in February 2020. In 2021, Wet'suwet'en chiefs, Canada, and the Province of British Columbia signed an agreement that it was hoped would provide a process for resolving future conflicts. See Meissner 2020; Simmons 2022.

56 See note 15 above.

REFERENCES

Arctic Council. 1996. "Declaration on the Establishment of the Arctic Council: Joint Communiqué of the Governments of the Arctic Countries on the Establishment of the Arctic Council." https://oaarchive.arctic-council.org/bitstream/handle/11374/85/EDOCS-1752-v2-ACMMCA00_Ottawa_1996_Founding_Declaration.PDF?sequence=5&isAllowed=y.

Armitage, A. 1995. *Comparing the Policy of Aboriginal Assimilation: Australia, Canada and New Zealand*. Vancouver: UBC Press.

Assembly of First Nations. 2018. *Affirming First Nations Rights, Title and Jurisdiction*. Ottawa: Assembly of First Nations. https://www.afn.ca/wp-content/uploads/2018/09/Affirming-FN-Rights-Title-and-Jurisdiction_EN.pdf.

BC Ministry of Agriculture. 2019. "Salmon Aquaculture in British Columbia." Press release, February 28. https://news.gov.bc.ca/factsheets/salmon-aquaculture-in-british-columbia.

BC Ministry of Indigenous Relations and Reconciliation. n.d. *Declaration on the Rights of Indigenous Peoples Act Action Plan: 2022–2027*. Victoria: BC Ministry of Indigenous Relations and Reconciliation. https://www2.gov.bc.ca/assets/gov/government/ministries-organizations/ministries/indigenous-relations-reconciliation/declaration_act_action_plan.pdf.

Bell, Jim. 2020. "Despite the Naysayers, Southern Labrador Inuit Rise Up to Claim Their Rights." *Nunatsiaq News*, March 5. https://nunatsiaq.com/stories/article/despite-the-naysayers-southern-labrador-inuit-rise-up-to-claim-their-rights/.

Berger, Thomas R. 1977. *Northern Frontier, Northern Homeland: The Report of the Mackenzie Valley Pipeline Inquiry*, vol. 1. Ottawa: Supply and Services Canada.

Canada–Newfoundland and Labrador Offshore Oil Petroleum Board. n.d. "Strategic Environmental Assessment (SEA)." https://www.cnlopb.ca/sea/.

Canadian Press. 2018. "Offshore Arctic Drilling Talks May Signal End of Ban: NWT Premier." *JWN*, October 5. https://www.jwnenergy.com/article/2018/10/5/offshore-arctic-drilling-talks-may-signal-end-ban-/.

Carpenter, Mary. 2017. "Lost Generations." *Canada's History*, April 1. https://www.pressreader.com/canada/canada-s-history/20170401/281500751056618.

CBC News. 2020. "Supreme Court Dismisses First Nations' Challenge against Trans Mountain Pipeline." *CBC News*, July 2. https://www.cbc.ca/news/canada/british-columbia/trans-mountain-pipeline-challenge-bc-first-nations-supreme-court-of-canada-1.5634232.

Centre for First Nations Governance. 2011. "Making First Nation Law: The Listuguj Mi'gmaq Fishery." https://fngovernance.org/making-first-nation-law-the-listuguj-migmaq-fishery-d26/.

CIRNA (Crown-Indigenous Relations and Northern Affairs Canada). 2015. "Peace and Friendship Treaties." https://www.rcaanc-cirnac.gc.ca/eng/1100100028589/1539608999656#wb-cont.

–. 2019. "Negotiations in Atlantic Canada." https://www.rcaanc-cirnac.gc.ca/eng/1100100028583/1529409875394.

–. 2020. "Recognition of Rights Discussion Tables." https://www.rcaanc-cirnac.gc.ca/eng/1511969222951/1529103469169.

–. 2021. *Northern Oil and Gas Annual Report 2020*. Gatineau, QC: CIRNA. https://www.rcaanc-cirnac.gc.ca/eng/1620937719262/1620937786832.

–. 2022. "Indigenous Peoples and Communities." https://www.rcaanc-cirnac.gc.ca/eng/1100100013785/1529102490303.

Coates, K. 2019. *The Marshall Decision at 20: Two Decades of Commercial Re-empowerment of the Mi'kmaq and Maliseet*. Ottawa: Macdonald-Laurier Institute. https://macdonaldlaurier.ca/files/pdf/20191015_Marshall_Decision_20th_Coates_PAPER_FWeb.pdf.

Cooke, A. 2020. "Mi'kmaw Fishermen Launch Self-Regulated Fishery in Saulnierville." *CBC News*, September 17. https://www.cbc.ca/news/canada/nova-scotia/mikmaw-fishermen-self-regulated-fishery-lower-saulnierville-1.5727920.

Cooke, M., R. Mitrou, D. Lawrence, E. Guimond, and D. Beavon. 2007. "Indigenous Well-Being in Four Countries: An Application of the UNDP's Human Development Index to Indigenous Peoples in Australia, Canada, New Zealand, and the United States." *BMC International Health and Human Rights* 7: 9. https://bmcinthealthhumrights.biomedcentral.com/articles/10.1186/1472-698X-7-9.

COSEWIC (Committee on the Status of Endangered Wildlife in Canada). 2018. "COSEWIC Assessment and Status Report on the Polar Bear (*Ursus maritimus*) in Canada 2018." https://www.canada.ca/en/environment-climate-change/services/species-risk-public-registry/cosewic-assessments-status-reports/polar-bear-2018.html.

Coté, C. 2010. *Spirits of Our Whaling Ancestors: Revitalizing Makah and Nuu-chah-nulth Traditions*. Vancouver: UBC Press.

Denny, S., A. Denny, E. Garden, and T. Paul. 2016. *Mn'tmu'k. Mi'kmaq Ecological Knowledge: Eastern Oysters in Unama'ki*. Cape Breton Island: Unama'ki Institute of Natural Resources. http://www.uinr.ca/wp-content/uploads/2016/11/Oyster-MEK-WEB-Spreads.pdf.

Department of Justice Canada. 2018. *Principles Respecting the Government of Canada's Relationship with Indigenous Peoples*. Ottawa: Department of Justice Canada. https://www.justice.gc.ca/eng/csj-sjc/principles.pdf.

DFO (Fisheries and Oceans Canada). 2007. *Recovery Strategy for Northern Abalone (*Haliotis kamtschatkana*) in Canada*. Vancouver: Fisheries and Oceans Canada. https://www.

sararegistry.gc.ca/virtual_sara/files/plans/rs_Northern_Abalone_0907_e.pdf.

–. 2016. *Evaluation of the Pacific Integrated Commercial Fisheries Initiative (PICFI).* Evaluation Directorate, Fisheries and Ocean Canada. Project Number 6B172. March.

–. 2019a. "Achievements." https://www.dfo-mpo.gc.ca/oceans/conservation/achievement-realisations/index-eng.html.

–. 2019b. "Closed Containment." https://www.dfo-mpo.gc.ca/aquaculture/programs-programmes/containment-eng.htm.

–. 2020. "Canada Joins Global Ocean Alliance with 21 Other Countries." Press release, July 9. https://www.canada.ca/en/fisheries-oceans/news/2020/07/canada-joins-global-ocean-alliance-advocates-for-protecting-30-per-cent-of-the-worlds-ocean-by-2030.html.

–. 2021a. "Implementing the Right to Fish in Pursuit of a Moderate Livelihood: Rebuilding Trust and Establishing a Constructive Path Forward." https://www.dfo-mpo.gc.ca/fisheries-peches/aboriginal-autochtones/moderate-livelihood-subsistence-convenable/surette-report-rapport-mar-2021-eng.html.

–. 2021b. "Minister Jordan Issues Statement on a New Path for First Nations to Fish in Pursuit of a Moderate Livelihood." Statement, March 3. https://www.canada.ca/en/fisheries-oceans/news/2021/03/minister-jordan-issues-statement-on-a-new-path-for-first-nations-to-fish-in-pursuit-of-a-moderate-livelihood.html.

–. 2022a. *Analysis of Commercial Fishing Licence, Quota and Vessel Values, as at December 31, 2021.* https://waves-vagues.dfo-mpo.gc.ca/library-bibliotheque/41059402.pdf.

–. 2022b. "Pikialasorsuaq (North Water Polynya)." https://www.dfo-mpo.gc.ca/oceans/management-gestion/pikialasorsuaq-eng.html.

–. 2023a. "Government of Canada and Coastal First Nations Announce Progress to Protect a Large Ecologically Unique Ocean Area Off the Pacific West Coast." https://www.dfo-mpo.gc.ca/oceans/aoi-si/tht-eng.html.

–. 2023b. "Reaching Canada's Marine Conservation Targets." https://www.dfo-mpo.gc.ca/oceans/conservation/plan/index-eng.html.

Duff, W. 1964. *The Indian History of British Columbia: The Impact of the White Man.* Victoria: Royal British Columbia Museum.

Ethier, Valerie. 2014. *Farmed Arctic Char, Salvelinus alpinus.* Monterey Bay Aquarium Seafood Watch. https://seafood.ocean.org/wp-content/uploads/2016/10/Char-Arctic-Farmed-Canada-Iceland-US.pdf.

FAO (Food and Agriculture Organization). 2018. *The State of World Fisheries and Aquaculture.* Rome: FAO. http://www.fao.org/state-of-fisheries-aquaculture.

First Nation Panel on Fisheries. 2004. *Our Place at the Table: First Nations in the B.C. Fishery.* http://fns.bc.ca/wp-content/uploads/2016/10/FNFishPanelReport0604.pdf.

GCRC (Geomatics and Cartographic Research Centre, Carleton University). 2020. "Residential Schools Land Memory Atlas." https://residentialschoolsatlas.org.

George, Jane. 2019. "Canada Submits Its Arctic Ocean Claim to the United Nations." *Nunatsiaq News,* May 24. https://nunatsiaq.com/stories/article/canada-submits-its-arctic-ocean-claim-to-the-united-nations/.

Gibbens, S. 2021. "The Arctic National Wildlife Refuge Just Got a Reprieve – but It's Not Safe Yet." *National Geographic,* June 3. https://www.nationalgeographic.com/environment/article/arctic-national-wildlife-refuge-oil-drilling-what-next.

Googoo, M. 2020. "Nova Scotia Chiefs Rejected $87-Million Offer from DFO, Want Moderate Livelihood Defined." *Ku'Ku'Kwes News,* October 5. http://kukukwes.com/2020/10/05/nova-scotia-chiefs-rejected-87-million-offer-from-dfo-want-moderate-livelihood-defined/.

Government of British Columbia. n.d.a. "Coastal First Nations." https://www2.gov.bc.ca/gov/content/environment/natural-resource-stewardship/consulting-with-first-nations/first-nations-negotiations/first-nations-a-z-listing/coastal-first-nations.

–. n.d.b. "Natural Gas Benefits Agreements." https://www2.gov.bc.ca/gov/content/environment/natural-resource-stewardship/consulting-with-first-nations/first-nations-negotiations/natural-gas-pipeline-benefits-agreements.

Government of Canada. 2019. "A Proposed New Impact Assessment System." https://www.canada.ca/en/services/environment/conservation/assessments/environmental-reviews/environmental-assessment-processes.html.

–. 2021. "Gwaii Haanas Gina 'Waadluxan KilGuhlGa Land-Sea-People Management Plan." https://parks.canada.ca/pn-np/bc/gwaiihaanas/info/consultations/gestion-management-2018.

Greaves, W., and W. Lackenbauer. 2019. *First Nations, LNG Canada, and the Politics of Anti-Pipeline Protests.* Calgary: Canadian Global Affairs Institute. https://d3n8a8pro7vhmx.cloudfront.net/cdfai/pages/4220/attachments/original/1561589071/First_Nations__LNG_Canada__and_the_Politics_of_Anti-Pipeline_Protests.pdf?1561589071.

Haisla Nation. 2019. "Haisla Nation and Seaspan Awarded LNG Canada Escort and Harbor Tugs Contract." Press

release, August 27. https://haisla.ca/haisla-nation-and
-seaspan-awarded-lng-canada-escort-and-harbor-tugs
-contract/.

Harris, D.C. 2001. *Landing Native Fisheries: Indian Reserves and Fishing Rights in British Columbia.* Vancouver: UBC Press.

–. 2008. *Fish, Law and Colonialism: The Legal Capture of Salmon in British Columbia.* Toronto: University of Toronto Press.

Havemann, P. 1999. "Indigenous Rights in the Political Jurisprudence of Australia, Canada and New Zealand: Parallel Chronologies." In *Indigenous Peoples' Rights in Australia, Canada, and New Zealand,* edited by P. Havemann, 22–64. Auckland, New Zealand: Oxford University Press.

Hill, C.J., R. Schuster, and J.R. Bennett. 2019. "Indigenous Involvement in the Canadian Species at Risk Recovery Process." *Environmental Science and Policy* 94: 220–26.

ICE (Indigenous Circle of Elders). 2018. *We Rise Together.* https://www.conservation2020canada.ca/s/PA234-ICE_Report_2018_Mar_22_web.pdf.

Jones, R. 2006. "Working Models for Collaborative Management." Unpublished report prepared for First Nation Marine Society. https://www.ceaa.gc.ca/050/documents_staticpost/cearref_21799/83896/Working_Models_for_Fisheries.pdf.

Jones, R., C. Rigg, and L. Lee. 2010. "Haida Marine Planning: First Nations as a Partner in Marine Conservation." *Ecology and Society* 15 (1): 12. http://www.ecologyandsociety.org/vol15/iss1/art12/.

Jones, R, C. Rigg, and E. Pinkerton. 2017. "Strategies for Assertion of Conservation and Local Management Rights: A Haida Gwaii Herring Story." *Marine Policy* 80: 154–67. https://doi.org/10.1016/j.marpol.2016.09.031.

Kinnear, L. 2007. "Contemporary Mi'kmaq Relationships between Humans and Animals: A Case Study of the Bear River First Nation Reserve in Nova Scotia." Master of Environmental Studies thesis, Dalhousie University. https://www.collectionscanada.gc.ca/obj/thesescanada/vol2/002/MR39166.PDF.

Knockwood, C. 2003. "The Mi'kmaq-Canadian Treaty Relationship: A 277-Year Journey of Rediscovery." In *Box of Treasures or Empty Box? Twenty Years of Section 35,* edited by A. Walkem and H. Bruce, 43–61. Penticton, BC: Theytus Books.

Kymlicka, W. 2007. "Multicultural Odysseys." *Ethnopolitics* 6 (4): 585–97. https://doi.org/10.1080/17449050701659789.

Maher, S. 2020. "Mi'kmaq Fishers Won at the Supreme Court. But They're Still Fighting for Their Livelihoods."

Maclean's, October 2. https://macleans.ca/news/canada/mikmaq-fishers-won-at-the-supreme-court-but-theyre-still-fighting-for-their-livelihoods/.

Mandell, L. 2018. "Our Interconnected Journey." *BC Studies* 200: 53–75.

McCarthy Tetrault. 2018. "Moving Towards Sustainability and Public Interest: Federal Government Introduces the Impact Assessment Act." https://www.mccarthy.ca/en/insights/blogs/canadian-era-perspectives/moving-towards-sustainability-and-public-interest-federal-government-introduces-impact-assessment-act.

–. 2019. "Impact Assessment Act." https://www.mccarthy.ca/en/insights/blogs/canadian-era-perspectives/impact-assessment-act.

McMillan, L.J., and K. Prosper. 2016. "Remobilizing *Netukulimk:* Indigenous Cultural and Spiritual Connections with Resource Stewardship and Fisheries Management in Atlantic Canada." *Reviews in Fish Biology and Fisheries* 26: 629–47. https://doi.org/10.1007/s11160-016-9433-2.

McRae, D.M., and P.H. Pearse. 2004. *Treaties and Transition: Towards a Sustainable Fishery on Canada's Pacific Coast.* Vancouver: Fisheries and Oceans Canada. https://waves-vagues.dfo-mpo.gc.ca/Library/280188.pdf.

Meissner, D. 2020. "Virtual Signing Ceremony Marks B.C.'s Wet'suwet'en Rights and Title Agreement." *Globe and Mail,* May 14. https://www.theglobeandmail.com/canada/article-virtual-signing-ceremony-marks-bcs-wetsuweten-rights-and-title/.

Miller, J.R. 1990. *Skyscrapers Hide the Heavens: A History of Indian-White Relations in Canada.* Toronto: University of Toronto Press.

Moore, M. 2016. "Justice and Colonialism." *Philosophy Compass* 11 (8): 446–61. https://onlinelibrary.wiley.com/doi/10.1111/phc3.12337.

MPPI (Marine Plan Partnership Initiative). 2015. *Haida Gwaii Marine Plan 2015.* http://www.haidanation.ca/wp-content/uploads/2017/05/MarinePlan_HaidaGwaii_WebVer_21042015-opt-1.pdf.

Neatby, N.L. 1984. "Exploration and History of the Canadian Arctic." In *Handbook of North American Indians,* vol. 5, *Arctic,* edited by David Damas. Washington, DC: Smithsonian Institution.

Newell, D. 1993. *Tangled Webs of History: Indians and the Law in Canada's Pacific Coast Fisheries.* Toronto: University of Toronto Press.

Northwest Institute. 2019. "Proposed Liquefied Natural Gas (LNG) Projects in Northern B.C." http://lngin northernbc.ca/images/uploads/documents/LNG_Tables_Sept10_2019.pdf.

Nunavut Wildlife Management Board. n.d. "Home." https://www.nwmb.com/en/.

Parks Canada. 2022a. "The Governments of Canada and Nunatsiavut Sign Memorandum of Understanding to Assess Feasibility of Establishing a New Protected Area along Northern Labrador Coast." Press release, February 23. https://www.canada.ca/en/parks-canada/news/2022/02/the-governments-of-canada-and-nunatsiavut-sign-memorandum-of-understanding-to-assess-feasibility-of-establishing-a-new-protected-area-along-norther.html.

–. 2022b. "Tallurutiup Imanga National Marine Conservation Area Inuit Impact and Benefit Agreement." https://parks.canada.ca/amnc-nmca/cnamnc-cnnmca/tallurutiup-imanga/entente-agreement.

–. 2022c. "Where Is Tallurutiup Imanga?" https://parks.canada.ca/amnc-nmca/cnamnc-cnnmca/tallurutiup-imanga/emplacement-location.

Pearse, P. 1982. *Turning the Tide: A New Policy for Canada's Pacific Fisheries. The Commission on Pacific Fisheries Policy Final Report.* Vancouver: Minister of Supply and Services Canada.

Pikialasorsuaq Commission. n.d. "Welcome to the Pikialasorsuaq Commission." http://www.pikialasorsuaq.org/en/.

–. 2017. *People of the Ice Bridge: The Future of the Pikialasorsuaq.* Ottawa: Inuit Circumpolar Council Canada. http://www.pikialasorsuaq.org/en/Resources/Reports/.

Pinkerton, E. 2007. "Integrating Holism and Segmentalism: Overcoming Barriers to Adaptive Co-Management between Agencies and Multi-Sector Bodies." In *Adaptive Co-Management, Learning and Multi-Level Governance,* edited by D. Armitage, F. Berkes, and N. Doubleday, 151–69. Vancouver: UBC Press.

Plowman, S. 2021. "'Stop Criminalizing Our Treaty Rights': National Chief of Assembly of First Nations Tells DFO on N.S. Wharf." *CTV News,* September 9. https://atlantic.ctvnews.ca/stop-criminalizing-our-treaty-rights-national-chief-of-assembly-of-first-nations-tells-dfo-on-n-s-wharf-1.5579464.

Prime Minister of Canada. 2016a. "Statement by the Prime Minister of Canada on Advancing Reconciliation with Indigenous Peoples." https://pm.gc.ca/en/news/statements/2016/12/15/statement-prime-minister-canada-advancing-reconciliation-indigenous.

–. 2016b. "United States-Canada Joint Arctic Leaders' Statement." https://pm.gc.ca/en/news/statements/2016/12/20/united-states-canada-joint-arctic-leaders-statement.

–. 2018. "Government of Canada to Create Recognition and Implementation of Rights Framework." https://pm.gc.ca/en/news/news-releases/2018/02/14/government-canada-create-recognition-and-implementation-rights.

–. 2019. "ARCHIVED – Minister of Fisheries, Oceans and the Canadian Coast Guard Mandate Letter." https://pm.gc.ca/en/mandate-letters/2019/12/13/archived-minister-fisheries-oceans-and-canadian-coast-guard-mandate.

Prosper, K., L.J. McMillan, A.A. Davis, and M. Moffitt. 2011. "Returning to *Netukulimk:* Mi'kmaq Cultural and Spiritual Connections with Resource Stewardship and Self-Governance." *International Indigenous Policy Journal* 2 (4). https://doi.org/10.18584/iipj.2011.2.4.7.

Pyne S.A., and D.R. Taylor, eds. 2019. *Cybercartography in a Reconciliation Community: Engaging Intersecting Perspectives.* Elsevier. https://doi.org/10.1016/B978-0-12-815343-7.21001-4.

Raddi, S., and N. Weeks. 1985. *The Prehistoric and Historic Utilization of Bowhead Whales in the Canadian Western Arctic: A Community-Based Study.* Toronto: World Wildlife Fund.

Rouhana, N.N. 2011. "Key Issues in Reconciliation Challenging Traditional Assumptions on Conflict Resolution and Power Dynamics." In *Intergroup Conflicts and Their Resolution: A Social Psychological Perspective,* edited by Daniel Bar-Tal, 291–314. New York: Psychology Press. https://doi.org/10.4324/9780203834091.

Royal Commission on Aboriginal Peoples. 1996. *Report of the Royal Commission on Aboriginal Peoples,* vols. 1–5. Ottawa: Supply and Services Canada.

Séguin, R. 2010. "Old Harry Oil and Gas Prospect Gets a New Lease on Life." *Globe and Mail,* October 21. https://www.theglobeandmail.com/news/politics/old-harry-oil-and-gas-prospect-gets-a-new-lease-on-life/article1215321/.

Shelley, K., S. Denny, and L. Fanning. 2016. "A Mi'kmaw Perspective on Advancing Salmon Governance in Nova Scotia, Canada: Setting the Stage for Collaborative Coexistence." *International Indigenous Policy Journal* 7 (3). https://ir.lib.uwo.ca/iipj/vol7/iss3/4.

Simmons, M. 2022. "Emails Reveal How the RCMP Changed Its Story about Arresting Journalists in Wet'suwet'en Raid." *CTV News,* May 9. https://bc.ctvnews.ca/emails-reveal-how-the-rcmp-changed-its-story-about-arresting-journalists-in-wet-suwet-en-raid-1.5894870.

Statistics Canada. 2019. "Aboriginal Population Profile, 2016 Census." https://www12.statcan.gc.ca/census-recensement/2016/dp-pd/abpopprof/index.cfm?Lang=E.

Sterritt, N.J. 2016. *Mapping My Way Home: A Gitxsan History.* Smithers, BC: Creekstone Press.

Thomson, Jimmy. 2015. "Clam Gardens Are Cultivating a New Look at Ancient Land Use." Hakai Institute, https://hakai.org/clam-gardens-are-cultivating-new-look-at-ancient-land-use/.

Thompson, L.C., and M.D. Kincade. 1990. "Languages." In *Handbook of North American Indians,* vol. 7, *Northwest Coast,* edited by W. Suttles. Washington, DC: Smithsonian Institution.

Trans Mountain. 2018. "43 Indigenous Groups Have Signed Agreements in Support of the Trans Mountain Expansion Project." https://www.transmountain.com/news/2018/43-aboriginal-groups-have-signed-agreements-in-support-of-the-trans-mountain-expansion-project.

Transport Canada. 2018. "Haida Nation and Transport Canada Identify New Potential Places of Refuge for Ships in Haida Gwaii." Press release, March 14. https://www.newswire.ca/news-releases/haida-nation-and-transport-canada-identify-new-potential-places-of-refuge-for-ships-in-haida-gwaii-676832573.html.

–. 2019a. "Government of Canada Partners with Indigenous Communities in Ocean Protection and Marine Safety Initiatives." Press release, July 24. https://www.canada.ca/en/transport-canada/news/2019/07/government-of-canada-partners-with-indigenous-communities-in-ocean-protection-and-marine-safety-initiatives.html.

–. 2019b. "The *Oil Tanker Moratorium Act* Receives Royal Assent." Press release, June 21. https://www.canada.ca/en/transport-canada/news/2019/06/the-oil-tanker-moratorium-act-receives-royal-assent.html.

–. 2022. "Oceans Protection Plan." https://tc.canada.ca/en/campaigns/oceans-protection-plan.

Truth and Reconciliation Commission of Canada. 2015. *Honouring the Truth, Reconciling for the Future: Summary of the Final Report of the Truth and Reconciliation Commission of Canada.* Toronto: James Lorimer and Company.

UNDRIP (*United Nations Declaration on the Rights of Indigenous Peoples*). 2007. https://www.un.org/development/desa/indigenouspeoples/declaration-on-the-rights-of-indigenous-peoples.html.

UNEP (United Nations Environment Programme). 2022. *Convention on Biological Diversity. Decision 15/4, Kunming-Montreal Global Biodiversity Framework.* https://www.cbd.int/doc/decisions/cop-15/cop-15-dec-04-en.pdf.

UNESCO (UNESCO-IOC/European Commission). 2021. *MSPglobal International Guide on Marine/Maritime Spatial Planning.* Paris: UNESCO. https://www.mspglobal2030.org/unesco-and-european-commission-launch-new-flagship-guide-on-msp/.

Walters, M. 2008. "The Jurisprudence of Reconciliation: Aboriginal Rights in Canada." In *The Politics of Reconciliation in Multicultural Societies,* edited by W. Kymlicka and B. Bashir, 165–91. Oxford and New York: Oxford University Press.

Winbourne, J., Haida Oceans Technical Team, and Haida Fisheries Program. 2011. "Herring." In *Haida Marine Traditional Knowledge Study,* vol. 3, *Focal Species Summary,* 41–75. https://haidamarineplanning.com/wp-content/uploads/2016/08/Haida_Marine_Traditional_Knowledge_vol3.pdf.

Withers, P. 2018. "Ottawa Considers Dozens More Marine Protected Areas in Maritimes." *CBC News,* May 10. https://www.cbc.ca/news/canada/nova-scotia/marine-protected-area-maps-maritimes-1.4658048.

Zurba, M., K.F. Beazley, E. English, and J. Buchmann-Duck. 2019. "Indigenous Protected and Conserved Areas (IPCAs), Aichi Target 11 and Canada's Pathway to Target 1: Focusing Conservation on Reconciliation. *Land* 8 (1): 1–20. https://doi.org/10.3390/land8010010.

PART 2

Changing Oceans

3

Rapid Changes across Canada's Oceans and Their Impact on Coastal Communities

Travis C. Tai and Juliano Palacios-Abrantes, with Juan Jose Alava Saltos, Natalie C. Ban, Andrea Bryndum-Buchholz, William W.L. Cheung, Simon Courtenay, Sarah Harper, Carie Hoover, Heike K. Lotze, Nadja S. Steiner, U. Rashid Sumaila, Nicolás Talloni-Álvarez, Charlotte K. Whitney, and Kristen Wilson

The sea around us is rapidly changing. For the last century, humans have fished global wild stocks to feed a growing population, extracted millions of gallons of oil from the seabed to fulfill socio-economic needs, shipped goods globally, and discarded waste into the ocean's blue waters (Halpern et al. 2019). In addition, the ocean has absorbed over 90% of the anthropogenic carbon dioxide released since the Industrial Revolution, heating up over 1°C over the last 50 years (IPCC 2019). Climate change has caused multiple shifts in ocean conditions, such as warming temperatures, decreasing salinities and pH, increasing harmful algal blooms, and expansion in the distribution of pathogenic agents, all of which have been associated with negative impacts on marine fauna and flora (Gao et al. 2018). Overall, these changes are affecting interdependent marine social-ecological systems. Unlike the large changes discussed in Chapter 4, rapid environmental changes are events of large magnitudes that occur on a short time frame of less than a year (i.e., weekly, monthly, seasonal) and are typically more localized in spatial scale (i.e., local, regional) (Okey et al. 2014). These changes are characterized by increasing interannual variability affecting the physical, biogeochemical, and ecological systems, with downstream effects on human societies (Riche, Johannesen, and Macdonald 2014).

Each coast in Canada has experienced different types of rapid change, with some commonalities. For example, events such as marine heatwaves have been experienced in all ocean basins and are expected to increase in frequency with global warming (Oliver et al. 2018; Smale et al. 2019) (Figure 3.2). Despite the differences in the type and magnitude of rapid changes across the three oceans, their impacts on Canada's ocean economic sectors – e.g., oil and gas, transportation, fisheries, coastal tourism – will be considerable, with far-reaching consequences for Canadians. Moreover, due to the close connection and interdependence between natural and social systems, such changes have repercussions for human societies that depend on marine resources. Coastal communities in Canada, especially Indigenous Peoples, have a strong bond with the ocean and its marine fauna (see Chapter 2). In many Indigenous societies, marine animals such as salmon, orcas, and bowhead whales represent values related to family, community, and protection, among others (Bennett et al. 2018). Indigenous Peoples have long applied their place-based knowledge to sustainably manage the natural resources within their traditional territories over generations (Ford et al. 2008; Pearce et al. 2015). Thus, rapid changes in the environment represent an increased source of pressure and have had, and will continue to have, direct impacts on the management of fish for food security, cultural ceremony, and local economies. However, our ability to predict long-term (slow) large change is much more advanced than for rapid change. The unpredictability and uncertainty of rapid change events increase the vulnerability to change for Canada's oceans, their living organisms, and dependent societies. Developing a better understanding of their potential impacts will enhance our capacity to prepare for and mitigate their effects on natural and human systems.

In this chapter, we explore some of the rapid environmental changes and their impacts across the three oceans bordering Canada: the Arctic, Atlantic, and Pacific. First, we review past and present observed rapid changes, followed by potential futures of rapid change. We provide an interdisciplinary approach to highlight cultural/phenomenological, scientific, and socio-economic impacts that these rapid changes will have on the human societies that depend on the ocean not only for food security and wealth but also for their way of life.

RAPID CHANGES ACROSS CANADA'S THREE OCEANS

Taking Stock

Canada's three oceans face multiple challenges caused by rapid global climate change and non-climatic anthropogenic stressors (e.g., natural resource overexploitation),

with each coast having a different geomorphology, climate, and oceanography that renders challenges from these stressors distinctive for different parts of Canada (Figure 3.1). While the resources along the Pacific coast have fluctuated through time, abrupt changes are more prevalent today, with an oil spill in 2016 in the traditional territory of the Heiltsuk Nation being a particularly poignant example of how quickly the marine environment can change, with long-lasting social, cultural, and ecological consequences (Alava 2017). In another example, a 2015 oil spill in the Pacific near Vancouver (Miller, Hotte, and Sumaila 2014) could have long-lasting effects on the area in terms of domestic and international tourism.

In the last 60 years, Canada's climate has warmed 1.7°C on average, double the global trend, with annual average air temperatures in northern regions warming by up to 2.3°C (Bush and Lemmen 2018). Such warming can increase the frequency of rapid changes, with downstream ecological and societal consequences. In the

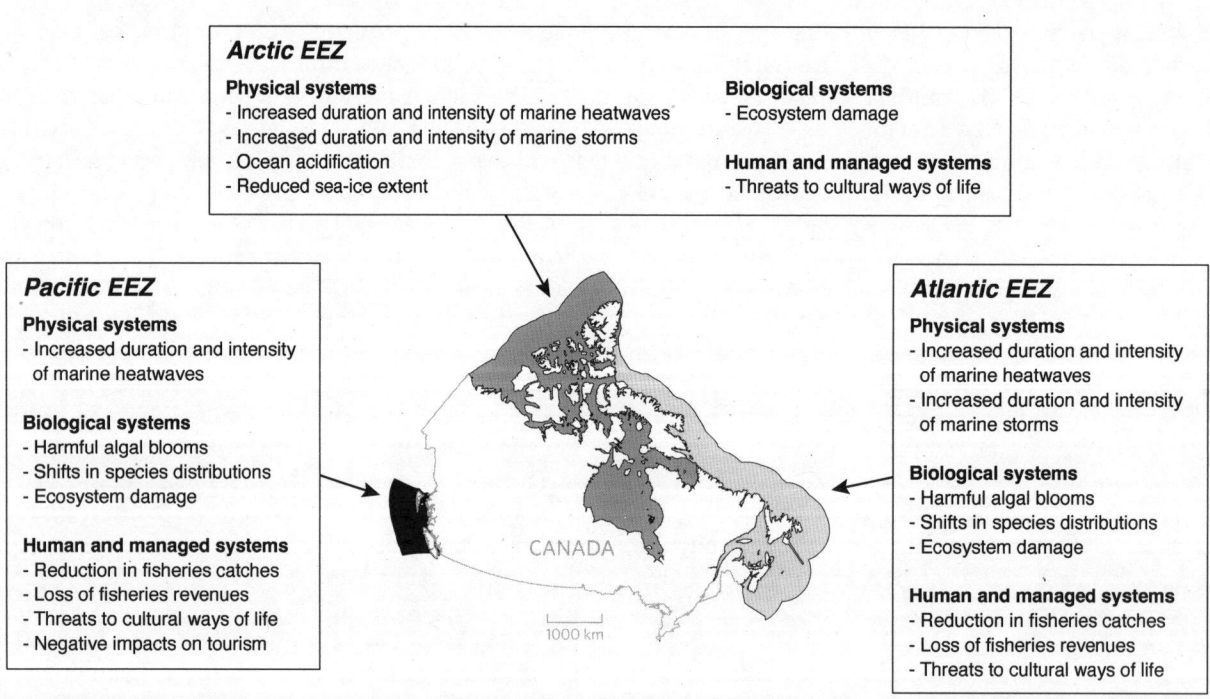

Figure 3.1 Impacts of rapid change across Canada's three exclusive economic zones. Map adapted from Sea Around Us (www.seaaroundus.org).

ocean, the 2012–15 marine heatwave known as the "Blob" in the Northeast Pacific had considerable impacts on marine ecosystems (Strasburger et al. 2016), affecting marine fisheries and livelihoods. The effects of the Blob may have contributed to very poor returns of sockeye salmon in 2019, which resulted in the closure of commercial fisheries and tighter restrictions on fishing by Indigenous Peoples. Indeed, poor or unpredictable returns have been characteristic for British Columbia salmon fisheries for longer than the past decade, but there have also been record returns (e.g., in 2010 for Fraser sockeye returns, and in 2018 for Alaskan sockeye salmon returns). These events increase unpredictability and the risk to fisheries. This widespread failure of salmon returns across British Columbia's coast has left many fishers "flat broke," as they had already invested in preparing and gearing up for the fishing season (Nair 2019). Many fishers hope for a climate change disaster relief fund, an important step for climate change mitigation and adaptation, but there has yet to be one established. Such a relief fund could provide aid when other unpredictable events occur, such as the 2019 landslide along a major migration portion of the Fraser River. For local Indigenous Peoples, such events have led to a lot of uncertainty, creating stress in communities and leading them to question this year's and future fishing seasons (Hennig 2019).

Examples of climate change impacts on the east coast of Canada include the considerable economic loss suffered by the lobster industry due to the 2012 marine heatwave in the Northwest Atlantic (Mills et al. 2013). Warmer temperatures resulted in earlier spring migration and increased moulting rates, extending the fishing season and leading to greater landings (Mills et al. 2013). Supply of lobster drastically exceeded market demand and the price collapsed. Prices per pound dropped by a dollar, i.e., a decrease of up to 50% of the price after spring and around 20% less than the year before (*CBC News* 2012, 2013). Furthermore, cross-border tension between Canada and the United States was created when Canadian lobster fishers and processors shut down two facilities and blocked a shipment of Maine lobsters to prevent further market flooding. Lobster fisheries, while most economically beneficial to the communities that fish them, are also important for Canada's economy as a whole, as they are typically one of the most lucrative fisheries in the world

Table 3.1

Notable rapid change events in Canada

Year	Event	Coast	Consequences
2012	Warm water mass	Atlantic	Phenology shift of phytoplankton bloom Change in lobster migration timing Lobster economy suffered with low prices
2013	Blob warm water	Pacific	Poor returns of sockeye salmon Mass mortality of Pacific cod and common murres Closure of commercial fisheries Tighter restrictions for fishing by Indigenous Peoples Disappearing humpback and fin whales
2016	Oil spill	Pacific	Ecosystem damage Loss of subsistence resources Loss of cultural way of life Loss of tourism
2019	Blob 2.0 warm water	Pacific	Increased harmful algal blooms Emergence of warm water species Poor salmon recruitment Fisheries closures

due to their extensive global market (Driscoll, Boyd, and Tyedmers 2015; Parker et al. 2018). Indigenous communities along the Atlantic coast rely on lobster fisheries and have protected rights to access these resources, important to their culture and critical to reconciliation efforts. These examples underscore the complexities of how abrupt rapid change events can disrupt physical, biological, social, cultural, and economic processes. The timing and magnitude of rapid change are inherently unpredictable, but documenting such events can lead to a better understanding of the potential consequences for Canada's oceans and their dependent communities (Table 3.1).

Future Scenarios

Climate change effects are expected to continue throughout the 21st century, resulting in an uncertain future

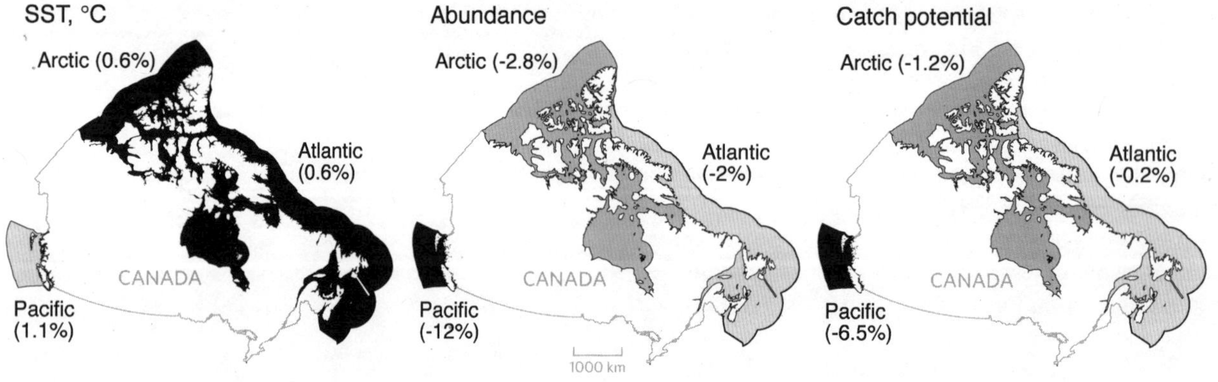

Figure 3.2 Projected impacts of marine heatwaves (MHW) on Canada's three oceans. This figure shows the projected effects of marine heatwaves in the 21st century that will add to the impacts from changes in decadal mean conditions under climate change. These effects include heatwaves indicated by sea surface temperature anomalies (relative to the decadal mean conditions), and changes in abundance and catch potential of exploited fishes and invertebrates. The projections are under the "no mitigation" (RCP8.5) greenhouse gas emissions scenario. Figure adapted from Cheung et al. 2021.

for Canada's oceans and their social-ecological health. Over the last century, humans have modified Earth's environment to such a degree that there is often no comparable data to project into the future. Moreover, it is impossible to know human behaviour with any precision; thus, we cannot foretell humanity's path into the future. Therefore, to plan ahead and adapt to potential social-ecological impacts of anthropogenic activities, scientists rely on understanding the baseline of past and current systems, projections based on the best available data, and the creation of future scenarios in order to understand future extremes at locally relevant scales (Bush and Lemmen 2018).

Future climate change scenarios suggest that Canada's climate will continue to warm regardless of which detailed scenario pans out. However, strong mitigation of greenhouse gases would result in an air temperature increase of 1.8°C (compared with 6.3°C with no mitigation[1]), decreasing the social-ecological impacts related to climate change (Bush and Lemmen 2018; Sumaila et al. 2019). The sea surface temperature (SST) of Canadian oceans is also expected to continue to warm over the 21st century, with substantial variability between seasons and regions (e.g., summer SST is expected to increase by 3°C and 4°C, in the Northeast Pacific and Arctic, respectively). Regardless of mitigative efforts, climate change will

continue to intensify extreme warm temperatures, increasing the frequency and severity of heatwaves on land (Greenan et al. 2018) and in the ocean (Frölicher, Fischer, and Gruber 2018) (Figure 3.2).

Globally, the probability of a marine heatwave by the end of the 21st century is projected to increase by a factor of 16 under a strong mitigation scenario (range 10–20) and 41 under a high-emission scenario (range 36–45) (Frölicher, Fischer, and Gruber 2018). This means that a future with high greenhouse gas emissions is projected to have one marine heatwave every three years, rather than every 100 years as in preindustrial times. Such events will cover larger areas (space), will last longer (time), and will bring warmer water (magnitude) (Frölicher, Fischer, and Gruber 2018). In addition, atmospheric and oceanic warming will result in reduction of sea-ice cover (mainly summer sea ice), leading to increased time and occurrence of extreme high-water-level events (wave action and larger storm surges) across Canada's Arctic and Atlantic coasts (Greenan et al. 2018). The oceans have absorbed over 90% of the carbon dioxide produced by humans to date (IPCC 2019), which has resulted in acidification of the marine environment (Bopp et al. 2013). In some regions of the Arctic, surface waters have already become corrosive due to low pH production. Projections following a business-as-usual scenario where no mitigation of greenhouse gases is made suggest that Arctic surface waters will become corrosive by mid-century (Feely, Doney, and Conley 2009).

All of these changes will have consequences for marine ecosystems. An increase in frequency and intensity of marine heatwaves threatens Canadian biodiversity (Smale et al. 2019); for example, projected increases in algal blooms will create hypoxic zones and lead to increasing mortality of local sessile populations unable to find refuge (Cavole et al. 2016), and elevated sea-ice melt will leave new areas for species to colonize, potentially creating a mismatch of coevolved species and unexpected trophic effects (Steiner et al. 2019). Due to the interdependence of social-ecological systems, rapid changes in oceanic conditions will have direct impacts on human communities along Canada's coasts, from losses in fisheries revenues due to marine heatwaves, to potential damage to tourism and to emerging fisheries from oil spills (Pinsky et al. 2018; Tai et al. 2019; Cheung and Frölicher 2020).

REGION-SPECIFIC RAPID CHANGE

Taking Stock

Arctic

The Arctic is rapidly changing, warming at twice the rate of global average temperatures in the past five years (2014–18) and exceeding all previous records since 1900 (Overland et al. 2018; Niemi et al. 2019). Seasonal temperature anomalies have also increased, with record high temperature in winter and autumn months of 2016 (Richter-Menge et al. 2016). Warming of sea surface temperatures have contributed to record-setting minimum sea-ice extents in the past 12 years (Osborne, Richter-Menge, and Jeffries 2018). Furthermore, regional variation in late-summer sea surface temperatures is linked to the variability in sea-ice retreat, air temperatures, and advection from adjacent Pacific and Atlantic Ocean waters (Osborne, Richter-Menge, and Jeffries 2018). Winter maximum sea ice set a record low in 2017 and was at its second-lowest in 2018, so there have now been four straight years with record low winter maximums (Perovich et al. 2017; National Snow and Ice Data Center 2018).

Rapid changes in ocean pH occur as a result of various seasonal events and can lead to localized acidification extremes (Steiner et al. 2015). Elevated atmospheric CO_2 intensifies ocean acidification due to the gas exchange at the air-sea interface. Changes in sea ice affect the amount of exposed open water area for exchange of CO_2. Localized upwelling events, river runoff, and biological activity are also major factors in ocean acidification. Colder waters already make the Arctic Ocean more susceptible to CO_2 uptake and acidification, and extreme acidification events could affect physiological performance and even survival of marine organisms (Steiner et al. 2015).

Ocean warming has affected hunting patterns of wildlife and humans due to changes in sea ice and sea state (Ford et al. 2008). This includes later formation and earlier breakup of sea ice with higher mobility and less stability, and increased sea state and wave action. Ultimately, it increases the dangers of sea-ice travel, alters the start and end of hunting seasons, and disrupts historical travel routes. Extreme events such as major storms, extensive snow or rain precipitation, and drought events can significantly affect the ecosystem. The number and intensity of Arctic winter storms have increased since 1976, potentially compromising ice growth (Graham et al. 2019). For example, in the Beaufort Sea strong wind events may cause earlier ice breakup and extensive open water regions, which can affect the timing of the beluga and bowhead whale migration into the Canadian Arctic Archipelago (e.g., Loseto et al. 2018). Interannual variability in snow and ice cover directly affects the light accessible to phytoplankton in the ocean and sea ice. Models suggest that this variability has led to increased interannual variability in ice algae and pelagic primary production over the last three decades (Steiner et al. 2019). In addition to temperature-related shifts, such changes in the lowest trophic level may directly affect the distribution and availability of key forage species.

Atlantic

The frequency and intensity of marine heatwaves are increasing in the western North Atlantic (Oliver et al. 2018); in particular, the Scotian Shelf and Bay of Fundy are a global hotspot for marine heatwaves and warming above the average global rate (Hobday and Pecl 2014; Smale et al. 2019). In 2012, the Scotian Shelf and Grand

Banks had their warmest SST, with annual offshore anomalies ranging from +1.7°C (±1.4 SD) to +2.5°C (±3.2 SD) relative to the 1981–2010 average (Hebert et al. 2013, 46). In 2016, the Gulf of St. Lawrence experienced its highest SST, where average November anomalies were +1.7°C (±2.2 SD), ranging from +0.8 to +2.2°C relative to 1985–2010 (Galbraith et al. 2017, 91). Record highs in deepwater temperature were also recorded between 2012 and 2016 throughout Atlantic Canada, particularly in August and September 2016, when Gulf of St. Lawrence temperatures at 300 m ranged from 0.53 to 0.85°C warmer than 1981–2010 climatology (Galbraith et al. 2017; Bernier, Jamieson, and Moore 2018; Hebert et al. 2018, 53). During a recent heatwave in the Gulf of Maine, some species, such as Atlantic cod (*Gadus morhua*), have shown strong declines or collapses (Pershing et al. 2015). Overall fish recruitment and projected biomass are generally negatively affected by warming waters and reduced primary production in Atlantic Canada (Britten, Dowd, and Worm 2016; Bryndum-Buchholz et al. 2019), but there are winners and losers, as some species are more vulnerable to warming than others (Shackell, Ricard, and Stortini 2014; Stortini et al. 2015).

Corresponding to record ocean temperatures and increased stratification in 2012, the spring phytoplankton bloom was earlier and shorter than the 1999–2012 average along the Scotian Shelf (Johnson et al. 2013, 42), Gulf of St. Lawrence (Plourde et al. 2014, 46), and the Newfoundland and Labrador Shelf (Pepin et al. 2013, 38), with a general shift in phytoplankton community structure to smaller species. Chlorophyll levels in the Bay of Fundy were higher than average in July and August 2012 due to the bloom of potentially toxic diatom and dinoflagellate species, and it is likely these blooms were widespread in the Gulf of Maine (Johnson et al. 2013). In addition, long-term declines in kelp populations in Atlantic Canada have resulted in a shift to turf and invasive seaweed species dominated rocky habitats in localized areas (Filbee-Dexter, Feehan, and Scheibling 2016). These declines are attributed to gradual increases in ocean temperatures, but a marine heatwave can result in further regime shifts to a turf-dominated ecosystem across a wider spatial scale (Wernberg et al. 2016).

Ocean chemistry has also become more variable. Rapid deoxygenation is already observed along the Canadian Atlantic coast, due to climate-related ocean circulation shifts in the northwest Atlantic, and is projected to become more pronounced with continued global warming (Claret et al. 2018). With ocean acidification, pH has been decreasing consistently since the 1990s (Bopp et al. 2013). While overall changes in pH are gradual, rapid changes in pH may occur as an indirect result of other environmental events (e.g., eutrophication, algal blooms, upwelling, and ocean currents).

Changes in marine storm intensity and frequency have also been recorded and have major implications on physical (e.g., temperature, stratification) and chemical (e.g., pH) characteristics. Since 1970, storm tracks have shifted 180 km northward in the North Atlantic Ocean, with a slight reduction in wind speed and wave heights within Atlantic Canada. Since 1958, there has also been an increase in extreme (air pressure <980 hectopascals [hPa]) marine storms during fall (October to December), with no change in other seasons (Wolf and Moser 2011).

Nutrient overabundance, particularly nitrogen, is one of the most serious environmental problems in the coastal waters of Prince Edward Island (e.g., Price et al. 2017; Foulon et al. 2019). The origin is principally agricultural, particularly the application of chemical fertilizers when growing potatoes. Nitrogen running off farmers' fields percolates down to the groundwater and ends up in estuaries, where the nutrients support blooms of annual opportunistic algae, including sea lettuce, *Ulva lactuca* (Coffin et al. 2018). These blooms tend to occur at the head of estuaries, where freshwater rivers and creeks meet the sea. Sea lettuce blooms degrade the estuarine environment in two ways. First, they shade out and kill eelgrass beds (*Zostera marina*), which play many important roles in estuarine ecology, including stabilization of sediments, nutrient cycling, and provision of habitat for marine biota; surveys show an estimated 40% reduction of eelgrass in estuaries of the southern Gulf of St. Lawrence. Second, when the large biomass of sea lettuce dies, its degradation by microbes uses up most or all of the dissolved oxygen in the water, causing hypoxic or anoxic conditions.

Pacific

Rapid changes on the Pacific coast of Canada include abrupt shifts in water temperatures, impacting marine ecosystems and people's livelihood (Talloni-Álvarez et al.

2019). The frequency and intensity of marine heatwaves is also increasing in the eastern North Pacific (Oliver et al. 2018). In late 2013, an unusual warm water mass – sea surface temperatures of 1°C to 4°C above average – was observed from Alaska to Baja California, linked to strong positive anomalies in sea-level pressure across the Pacific Northwest (Peterson et al. 2016). The Blob persisted for at least three years, triggering unexpected changes in marine productivity, ecosystems, marine species, and fisheries (Bond et al. 2015). Some of these changes included unprecedented harmful algal bloom, stranding of marine mammals and seabirds, and shifts in distribution and abundance of marine species (Cavole et al. 2016). Episodes of harmful algal bloom in 2015 were among the largest and most severe recorded in the Northeast Pacific (Peterson et al. 2016). Extremely low chlorophyll levels were also documented in the region of warm anomalies, associated with suppressed nutrient transports into the mixed layer (Whitney 2015), while numerous warm water species were sighted further north than ever before (Cavole et al. 2016). Ecosystem impacts triggered by the Blob resulted in reduction of prey availability and fish size, poor recruitment of Pacific salmon, and the closure of several fisheries (commercial and recreational) in the

OIL SPILLS IMPACTS ON MARINE ECOSYSTEMS AND HUMAN COMMUNITIES

Oil spills are identified as the major threat to sensitive marine ecosystems and critical habitats in the Pacific coast and marine regions of Canada.

The potential short- and long-term impacts across multiple sectors of rapid episodic events such as oil spills is succinctly explained in an interview with professor of fisheries economics U. Rashid Sumaila at the University of British Columbia (*CBC News* 2015): "Tourism is a lot about perception, so once people start to think that our waters are polluted with toxic oil, that can really affect [the tourism industry]." He added that "perceptions can hang on for a long time" (Sumaila et al. 2012; Miller and Sumaila 2015; Alava 2017). On October 13, 2016, the *Nathan E. Stewart* tug towing a barge through Heiltsuk territory on the Central Coast of British Columbia ran aground, spilling diesel and other pollutants on and around the ancient Village of Q'vúqvai in Gale Passage, near Bella Bella (Heiltsuk Tribal Council 2017). The rich ecosystem found there has long been an important harvesting site for the Heiltsuk, with many species caught there for food, social, and ceremonial (FSC) and commercial purposes, including clams, red sea urchin, sea cucumber, salmon, and herring spawn on kelp (SOK).

Prior to this incident, the area was relatively ecologically intact, but after the spill, it was closed to all fishing for an indeterminate period. This reduced the availability of FSC and commercial resources available to the Heiltsuk, and limited opportunities for teaching effective stewardship through traditional cultural practices. The spill had a devastating effect on the community, with potentially long-lasting impacts that could have been mitigated through a more effective marine spill response (Eykelbosh et al. 2018). However, shipping along Canada's Pacific coast has become a contentious topic, with increased prospects for shipping of oil. In November 2016, the Canadian government approved the Kinder Morgan Trans Mountain Pipeline Expansion project, which would carry nearly a million barrels of oil from Alberta to British Columbia (David Suzuki Foundation 2016; Pembina Institute 2016). In 2018, the development was halted by the Federal Court of Appeal as it broke the law, violated the rights of Indigenous Peoples, and failed to include the negative effects of a sevenfold increase in tanker traffic on the marine environment of killer whales (*Orcinus orca*), the most endangered populations of marine mammals in Canada (Alava 2017; MacDuffee and Young 2018). Nonetheless, on June 18, 2019, Canada's prime minister approved the expansion of the controversial pipeline project in British Columbia (Austen 2019). Ironically, the approval occurred a day after the House of Commons declared a national climate emergency after the national climate change report suggested that Canada was warming twice as fast as the world average (Jackson 2019).

Northwest Pacific coast due to the toxic *Pseudo-nitzschia* bloom (Cavole et al. 2016). As well, a die-off of sea scallops on the east coast of Vancouver Island during the summer of 2013 was associated with an unusually cold water mass from coastal upwelling, in combination with variable pH and deoxygenation (Gao et al. 2018). Pacific oysters in British Columbia have also been exposed to mass mortalities during recent summers, with losses for adult oysters ranging from 50 to 90% on intertidal farms and 25 to 100% in suspended culture sites, representing losses over Cdn$100 million (DFO 2017).

However, climate change is not the only threat. Canada's Pacific Ocean has seen dramatic changes over the past century, related to socio-economic factors such as shipping and transportation, pulp mill discharges, and mining and municipal wastewater effluents (Alava 2019), as well as the impacts of colonialism, the commercialization of marine species, changes to marine management, and the interrelationship among these. Such changes are particularly important in the coastal waterways of British Columbia, where we find the large urban centres of Metro Vancouver (~2.8 million people) and Victoria (~420,000 people) (Statistics Canada 2023), as well as within the traditional territories of Indigenous Peoples distributed along the West Coast (see Chapter 2). Canada's Pacific Ocean is thus substantially changed from what it looked like 100 years ago, even without considering climate change. The commercialization and subsequent mismanagement of fisheries has changed the Northeast Pacific Ocean's ecological communities. Serial depletions include cetaceans and pinnipeds due to whaling (Trites 2007), sea otters during the fur trade (Gregr et al. 2008), and a variety of fishes and invertebrates due to commercialization (Pauly et al. 2001). Continued reliance on petroleum products has many costs, including the potential to rapidly alter the marine environment in the case of an oil spill (see text box titled "Oil spills impacts on marine ecosystems and human communities").

Future Scenarios for Each Region

Arctic

Rapid change events in the Arctic include extreme seasonal temperature anomalies and acidification extremes. These will likely increase in frequency, while the timing of seasonal ice formation and breakup will become less reliable and predictable (Steiner et al. 2015). Furthermore, climate change may increase the frequency of marine fog, wind, and storminess. Reduced sea-ice cover and increased storminess will likely interact in more areas along the ice edge, potentially increasing hazards and less stable sea ice for animals and humans.

Northern communities in the Arctic have shown concern for the rapid change events that put their livelihoods and way of life at risk. Uncertainty in resource availability, along with decreased reliability of safety for traditional hunting methods, means that these are compromised. Inuit communities are dependent on marine ecosystems, obtaining 40% of their caloric intake from marine resources (Watts et al. 2017); up to 80% is derived from marine mammals alone. However, distribution and abundance of marine species are projected to change under high climate change scenarios, which may alter marine ecosystem composition and functioning. Models project that there will be an increase in localized invasions of subpolar species and subsequently an overall increase in fisheries catch and revenue potential (Hoover, Pitcher, and Christensen 2013; Lam, Cheung, and Sumaila 2014; Suprenand, Ainsworth, and Hoover 2018; Tai et al. 2019). Some key ecosystem Arctic species, such as the Arctic cod (*Boreogadus saida*), are at their northern limit and are facing range contraction from the south as environmental conditions are less suitable (Steiner et al. 2019). Arctic cod are highly abundant in Arctic marine ecosystems and are an important link for energy transfer between trophic levels. Models generally predict long-term responses to climatological changes but lack the ability to predict responses to increased variability and extreme events. Responses of Arctic cod to extreme events could affect growth, fecundity, development, or mortality, which could lead to drastic changes for the ecosystem and organisms directly important for northern communities.

The IPCC special report on extreme events (Field et al. 2012) emphasizes the particular relevance of extreme climate events on society and ecosystems due to their potentially severe impacts. Model simulations suggest increases in daily maximum near-surface temperatures (decrease in minimum temperature) in the future, with an increase in the duration of warm periods (decrease in duration of cold periods), and increases in extreme precipitation events in northern high latitudes (Sillmann et al. 2013).

Atlantic

In the North Atlantic Ocean, including the Canadian Atlantic economic exclusion zone (EEZ), SST is projected to consistently increase above the global average under both low-emission and business-as-usual scenarios in the next couple of decades (Bryndum-Buchholz et al. 2019), while the intensity (how hot), spatial extent (how large), frequency (how often), and duration (how long) of marine heatwaves is projected to continue to increase by 2100 (Greenan et al. 2018). Projections of sea-level rise under a high climate change scenario suggest over 15 cm by the year 2100 along the east coast of Canada due to fluid dynamic processes alone (e.g., thermal expansion).

By the end of the 21st century, common seaweed and seagrass species in Atlantic Canada will have greater exposure to water temperatures near the warm limit of their thermal niche in Atlantic Canada, making them more susceptible to marine heatwaves (Wilson and Lotze 2019; Wilson, Skinner, and Lotze 2019). Laboratory experiments involving Atlantic Canada rockweed and kelp populations have shown that their susceptibility to marine heatwaves is dependent on the duration and magnitude of the heatwave (Simonson, Scheibling, and Metaxas 2015; Wilson et al. 2015; Kay et al. 2016). Similar changes are also projected for highly valuable species in a high climate change scenario. For example, overall abundance of Atlantic lobster in Canadian waters is projected to decrease, with decreases at lower latitudes and increases at higher latitudes and their distribution shifts north (Tai, Harley, and Cheung 2018). As seen in the 2012 heatwave, changes in catch can have many lasting consequences for communities and the national economy.

Rapid melting of the Greenland Ice Sheet may also have severe consequences for ocean circulation and global climate processes (IPCC 2019). While the melting of this ice sheet may be a relatively rapid event, it has much greater large-scale changes, potentially affecting El Niño–Southern Oscillation characteristics, impacts on the Amazonian rainforest, and even shrinking the West Antarctic Ice Sheet. Many of these cause-and-effect relationships have yet to be quantitatively resolved due to the uncertainty of the initial responses of the Greenland Ice Sheet to warming and its downstream effects on Atlantic circulation processes.

Pacific

Despite a projected increase in SST in all seasons in the Canadian Pacific (Wolf and Moser 2011), there is low certainty about the drivers of extreme events in British Columbia, mainly due to very few studies examining these issues in the region (Gao et al. 2018). As a consequence, the ecological repercussions are still not clear. For example, some have hypothesized that the high temporal and spatial heterogeneity of the oceanographic transition zone in this region may make the biota more resistant and resilient to climate change impacts (Okey et al. 2014), but a recent study found limited evidence of potential climate refugia along the BC coast (Ban et al. 2016). The lack of scientific knowledge regarding different processes can be supported by Indigenous Ecological Knowledge (IEK). "I think it [adaptation] has to come from the community ... I really believe that people should be empowered to know that they can make those changes" (personal communication by anonymous informant, Nuxalk Nation, July 2018).

Experiential learning of Indigenous knowledge has persisted in many Indigenous Peoples in British Columbia through cultural revitalization programs (e.g., Rediscovery youth camps; Supporting Emerging Aboriginal Stewards [SEAS] programs), which provide hopeful examples of how Indigenous Peoples may reclaim self-determination and lead their own adaptation pathways (Turner and Spalding 2013). Some Indigenous Peoples of the Central Coast frame the future in a positive light, highlighting that Indigenous Peoples and knowledge will continue to adapt to a changing climate.

MOVING FORWARD WITH RAPID OCEAN CHANGES FOR CANADA: VISIONING

Canada's Arctic, Atlantic, and Pacific Oceans have all been recently impacted by rapid changes, from marine heatwaves with fisheries repercussions, to harmful algal bloom events affecting biodiversity, to extreme storms threatening coastal structures, and to non-climatic events such as oil spills and sediment runoffs (Boldt, Javorski, and Chandler 2019; Niemi et al. 2019). Moreover, results from global climate change models indicate that rapid change events not only will continue through the 21st century but in many cases will intensify, regardless of which future climate change scenario is considered, with greater consequences to social-ecological systems if countries do not mitigate greenhouse gases. Impacts of rapid changes have already had consequences on social-ecological systems. Thus, it is imperative to protect Canada's oceans, critical habitats, and coastal zones; rapid changes from marine habitat degradation, pollution risks, and climate change urgently need to be addressed.

Implementing precautionary actions would align with international instruments, such as the United Nations Sustainable Development Goal to conserve and sustainably use the oceans, seas, marine resources, and fisheries (SDG 14); the *Stockholm Convention on Persistent Organic Pollutants*; the *Minamata Convention on Mercury*; and the 2015 *Paris Agreement* under the United Nations Framework Convention on Climate Change. Efforts like Canada's Climate Change Report (Bush and Lemmen 2018) are a great first step in understanding the magnitude of climate change impacts and should be continually updated as new research is developed. Initiatives that support the compilation and management of data and metadata are also important to advance climate, ecosystem, and other modelling approaches needed to support ocean policy, as these are very data-intensive, requiring knowledge and data across disciplines, scales, and knowledge types (Tittensor et al. 2018). The compilation of metadata in a single repository is a key first step toward an integrated database of ocean research where data can be easily found and shared (Cisneros-Montemayor et al. 2016).

Allied to this, there is a need to maintain existing programs while creating new marine observation programs to feed models that predict extreme events such as the Blob. A pilot of such a model has been created by the US National Oceanic and Atmospheric Agency (NOAA), which has helped identify a new marine heatwave threatening the region from the coast of Alaska to California. While this tool is still in development, scientists are producing a range of indices that could help forecast future marine heatwaves along the Pacific coast (Integrated Ecosystem Assessment 2023). Overall, while global trends can somehow inform at regional scales, future research at the regional scale (e.g., ocean basin) focused on rapid extremes rather than trends can shed light on the future implications of a rapidly changing ocean for the social-ecological systems of the Canadian coast.

Scientific research alone will be insufficient to account for, and respond to, rapid changes in Canada's oceans. Past and ongoing European colonization of coastal regions resulted in rapid and drastic changes in Indigenous management practices because they were criminalized (Harris 2008; Ommer 2007). Indigenous Peoples were forcibly relocated and their populations declined precipitously due to smallpox and other diseases; together with declines of marine species due to commercialization, these contributed to changed access to Indigenous Peoples' management practices (Harris 2008; Ommer 2007). To identify and prioritize adaptive actions most relevant to a community, it is important to explore perceptions from individuals with first-hand lived experience (Marshall and Marshall 2007; Petheram et al. 2010; Wolf and Moser 2011). Therefore, addressing climate change adaptation requires not only the best science available but also recognition of the diverse and cumulative challenges facing Indigenous Peoples worldwide (Huntington et al. 2019) to provide insights for proactive strategies to support their ways of life. Only with the integration of different types of knowledge, such as Indigenous ecological knowledge, will society achieve sustainable use of the ocean (Huntington et al. 2019). Due to their long-term place-based knowledge, Indigenous Peoples have an intrinsic capacity to respond to novel environmental changes by adapting their way of life accordingly. Certainly, Indigenous Peoples have experienced and persisted through environmental change over long time scales and developed cultural practices and knowledge related to invasive species (Reo et al. 2017), floods (Brown and Brown 2009; Horne 2012), and changes in weather patterns (Cunsolo Willox et al. 2012). Those with lived experience can help to interpret model outputs and provide advice to policy-makers to achieve adaptive, collaborative, ecosystem-based management strategies.

Finally, policies have to be informed and based on the best available science and must be inclusive of all levels of society. Only through a truly equitable process that respects the values of different groups can sustainable ocean development be achieved (Cisneros-Montemayor et al. 2019).

NOTE

1 In this particular example, the "strong mitigation" and "no mitigation" scenarios refer to the Representative Concentration Pathway (RCP) 2.6 and RCP8.5 scenarios.

REFERENCES

Alava, J.J. 2017. "Pipelines Imperil Canada's Ecosystem." *Science* 355 (6321): 140.

–. 2019. "Legacy and Emerging Pollutants in Marine Mammals' Habitat from British Columbia: Management Perspectives for Sensitive Marine Ecosystems." In *Stewarding the Sound: The Challenge of Managing Sensitive Coastal Ecosystems,* edited by Leah Bendell, Patricia Gallaugher, Laurie Wood, and Shelley McKeachie, 87–114. Boca Raton, FL: CRC Press, Taylor and Francis Group.

Austen, I. 2019. "Canada Approves Expansion of Controversial Trans Mountain Pipeline." *New York Times,* June 18. https://www.nytimes.com/2019/06/18/world/canada/trudeau-trans-mountain-pipeline.html.

Ban, S.S., H.M. Alidina, T.A. Okey, R.M. Gregg, and N.C. Ban. 2016. "Identifying Potential Marine Climate Change Refugia: A Case Study in Canada's Pacific Marine Ecosystems." *Global Ecology and Conservation* 8: 41–54.

Bennett, N.J., M. Kaplan-Hallam, G. Augustine, N. Ban, D. Belhabib, I. Brueckner-Irwin, A. Charles, et al. 2018. "Coastal and Indigenous Community Access to Marine Resources and the Ocean: A Policy Imperative for Canada." *Marine Policy* 87: 186–93.

Bernier, R.Y., R.E. Jamieson, and A.M. Moore, eds. 2018. *State of the Atlantic Ocean Synthesis Report.* Canadian Technical Report of Fisheries and Aquatic Sciences 3167. Moncton: Fisheries and Oceans Canada, Gulf Region.

Boldt, J.L., A. Javorski, and P.C. Chandler, eds. 2019. *State of the Physical, Biological and Selected Fishery Resources of Pacific Canadian Marine Ecosystems in 2019.* Sidney, BC: Canada Institute of Ocean Sciences.

Bond, N.A., M.F. Cronin, H. Freeland, and N. Mantua. 2015. "Causes and Impacts of the 2014 Warm Anomaly in the NE Pacific." *Geophysical Research Letters* 42: 3414–20.

Bopp, L., L. Resplandy, J.C. Orr, S.C. Doney, J.P. Dunne, M. Gehlen, P. Halloran, et al. 2013. "Multiple Stressors of Ocean Ecosystems in the 21st Century: Projections with CMIP5 Models." *Biogeosciences* 10 (10): 6225–45. https://pure.mpg.de/rest/items/item_1841117/component/file_1841119/content.

Britten, G.L., M. Dowd, and B. Worm. 2016. "Changing Recruitment Capacity in Global Fish Stocks." *Proceedings of the National Academy of Sciences* 113 (1): 134–39.

Brown, F., and Y.K. Brown. 2009. *Staying the Course, Staying Alive: Coastal First Nations Fundamental Truths: Biodiversity, Stewardship and Sustainability.* Victoria, BC: Biodiversity BC.

Bryndum-Buchholz, A., D.P. Tittensor, J.L. Blanchard, W.W.L. Cheung, M. Coll, E.D. Galbraith, S. Jennings, O. Maury, and H.K. Lotze. 2019. "Twenty-First-Century Climate Change Impacts on Marine Animal Biomass and Ecosystem Structure across Ocean Basins." *Global Change Biology* 25 (2): 459–72.

Bush, E., and D.S. Lemmen, eds. 2018. *Canada's Climate Change Report.* Ottawa: Government of Canada. https://changingclimate.ca/CCCR2019/.

Cavole, L.M., A.M. Demko, R.E. Diner, A. Giddings, I. Koester, C.M.L.S. Pagniello, M.-L. Paulsen, et al. 2016. "Biological Impacts of the 2013–2015 Warm-Water Anomaly in the Northeast Pacific: Winners, Losers, and the Future." *Oceanography* 29 (2): 273–85.

CBC News. 2012. "Lobster Prices Way Down since Spring." *CBC News,* August 24. https://www.cbc.ca/news/canada/prince-edward-island/lobster-prices-way-down-since-spring-1.1214072.

–. 2013. "Lobster Protest Continues over Low Prices." *CBC News,* May 10. https://www.cbc.ca/news/canada/new-brunswick/lobster-protest-continues-over-low-prices-1.1408457.

–. 2015. "Vancouver Oil Spill Could Have Far Reaching Impacts: Expert." *CBC News,* April 10. https://www.cbc.ca/news/canada/british-columbia/vancouver-oil-spill-could-have-far-reaching-impacts-expert-1.3027976.

Cheung, W.W.L., and T.L. Frölicher. 2020. "Marine Heatwaves Exacerbate Climate Change Impacts for Fisheries in the Northeast Pacific." *Scientific Reports* 10 (1): 1–10. https://www.nature.com/articles/s41598-020-63650-z.

Cheung, W.W.L., T.L. Frölicher, V.W.Y. Lam, M.A. Oyinlola, G. Reygondeau, U.R. Sumaila, T.C. Tai, L.C.L. Teh, and C.C.C. Wabnitz. 2021. "Marine Heatwaves Amplify the Impacts of Climate Change on Fish and Fisheries." *Science Advances* 7: eabh0895. https://doi.org/10.1126/sciadv.abh0895.

Cisneros-Montemayor, A.M., W.W.L. Cheung, K. Bodtker, L. Teh, N. Steiner, M. Bailey, C. Hoover, and U.R. Sumaila. 2016. "Towards an Integrated Database on Canadian Ocean Resources: Benefits, Current States, and Research Gaps." *Canadian Journal of Fisheries and Aquatic Sciences* 74 (1): 65–74.

Cisneros-Montemayor, A.M., M. Moreno-Báez, M. Voyer, E.H. Allison, W.W.L. Cheung, M. Hessing-Lewis, M.A.

Oyinlola, G.G. Singh, W. Swartz, and Y. Ota. 2019. "Social Equity and Benefits as the Nexus of a Transformative Blue Economy: A Sectoral Review of Implications." *Marine Policy* 109: 103702.

Claret, M., E.D. Galbraith, J.B. Palter, D. Bianchi, K. Fennel, D. Gilbert, and J.P. Dunne. 2018. "Rapid Coastal Deoxygenation Due to Ocean Circulation Shift in the Northwest Atlantic." *Nature Climate Change* 8 (10): 868–72.

Coffin, M.R.S., S.C. Courtenay, K.M. Knysh, C.C. Pater, and M.R. van den Heuvel. 2018. "Impacts of Hypoxia on Estuarine Macroinvertebrate Assemblages across a Regional Nutrient Gradient." *FACETS* 3 (1): 23–44.

Cunsolo Willox, A., S.L. Harper, J.D. Ford, K. Landman, K. Houle, and V.L. Edge. 2012. "'From This Place and of This Place': Climate Change, Sense of Place, and Health in Nunatsiavut, Canada." *Social Science and Medicine* 75 (3): 538–47.

David Suzuki Foundation. 2016. "Expanding Pipelines Now Doesn't Make Environmental or Economic Sense." Press release, November 29. https://davidsuzuki.org/press/expanding-pipelines-now-doesnt-make-environmental-economic-sense/.

DFO (Fisheries and Oceans Canada). 2017. "Seasonal Mortality in Pacific Oysters in Baynes Sound: Effect of Environmental Variables, Spawning, and Pathogens." https://www.dfo-mpo.gc.ca/aquaculture/rp-pr/acrdp-pcrda/projects-projets/17-1-P-05-eng.html.

Driscoll, J., C. Boyd, and P. Tyedmers. 2015. "Life Cycle Assessment of the Maine and Southwest Nova Scotia Lobster Industries." *Fisheries* 172: 385–400.

Eykelbosh, A., C.M. Slett, P. Wilson, and L.P.E. Health. 2018. "Supporting Indigenous Communities during Environmental Public Health Emergencies." *Environmental Health Review* 61 (1): 9–11.

Feely, R.A., S.C. Doney, and S.R. Conley. 2009. "Ocean Acidification: Present Conditions and Future Changes in a High-CO_2 World." *Oceanography* 22 (4): 36–47.

Field, C.B., V. Barros, T.F. Stocker, D. Qin, D.J. Dokken, K.L. Ebi, M.D. Mastrandrea, et al., eds. 2012. *Managing the Risks of Extreme Events and Disasters to Advance Climate Change Adaptation.* Cambridge: Cambridge University Press.

Filbee-Dexter, K., C.J. Feehan, and R.E. Scheibling. 2016. "Large-Scale Degradation of a Kelp Ecosystem in an Ocean Warming Hotspot." *Marine Ecology Progress Series* 543: 141–52.

Ford, J.D., B. Smit, J. Wandel, M. Allurut, K. Shappa, H. Ittusarjuat, and K. Qrunnut. 2008. "Climate Change in the Arctic: Current and Future Vulnerability in Two Inuit Communities in Canada." *Geographical Journal* 174 (1): 45–62.

Foulon, É., A.N. Rousseau, G. Benoy, and R.L. North. 2019. "A Global Scan of How the Issue of Nutrient Loading and Harmful Algal Blooms Is Being Addressed by Governments, Non-Governmental Organizations, and Volunteers." *Water Quality Research Journal* 55 (1): 1–23.

Frölicher, T.L., E.M. Fischer, and N. Gruber. 2018. "Marine Heatwaves under Global Warming." *Nature* 560 (7718): 360–64.

Galbraith, P.S., J. Chassé, C. Caverhill, P. Nicot, D. Gilbert, B. Pettigrew, D. Lefaivre, D. Brickman, L. Devine, and C. Lafleur. 2017. *Physical Oceanographic Conditions in the Gulf of St. Lawrence in 2016.* DFO Canadian Science Advisory Secretariat Research Document 2017/044. Ottawa: Fisheries and Oceans Canada. https://waves-vagues.dfo-mpo.gc.ca/Library/40613677.pdf.

Gao, Y., R.A. Svec, J. Morgan, and D.L. Dettman. 2018. "Isotopic Records on the Massive Death of Sea Scallops in Vancouver Island of Canada." *Applied Geochemistry* 97: 256–62.

Graham, R.M., P. Itkin, A. Meyer, A. Sundfjord, G. Spreen, L.H. Smedsrud, G.E. Liston, et al. 2019. "Winter Storms Accelerate the Demise of Sea Ice in the Atlantic Sector of the Arctic Ocean." *Scientific Reports* 9 (1): 9222.

Greenan, B.J.W., T.S. James, J.W. Loder, P. Pepin, K. Azetsu-Scott, D. Ianson, R.C. Hamme, et al. 2018. "Changes in Oceans Surrounding Canada." In *Canada's Climate Change Report,* edited by E. Bush and D.S. Lemmen, 343–423. Ottawa: Government of Canada.

Gregr, E.J., L.M. Nichol, J.C. Watson, J.K.B. Ford, and G.M. Ellis. 2008. "Estimating Carrying Capacity for Sea Otters in British Columbia." *Journal of Wildlife Management* 72 (2): 382–88.

Halpern, B.S., M. Frazier, J. Afflerbach, J.S. Lowndes, F. Micheli, C.O.X. Hara, C. Scarborough, and K.A. Selkoe. 2019. "Recent Pace of Change in Human Impact on the World's Ocean." *Scientific Reports* 9: 11609.

Harris, D.C. 2008. *Fish, Law, and Colonialism: The Legal Capture of Salmon in British Columbia.* Toronto: University of Toronto Press.

Hebert, D., R. Pettipas, D. Brickman, and M. Dever. 2013. *Meteorological, Sea Ice and Physical Oceanographic Conditions on the Scotian Shelf and in the Gulf of Maine during 2012.* DFO Canadian Science Advisory Secretariat

Research Document 2013/058. Ottawa: Fisheries and Oceans Canada.

–. 2018. *Meteorological, Sea Ice and Physical Oceanographic Conditions on the Scotian Shelf and in the Gulf of Maine during 2016*. DFO Canadian Science Advisory Secretariat Research Document 2018/016. Ottawa: Fisheries and Oceans Canada.

Heiltsuk Tribal Council. 2017. *Investigation Report: The 48 Hours after the Grounding of the* Nathan E. Stewart *and Its Oil Spill*. Bella Bella, BC: Heiltsuk Tribal Council. https://heiltsuknation.ca/documents/investigation-report-the-48-hours-after-the-grounding-of-the-nathan-e-stewart-and-its-oil-spill/.

Hennig, C. "'What Fishing Season?': Local First Nations Worry about State of Fishing in Fraser River." *CBC News,* August 17. https://www.cbc.ca/news/canada/british-columbia/what-fishing-season-local-first-nations-worried-1.5251253.

Hobday, A.J., and G.T. Pecl. 2014. "Identification of Global Marine Hotspots: Sentinels for Change and Vanguards for Adaptation Action." *Reviews in Fish Biology and Fisheries* 24 (2): 415–25.

Hoover, C., T.J. Pitcher, and V. Christensen. 2013. "Effects of Hunting, Fishing and Climate Change on the Hudson Bay Marine Ecosystem: II. Ecosystem Model Future Projections." *Ecological Modelling* 264: 143–56.

Horne, J. 2012. "WSANEC: Emerging Land or Emerging People." *Arbutus Review* 3 (2): 6–19.

Huntington, H.P., M. Carey, C. Apok, B.C. Forbes, S. Fox, L.K. Holm, A. Ivanova, J. Jaypoody, G. Noongwook, and F. Stammler. 2019. "Climate Change in Context: Putting People First in the Arctic." *Regional Environmental Change* 19 (4): 1217–23.

Integrated Ecosystem Assessment. 2023. "The California Current Marine Heatwave Tracker – Blobtracker." https://www.integratedecosystemassessment.noaa.gov/regions/california-current/cc-projects-blobtracker.

IPCC (Intergovernmental Panel on Climate Change). 2019. *IPCC Special Report on the Ocean and Cryosphere in a Changing Climate: Summary for Policymakers*. Cambridge: Cambridge University Press.

Jackson, H. 2019. "National Climate Emergency Declared by House of Commons." *Global News,* June 17. https://globalnews.ca/news/5401586/canada-national-climate-emergency/.

Johnson, C., G. Harrison, B. Cassault, J. Spry, W. Li, and E. Head. 2013. *Optical, Chemical, and Biological Oceanographic Conditions on the Scotian Shelf and in the Eastern Gulf of Maine in 2012*. DFO Canadian Science Advisory Secretariat Research Document 2013/070. Ottawa: Fisheries and Oceans Canada.

Kay, L.M., A.L. Schmidt, K.L. Wilson, and H.K. Lotze. 2016. "Interactive Effects of Increasing Temperature and Nutrient Loading on the Habitat-Forming Rockweed *Ascophyllum nodosum*." *Aquatic Botany* 133: 70–78.

Lam, V.W.Y., W.W.L. Cheung, and U.R. Sumaila. 2014. "Marine Capture Fisheries in the Arctic: Winners or Losers under Climate Change and Ocean Acidification?" *Fish and Fisheries* 17 (2): 335–57.

Loseto, L.L., C. Hoover, S. Ostertag, D. Whalen, T. Pearce, J. Paulic, J. Iacozza, and S. MacPhee. 2018. "Beluga Whales (*Delphinapterus leucas*), Environmental Change and Marine Protected Areas in the Western Canadian Arctic." *Estuarine, Coastal and Shelf Science* 212: 128–37. https://www.sciencedirect.com/science/article/abs/pii/S027277 1417311319.

MacDuffee, M., and J. Young. 2018. "Salish Sea Orcas Need Immediate Actions to Survive." David Suzuki Foundation, https://davidsuzuki.org/expert-article/salish-sea-orcas-need-immediate-actions-to-survive/.

Marshall, N.A., and P.A. Marshall. 2007. "Conceptualizing and Operationalizing Social Resilience within Commercial Fisheries in Northern Australia." *Ecology and Society* 12 (1).

Miller, D., and U.R. Sumaila. 2015. "Doubly Lucky: Economic Impact of the English Bay Bunker Oil Spill of Vancouver, BC." UBC Fisheries Centre Working Paper Series 2015-101, University of British Columbia, Vancouver.

Miller, D.D., N. Hotte, and U.R. Sumaila. 2014. "Mandating responsible flagging practices as a strategy for reducing the risk of coastal oil spills." *Marine Pollution Bulletin* 81 (1): 24–26. https://www.sciencedirect.com/science/article/abs/pii/S0025326X14000149.

Mills, K.E., A.J. Pershing, C.J. Brown, Y. Chen, F.-S. Chiang, D.S. Holland, S. Lehuta, et al. 2013. "Fisheries Management in a Changing Climate: Lessons from the 2012 Ocean Heat Wave in the Northwest Atlantic." *Oceanography* 26 (2): 191–95.

Nair, R. "'They're Flat Broke': Salmon Fishermen Demand Disaster Relief for Failed Season." *CBC News,* August 21. https://www.cbc.ca/news/canada/british-columbia/salmon-fishermen-disaster-relief-1.5255217.

National Snow and Ice Data Center. 2018. "Arctic Sea Ice Maximum at Second Lowest in the Satellite Record." Press release, March 23. https://nsidc.org/news-analyses/news

-stories/arctic-sea-ice-maximum-second-lowest-satellite
-record.

Niemi, A., S. Ferguson, K. Hedges, H. Melling, C. Michel, B. Ayles, K. Azetsu-Scott, et al. 2019. *State of the Arctic Ocean Report 2019*. Canadian Technical Report of Fisheries and Aquatic Sciences. Fisheries and Oceans Canada.

Okey, T.A., H.M. Alidina, V. Lo, and S. Jessen. 2014. "Effects of Climate Change on Canada's Pacific Marine Ecosystems: A Summary of Scientific Knowledge." *Reviews in Fish Biology and Fisheries* 24 (2): 519–59.

Oliver, E.C.J., M.G. Donat, M.T. Burrows, P.J. Moore, D.A. Smale, L.V. Alexander, J.A. Benthuysen, et al. 2018. "Longer and More Frequent Marine Heatwaves over the Past Century." *Nature Communications* 9: 1–12.

Ommer, R.E. 2007. *Coasts Under Stress: Restructuring and Social-Ecological Health*. Montreal and Kingston: McGill-Queen's University Press.

Osborne, E., J. Richter-Menge, and M. Jeffries, eds. 2018. *Arctic Report Card 2018*. http://www.arctic.noaa.gov/Report-Card.

Overland, J.E., E. Hanna, I. Hanssen-Bauer, S.-J. Kim, J.E. Walsh, M. Wang, U.S. Bhatt, and R.L. Thoman. 2018. "Surface Air Temperature." In *Arctic Report Card 2018*. https://arctic.noaa.gov/Report-Card.

Parker, R.W.R., J.L. Blanchard, C. Gardner, B.S. Green, K. Hartmann, P.H. Tyedmers, and R.A. Watson. 2018. "Fuel Use and Greenhouse Gas Emissions of World Fisheries." *Nature Climate Change* 8 (4): 333–37.

Pauly, D., M.L. Palomares, R. Froese, P. Sa-a, M. Vakily, D. Preikshot, and S. Wallace. 2001. "Fishing Down Canadian Aquatic Food Webs." *Canadian Journal of Fisheries and Aquatic Sciences* 58 (1): 51–62.

Pearce, T., J. Ford, A.C. Willox, and B. Smit. 2015. "Inuit Traditional Ecological Knowledge (TEK): Subsistence Hunting and Adaptation to Climate Change in the Canadian Arctic." *Arctic* 68 (2): 233–13.

Pembina Institute. 2016. "Pipeline Approvals Highlight Need for a Complete Pan-Canadian Plan." Press release, November 29. https://www.pembina.org/media-release/pipeline-approvals-highlight-need-for-a-complete-pan-canadian-plan.

Pepin, P., G. Maillet, S. Fraser, T. Shears, and G. Redmond. 2013. *Optical, Chemical, and Biological Oceanographic Conditions on the Newfoundland and Labrador Shelf during 2011–12*. DFO Canadian Science Advisory Secretariat Research Document 2013/051. Ottawa: Fisheries and Oceans Canada.

Perovich, D., W. Meier, M. Tschudi, S. Farrell, S. Hendricks, S. Gerland, C. Haas, et al. 2017. "Sea Ice." In *Arctic Report Card 2017*. http://www.arctic.noaa.gov/Report-Card.

Pershing, A.J., M.A. Alexander, C.M. Hernandez, L.A. Kerr, A. Le Bris, K.E. Mills, J.A. Nye, et al. 2015. "Slow Adaptation in the Face of Rapid Warming Leads to Collapse of the Gulf of Maine Cod Fishery." *Science* 350 (6262): 809–12.

Peterson, W., N. Bond, and M. Robert. 2016. "The Blob (Part Three): Going, Going, Gone?" *PICES Press* 24: 46–48.

Petheram, L., K.K. Zander, B.M. Campbell, C. High, and N. Stacey. 2010. "'Strange Changes': Indigenous Perspectives of Climate Change and Adaptation in NE Arnhem Land (Australia)." *Global Environmental Change* 20 (4): 681–92.

Pinsky, M.L., G. Reygondeau, R. Caddell, J. Palacios-Abrantes, J. Spijkers, and W.W.L. Cheung. 2018. "Preparing Ocean Governance for Species on the Move." *Science* 360 (6394): 1189–91.

Plourde, S., M. Starr, L. Devine, J.-F. St-Pierre, L. St-Amand, P. Joly, and P.S. Galbraith. 2014. *Chemical and Biological Oceanographic Conditions in the Estuary and Gulf of St. Lawrence during 2011 and 2012*. DFO Canadian Science Advisory Secretariat Research Document 2014/049. Ottawa: Fisheries and Oceans Canada.

Price, A.M., M.R.S. Coffin, V. Pospelova, J.S. Latimer, and G.L. Chmura. 2017. "Effect of Nutrient Pollution on Dinoflagellate Cyst Assemblages across Estuaries of the NW Atlantic." *Marine Pollution Bulletin* 121 (1–2): 339–51.

Reo, N.J., K.P. Whyte, D. McGregor, M.P. Smith, and J.F. Jenkins. 2017. "Factors That Support Indigenous Involvement in Multi-Actor Environmental Stewardship." *AlterNative: An International Journal of Indigenous Peoples* 13 (2): 58–68.

Riche, O., S.C. Johannessen, and R.W. Macdonald. 2014. "Why Timing Matters in a Coastal Sea: Trends, Variability and Tipping Points in the Strait of Georgia, Canada." *Journal of Marine Systems* 131: 36–53.

Richter-Menge, J., J.E. Overland, and J.T. Mathis, eds. 2016. *Arctic Report Card 2016*. http://www.arctic.noaa.gov/Report-Card.

Shackell, N.L., D. Ricard, and C. Stortini. 2014. "Thermal Habitat Index of Many Northwest Atlantic Temperate Species Stays Neutral under Warming Projected for 2030 but Changes Radically by 2060." *PLOS One* 9 (3): e90662.

Sillmann, J., V.V. Kharin, F.W. Zwiers, X. Zhang, and D. Bronaugh. 2013. "Climate Extremes Indices in the CMIP5 Multimodel Ensemble: Part 2. Future Climate Projections."

Journal of Geophysical Research: Atmospheres 118 (6): 2473–93.

Simonson, E.J., R.E. Scheibling, and A. Metaxas. 2015. "Kelp in Hot Water: I. Warming Seawater Temperature Induces Weakening and Loss of Kelp Tissue." *Marine Ecology Progress Series* 537: 89–104.

Smale, D.A., T. Wernberg, E.C.J. Oliver, M. Thomsen, B.P. Harvey, S.C. Straub, M.T. Burrows, et al. 2019. "Marine Heatwaves Threaten Global Biodiversity and the Provision of Ecosystem Services." *Nature Climate Change* 9 (4): 306–12.

Statistics Canada. 2023. "Table 17-10-0135-01: Population Estimates, July 1, by Census Metropolitan Area and Census Agglomeration, 2016 Boundaries." https://doi.org/10.25318/1710013501-eng.

Steiner, N., K. Azetsu-Scott, J. Hamilton, K. Hedges, X. Hu, M.Y. Janjua, D. Lavoie, et al. 2015. "Observed Trends and Climate Projections Affecting Marine Ecosystems in the Canadian Arctic." *Environmental Reviews* 23 (2): 191–239.

Steiner, N., W.W. Cheung, A.M. Cisneros-Montemayor, H. Drost, H. Hayashida, C. Hoover, J. Lam, et al. 2019. "Impacts of the Changing Ocean–Sea Ice System on the Key Forage Fish Arctic Cod (*Boreogadus saida*) and Subsistence Fisheries in the Western Canadian Arctic: Evaluating Linked Climate, Ecosystem and Economic (CEE) Models." *Frontiers in Marine Science* 6 (179).

Stortini, C.H., N.L. Shackell, P. Tyedmers, and K. Beazley. 2015. "Assessing Marine Species Vulnerability to Projected Warming on the Scotian Shelf, Canada." *ICES Journal of Marine Science* 72 (6): 1731–43.

Strasburger, W.W., J.H. Moss, K.A. Siwicke, and E.M. Yasumiishi. 2016. *Results from the Eastern Gulf of Alaska Ecosystem Assessment, July through August 2016*. NOAA Technical Memorandum No. NMFS-AFSC-363. Washington, DC: National Oceanic and Atmospheric Administration.

Sumaila, U.R., A.M. Cisneros-Montemayor, A. Dyck, L. Huang, W.W.L. Cheung, J. Jacquet, K. Kleisner, et al. 2012. "Impact of the Deepwater Horizon Well Blowout on the Economics of US Gulf Fisheries." *Canadian Journal of Fisheries and Aquatic Sciences* 69 (3): 499–510.

Sumaila, U.R., T.C. Tai, V.W.Y. Lam, W.W.L. Cheung, M. Bailey, A.M. Cisneros-Montemayor, O.L. Chen, and S.S. Gulati. 2019. "Benefits of the Paris Agreement to Ocean Life, Economies, and People." *Science Advances* 5 (2): eaau3855.

Suprenand, P.M., C.H. Ainsworth, and C. Hoover. 2018. "Ecosystem Model of the Entire Beaufort Sea Marine Ecosystem: A Temporal Tool for Assessing Food-Web Structure and Marine Animal Populations from 1970 to 2014." Marine Science Faculty Publications 261, University of South Florida. https://scholarcommons.usf.edu/msc_facpub/261/.

Tai, T.C., C.D.G. Harley, and W.W.L. Cheung. 2018. "Comparing Model Parameterizations of the Biophysical Impacts of Ocean Acidification to Identify Limitations and Uncertainties." *Ecological Modelling* 385: 1–11.

Tai, T.C., N. Steiner, C. Hoover, W.W.L. Cheung, and U.R. Sumaila. 2019. "Evaluating Present and Future Potential of Arctic Fisheries in Canada." *Marine Policy* 108: 103637.

Talloni-Álvarez, N.E., U.R. Sumaila, P. Le Billon, and W.W.L. Cheung. 2019. "Climate Change Impact on Canada's Pacific Marine Ecosystem: The Current State of Knowledge." *Marine Policy* 104: 163–76.

Tittensor, D.P., T.D. Eddy, H.K. Lotze, E.D. Galbraith, W.W.L. Cheung, M. Barange, J.L. Blanchard, et al. 2018. "A Protocol for the Intercomparison of Marine Fishery and Ecosystem Models: Fish-MIP v1.0." *Geoscientific Model Development* 11 (4): 1421–42. https://www.geosci-model-dev.net/11/1421/2018/.

Trites, A.W. 2007. "Killer Whales, Whaling, and Sequential Megafaunal Collapse in the North Pacific: A Comparative Analysis of the Dynamics of Marine Mammals in Alaska and British Columbia Following Commercial Whaling." *Marine Mammal Science* 23: 751–65.

Turner, N., and P. Spalding. 2013. "'We Might Go Back to This': Drawing on the Past to Meet the Future in Northwestern North American Indigenous Communities." *Ecology and Society* 18 (4): 29.

Watts, P., K. Koutouki, S. Booth, and S. Blum. 2017. "Inuit Food Security in Canada: Arctic Marine Ethnoecology." *Food Security* 9 (3): 421–40.

Wernberg, T., S. Bennett, R.C. Babcock, T. de Bettignies, K. Cure, M. Depczynski, F. Dufois, et al. 2016. "Climate-Driven Regime Shift of a Temperate Marine Ecosystem." *Science* 353 (6295): 169–72.

Whitney, F.A. 2015. "Anomalous Winter Winds Decrease 2014 Transition Zone Productivity in the NE Pacific." *Geophysical Research Letters* 42 (2): 428–31.

Wilson, K.L., L.M. Kay, A.L. Schmidt, and H.K. Lotze. 2015. "Effects of Increasing Water Temperatures on Survival and Growth of Ecologically and Economically Important Seaweeds in Atlantic Canada: Implications for Climate Change." *Marine Biology* 162 (12): 2431–44.

Wilson, K.L., and H.K. Lotze. 2019. "Projected Range Shift of Eelgrass *Zostera marina* in the Northwest Atlantic with Climate Change." *Marine Ecology Progress Series* 620: 47–62.

Wilson, K.L., M.A. Skinner, and H.K. Lotze. 2019. "Projected 21st-Century Distribution of Canopy-Forming Seaweeds in the Northwest Atlantic with Climate Change." *Diversity and Distributions* 25 (4): 582–602.

Wolf, J., and S.C. Moser. 2011. "Individual Understandings, Perceptions, and Engagement with Climate Change: Insights from In-Depth Studies across the World." *Wiley Interdisciplinary Reviews: Climate Change* 2 (4): 547–69.

4

Large Changes in Canada's Oceans and Their Impacts on Ecosystems and Fisheries

Nadja S. Steiner, Andrea Bryndum-Buchholz, William W.L. Cheung, Amber M. Holdsworth, Heike K. Lotze, Nidhi Nagabhatla, Sarah L. Newell, Thomas A. Okey, Juliano Palacios- Abrantes, U. Rashid Sumaila, Travis C. Tai, Kristen L. Wilson, and the Community of Chesterfield Inlet

Canada's marine environments are undergoing large-scale changes defined by shifts and trends in their mean physical or chemical states over decadal time scales or large spatial scales (ocean basin–wide). These changes affect biological communities and human societies through impacts on marine ecosystems, subsistence practices, especially those of Indigenous Canadians, commercial and recreational fisheries, ecotourism, and other ecosystem services.

Climate models and regional downscaling approaches are used to assess these trends and projections. Recent national assessments for Canada's three oceans highlight increased atmospheric and ocean warming (especially in the Arctic), modified ocean stratification affecting the nutrient supply for marine species, regionally varying changes in primary production, decreasing oxygen levels, advancing ocean acidification (OA) (most prominently in the Arctic), decreasing sea-ice extent and thickness, modified sea-ice characteristics and storm waves, and increasing, climate change–related extreme events. These observed trends are projected to continue into the future with continued greenhouse gas emissions (Steiner et al. 2015; Christian and Foreman 2013; Loder et al. 2013). Differences among future emissions scenarios become apparent only after several decades, as upcoming changes are already determined by past emissions. On decadal time scales, natural variability is expected to equal long-term trends in magnitude, complicating the detection of long-term trends in recorded observations. Furthermore, local effects (e.g., topography-influenced ocean/atmosphere circulation) might mask trends locally. These results are reinforced in *Canada's Changing Climate Report* (Bush and Lemmen 2019), the *IPCC Special Report on the Ocean and Cryosphere in a Changing Climate* (Pörtner et al. 2019), and recent Arctic Monitoring and Assessment Programme (AMAP) reports (AMAP 2018a, 2018b, 2018c). The *IPCC Special Report* (Pörtner et al. 2019) indicates that many marine species have undergone shifts in geographic ranges, seasonality, composition, and abundance, with cascading impacts on ecosystem structure and functioning. This has impacted ecosystems and their services with regionally diverse (positive and negative) outcomes but negative consequences for health and well-being generally and for Indigenous Peoples dependent on fisheries.

The global assessment of biodiversity and ecosystem services by the Intergovernmental Science-Policy Platform on Biodiversity and Ecosystem Services (IPBES) (2019) indicates that 66% of the global ocean area is experiencing increasing cumulative human impacts, including overexploitation of marine species, land- and sea-based pollution, and land- and sea-use changes. It states that sustaining and conserving fisheries, marine species, and ecosystems can be achieved through coordinated interventions on land, in freshwater and oceans, and through multi-level coordination across stakeholders. The regional assessment report for the Americas describes how human-induced changes (e.g.,

COUNTRY FOOD IN CHESTERFIELD INLET, NUNAVUT

Chesterfield Inlet is a small coastal Inuit community in Hudson Bay, Nunavut (population ~400). Community members practise traditional methods of harvesting local wildlife and other country food for consumption. Sharing country food is an important practice that supports the food security of Elders and families without a hunter. Elders advocate for sharing rather than selling country food, as it would be too expensive for many community members and because this practice promotes community well-being. For thousands of years, Inuit Traditional Knowledge (Inuit Qaujimajatuqangit, IQ) has helped keep them safe while on the land and guided harvesting methods. However, community members are noticing that climate change is affecting their ability to use IQ to harvest safely and effectively. Sea ice is thinner, forming later in the fall, breaking up earlier in the spring, and putting hunters' travel at risk. Snow quality and wind direction have changed, making it a challenge to build shelters or navigate back to the community. Locals are also noticing changes in wildlife, and different varieties of plants that are taller than those typically seen. The concern about parasite infections in walrus has increased, resulting in the need to test fresh meat before consumption. Caribou and seal harvests are becoming inconsistent due to increased ship traffic with less ice. Given the importance of sharing country food, any decrease in quantity and quality of harvested species will impact the entire community (Newell et al. 2020).

pollution, invasive species) increasingly influence and amplify the impacts of climate change–related drivers on livelihoods and economies (IPBES 2018).

Changes in Canada's oceans, and their consequences for ecosystems and fisheries, are highlighted in this chapter through a three-lens perspective: (1) local/Indigenous views of large changes through quotes and storytelling; (2) scientific perspectives through evaluation of observed and modelled trends and projections; and (3) impacts on human communities that depend on marine resources. Changes vary regionally due to differences in human uses and environmental conditions in Canada's Pacific, Arctic, and Atlantic regions.

LARGE CHANGES ACROSS CANADA'S THREE OCEANS

Taking Stock

The distribution and abundance of marine species are linked to environmental conditions. Hence, changes in those conditions will directly and indirectly affect species composition, ecosystem function, and fisheries. While modelling approaches are advancing rapidly, the complexity of physical and biological dynamics and their interactions with human activities in Canada's marine ecosystems can confound projections of ecological outcomes.

Changing Temperatures and Stratification

The impact of ocean warming on the physiology of ectothermic marine species (i.e., animals that are dependent on external sources of body heat, such as fish, copepods, and bivalves) is of particular concern due to their complete dependence on habitat for thermoregulation. Temperature drives metabolism and basic performance, including development, growth, reproduction, and survival (e.g., Pörtner and Farrell 2008). Marine species are acclimated to optimum temperature ranges, and their distributions are shaped by temperature, prey availability, and predators. The distribution of marine primary producers (phytoplankton), which fuel the marine food chain, is highly dependent on environmental conditions that affect light and nutrient availability. Key nutrient supply is provided through mixing (e.g., areas with high wind input or strong tidal currents) and upwelling from the deeper ocean. Increased stratification due to warming or freshwater input (via sea-ice melt, precipitation, runoff) can reduce this nutrient supply.

Advancing Ocean Acidification

Fennel and colleagues (2019) reviewed the carbon flux for the North American coastal waters, indicating those as an overall sink of atmospheric carbon dioxide with an estimated 160 teragrams of carbon per year (Tg C yr^{-1}) uptake for the North American exclusive economic zone (EEZ) (59 Tg C yr^{-1} anthropogenic contribution). The estimated North American EEZ uptake (representing ~4% of the global ocean surface area) amounts to 6.4% of the global ocean uptake of atmospheric CO$_2$ of 2,500 Tg C yr^{-1}, indicating about 50% higher uptake than the global average. Over half of this uptake is occurring in the Labrador Sea, Gulf of Alaska, and Bering Sea, which account for 93 Tg C yr^{-1} (26 Tg C yr^{-1} anthropogenic contribution), 58% of the total uptake, while making up only 29% of the combined EEZ area. The Arctic and Subarctic, mid-latitude Atlantic, and mid-latitude Pacific account for 104, 62, and –3.7 Tg C yr^{-1}, respectively. The direct effect of rising atmospheric CO$_2$ levels alone will likely amplify coastal CO$_2$ uptake and carbon export, at least until the ocean's buffering capacity is significantly reduced. The extent of this increase will depend on the rate of the atmospheric CO$_2$ rise, the residence time of shelf waters, the carbon content of open-ocean source waters supplied to coastal regions, and indirect effects due to changes in atmosphere-ocean interactions (e.g., sea-ice cover, wind, and heat fluxes).

A major concern related to enhanced CO$_2$ uptake is ocean acidification, which can further affect the physiology, growth, and life cycles of marine organisms, specifically calcifiers, and can trigger cascading ecosystem effects (AMAP 2018c). The saturation state (Ω) of calcium carbonate (CaCO$_3$) in seawater, in the form of aragonite and calcite, is reduced with increases in dissolved CO$_2$ concentrations. When the saturation state is reduced to the undersaturated level ($\Omega < 1.0$), shells and skeletons of calcifiers begin corroding and dissolving, and the formation of new ones are impeded. Decreased saturation states can also have adverse effects on metabolism, respiration, immune response, and the development of different life stages of marine fish and invertebrate species (AMAP 2018c).

Changing Oxygen Levels

Ocean warming reduces the amount of soluble oxygen in the ocean while decreasing ocean-atmosphere exchange due to increased ocean stratification. The resulting oxygen loss (deoxygenation) is a major consequence of climate change and is exacerbated by other aspects of global change. An average global loss of 2% or more has been recorded in the open ocean over the past 50–100 years, with greater losses in some intermediate waters (100–600 m), including in the North Pacific. In coastal systems, deoxygenation is further accelerated by increasing nutrient delivery from watersheds (Levin 2018; Breitburg et al. 2018). "Hypoxia" refers to low or depleted oxygen and is often associated with the overgrowth and decomposition of certain species of algae. Oxygen minimum zones have expanded, and areas with hypoxia have increased significantly since the 1950s (Breitburg et al. 2018), reducing habitat for marine species. Changes in respiration, circulation, nutrient inputs, and possibly methane release all contribute to oxygen loss, often indirectly through biological processes, and even small changes in oxygenation can have significant biological effects (Levin 2018). Global deoxygenation (with regional variability) is projected to continue for the next century under most emissions scenarios.

Changing Sea Ice

> [Sea ice is] a lot thinner than it used to be. Back then [when the participant was young] there were 20 feet thick [ice floes] ... some of them thick enough to last the whole summer till winter, but not any more. They're like only 8 feet, 10 feet thick. And a lot of [ice is] too thin, a lot thinner than it used to be, melts faster ... It's a lot more scary to hunt ... It's getting harder to go down to the floe edge without gear or anything, never know when you gonna hit a soft spot ... Back then ... you would never see the floe edge but now you can see the floe edge all year long. (Community member, Chesterfield Inlet)

Changes in sea ice directly impact on-ice travel for humans and animals, the exchange of heat and gases between atmosphere and ocean, and the marine ecosystem (Lannuzel et al. 2020). Sea ice provides a habitat and nursery for zooplankton, fish, and marine mammal species, with bottom ice algae providing food and fuelling the benthic system. Sea-ice retreat may reduce ice algae habitat (Tedesco, Vichi, and Scoccimarro 2019) but increase oceanic phytoplankton production (Vancoppenolle et al. 2013). Impacts on the food chain might positively or negatively affect the body condition of higher–trophic level species (Harwood et al. 2015), leading to winners and losers. Changes in ice breakup and freeze-up impact iconic Arctic species directly, by affecting either their pupping or nursery platforms. Changes in sea-ice presence also affect their migration pathways on ice (polar bears) or along ice (e.g., beluga whales), refuge from predation, and accessibility to food (e.g., Loseto et al. 2018). Sea-ice retreat accelerates OA in the Arctic, through both surface water dilution and increased air-ocean interaction (AMAP 2018c; Steiner et al. 2014, 2015), and allows for enhanced wave action, coastal erosion, and sediment input into the ocean (e.g., AMAP 2018a, 2018b). Enhanced melt also affects ocean stratification in the Arctic and North Atlantic outflow regions.

Changes in Ocean Usage and Multiple Stressors

Coastal habitats, including estuaries and deltas, are critical for marine biota, foraging mammals (e.g., Figure 4.1), and regional economies (e.g., the lobster fisheries in Atlantic Canada, Figure 4.2), and have been severely affected by sea-use changes (e.g., coastal development, aquaculture, seafloor trawling) and land-use changes (e.g., land clearance, urban sprawl along coastlines, erosion, pollution), as described in the IPBES global assessment report (IPBES 2019). The report further states that ocean mining has expanded since the early 1980s and sea-ice retreat will likely allow expansion into Arctic regions and increased marine shipping (IPBES 2019). Over 40% of the global ocean area was strongly affected by multiple drivers in 2008, and 66% was experiencing increasing cumulative impacts in 2014; only 3% was described as free from human pressure. These changes add stressors to the environmental changes described above. Pörtner and Farrell (2008) suggest that most marine species perform best within a window of optimal thermal conditions. The additional energy a species requires to acclimate to shifting conditions can reduce its optimum performance

Figure 4.1 Grizzly bear foraging at the rocky shores of the Pacific Great Bear Rainforest, British Columbia | Photo: Finn Steiner

▶ *Figure 4.2* Peggy's Cove, Nova Scotia. A lobster feast in Peggy's Cove is a key tourist attraction in this Atlantic region. | Photo: iStock

window (e.g., Drost, Carmack, and Farrell 2016). Additional stressors (e.g., deoxygenation, acidification, nutrient stress) may shrink the thermal window and limit the thermal range of optimal performance and acclimation potential (Pörtner and Farrell 2008). Identifying these physiological limits in marine species is key to understanding their ability and way of response to multiple stressors, including shifting spatial distributions (Steiner, Drost, and Hunter 2018). Human habitation, overfishing, industrial and agricultural activities, increased shipping, noise, pollution, anthropogenic contaminants, plastics, altered food webs, and the introduction of invasive species represent yet more stressors on the marine ecosystem (e.g., CAFF 2013), which can exacerbate the impacts of oceanographic changes on biological communities and human societies of Canada's oceans (Newell 2018). The *Adaptation Actions for a Changing Arctic* reports highlight that climate warming, combined with changes in the natural and social-ecological systems, is causing cascading effects on marine ecosystems and society, with substantial impacts on human health and quality of life, particularly through the impacts on food resources, shipping, fisheries expansion, and oil and gas exploration (AMAP 2018a, 2018b).

Extreme Events

> *Prince Rupert residents only had one dry day*
> *during the entire month of August and got a third*
> *of their annual rainfall in three summer months.*
> (CBC News *2020*)

Extreme climate events are of high relevance to social-ecological systems due to their potentially severe impacts (e.g., Pörtner et al. 2019). Extreme events, discussed in Chapter 3, tend to be of short duration and are projected to increase throughout the century. However, super-imposing potentially intensifying (in magnitude and frequency) rapid changes on longer-term trends (i.e., large changes) can lead to significantly larger extreme events (heatwaves, storm surges, flooding, sea-ice loss, perma-frost thaw, etc.) than those experienced in any particular region before. This may lead to the need for a much higher level of preparedness in affected communities.

Socio-Economic Impacts

> *The chars are really late going downstream*
> *this year. We usually catch like 60, 70 a day,*
> *I got only 6 this year. (Community member,*
> *Chesterfield Inlet)*

Human uses of the ocean are complex, shaped by (1) physical characteristics of the ocean (its currents and distribution of marine life); (2) terrestrial conditions that have influenced the locations of human settlements; (3) economic pressures; and (4) societal rules (developed to control human activities, including national legisla-tion, the law of the sea, international agreements on human uses of the land and sea). Commercial fishing is a key contributor to the Canadian economy, with around $7 billion in fish and seafood products exported annually (DFO 2019). Traditional subsistence fisheries in the Arctic are still a key part of Inuit life (Figure 4.3; see also the text

Figure 4.3 Elder Andre Nuliajuk Tautu hunting on the sea ice near Chesterfield Inlet, Nunavut | Photo: Doris Tautu

box "Country food in Chesterfield Inlet, Nunavut"). Current subsistence catch across the Canadian Arctic represents only 1% of the total, but it is much higher in regions with few or nonexistent commercial fisheries (e.g., Beaufort Sea, 80%; Hudson Bay, 55%), where changes in marine resources will directly impact food security (Tai et al. 2019). Models have projected the sustainable fisheries potential in Canada's Arctic to strongly increase under the no-mitigation (high-emission) Representative Concentration Pathway (RCP) 8.5, with moderate increases for the high-mitigation (low-emission) RCP2.6 (Tai et al. 2019). However, many species responses are uncertain, and Arctic ecosystems may not be ecologically suitable for subpolar species or resilient to increased exploitation, suggesting a cautionary interpretation of such results and the need to take adequate measures to

ensure sustainability and consideration of ecological, social, cultural, and economic impacts (Tai et al. 2019). Including local information, experiences, and responses in assessments and management decisions is a key component of these considerations (e.g., see the text box "Understanding coastal ocean ecosystem changes with the Local Environmental Observer [LEO] Network"). Globally, severe impacts to ocean ecosystems are illustrated by loss of one-third of fish stocks, classified as overexploited and >50% ocean area being subjected to industrial fishing (IPBES 2019). Impacts of these phenomena vary geographically, with many fish populations projected to move poleward due to ocean warming (Pörtner et al. 2019).

Future Climate Scenarios

Recent assessment reports highlight past trends and future projections of ocean conditions internationally (e.g., Pörtner et al. 2019), nationally (Bush and Lemmen 2019), and regionally for the Pacific (Christian and Foreman 2013), Arctic (Steiner et al. 2015; AMAP 2018a, 2018b, 2018c), and Atlantic (Loder and Van Der Baaren 2013). Projections indicate significant increases in air temperature (0–3°C in summer and 3–7°C in winter over the next 50 years), slight increases in precipitation and snow depth, intensification of extreme events (hot/cold spells, marine heatwaves, extreme precipitation), increases in storm strength and size, and slight increases in wave height and wind speed. Arctic-wide decreases of multiyear ice and increases in open water area and duration are projected to continue into the future. This will allow storm waves to have greater impacts on coasts, enhancing erosion and land-sea sediment transport. Observed trends in ocean properties are projected to continue. Ocean acidification is projected to increase, particularly

▶ *Figure 4.4* Simulated contemporary (2006–25, left column) and future (2066–85, middle column) climatologies for annual mean sea surface temperature (°C), (a–c); sea-ice concentration (%), (d–f); integrated primary production (PP, mmol-C m^{-2} d^{-1}), (g–i); pH, (j–l); and differences (future minus contemporary, right column), from an ensemble of six earth system model simulations (IPSL-CM5A-LR, GFDL-ESM2M, CanESM2, HadGEM2-ES, MPI-ESM-LR, MIROC-ESM; for details, see summary in Steiner et al. 2014) in response to the high-emission scenario (RCP8.5).

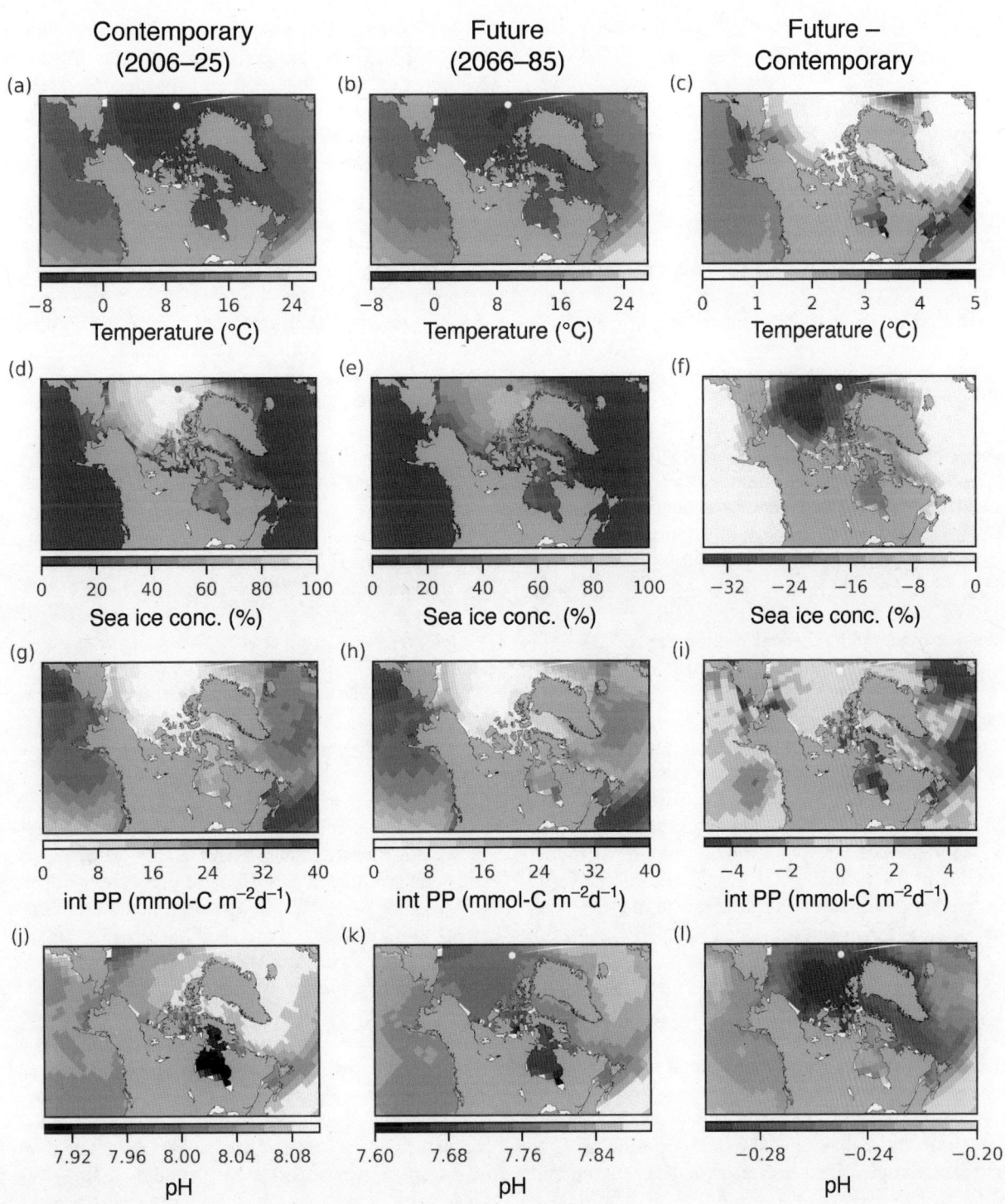

Contemporary (2006–25)

Future (2066–85)

Future – Contemporary

(a) (b) (c)
Temperature (°C)

(d) (e) (f)
Sea ice conc. (%)

(g) (h) (i)
int PP (mmol-C m^{-2}d^{-1})

(j) (k) (l)
pH

throughout the Canadian Arctic. Figure 4.4 shows simulated contemporary (2006–25) and future (2066–85) climatologies for annual mean sea surface temperature, (a–c); sea-ice concentration, (d–f); integrated primary production (PP), (g–i); pH (j–l); and differences (future minus contemporary), from an ensemble of six earth system model (ESM) simulations (see figure caption) for the RCP8.5 scenario. Results indicate sea surface temperature changes of 0–5°C, with lowest changes in areas ice-covered for most of the year. Significant ice retreat is shown in the Canada Basin and Beaufort Sea (>25%). These changes are more pronounced in summer (up to 60%) (Steiner et al. 2015). Integrated PP is suggested to decrease in much of Canada's Pacific and Atlantic Oceans, where stronger stratification reduces the upper ocean nutrient supply. Increases are projected for the Arctic, where the increase in light has a larger impact on phytoplankton growth. For zooplankton, projections suggest consistent declines until the 2040s under RCP2.6 and RCP8.5, with subsequent stabilization under RCP2.6 and continued declines under RCP8.5 (~ −14% ± 3% relative to 1990–99) (Kwiatkowski, Aumont, and Bopp 2019). Consistent decreases in pH highlight the progression of OA with accelerated rates (pH and Ω) in the Arctic (Steiner et al. 2014; AMAP 2018c).

Climate Model Uncertainty

Uncertainty in climate models (including ESMs) has been linked to three main sources: (1) natural internal variability, which is intrinsic to the climate system; (2) emission uncertainty; and (3) uncertainty related to the climate system response to the emissions (model uncertainty) (Kirtman et al. 2013). Emission-related uncertainty is estimated using the spread of projections for different emission scenarios (e.g., RCPs), and the spread among different models is used to estimate model response uncertainty. Due to their internal variability, however, ESMs cannot reliably simulate interannual and decadal variability, which might overlay and mask or enhance a longer-term trend; Arctic sea ice is a key example (Swart et al. 2015).

Projections of Arctic primary production show inconsistent results, mainly related to the opposing impacts of increased light versus decreased nutrient supply, and respective representations in models, but also due to limited understanding and parameterization of ecosystem functions and processes (e.g., Vancoppenolle et al. 2013; Steiner et al. 2015). Differences in projected ice retreat in the ESMs further affect light availability and exchange processes at the ocean surface, increasing model uncertainty. Many processes that are responsible for mixing and nutrient supply are smaller-scale processes (e.g., coastal upwelling, local mixing) and require higher-resolution models. However, dynamical or statistical downscaling procedures may introduce additional uncertainties. Species distribution and food web models forced by ESMs such as those used in Fish-MIP inherit these uncertainties while introducing additional uncertainties due to limited knowledge and parameterization of higher trophic level (HTL) species complexity and physiological responses to multiple stressors.

REGION-SPECIFIC LARGE CHANGES

> *I live[d] here all my life in Chester, born and raised. I noticed the climate changed a lot, since the '60s … Now we have a lot faster ice breaking up and [the snow and ice are] melting a lot faster. It's freezing a lot later, too. Like, in the '60s we used to be able to cross the inlet almost at the beginning of November. But nowadays you have to wait till almost beginning of January to be able to cross, that's how late it freezes up now. (Community member, Chesterfield Inlet)*

Modelling approaches are important tools for hindcasting historical, and projecting future, changes in marine species and ecosystems. For example, species distribution and habitat suitability models use current species distributions and species thresholds to estimate the environmental comfort zone of a species of interest, then assesses shifts in these environmental conditions (from climate models) and the species' capacity to adjust to those shifts. Figure 4.5 shows results from the dynamic bioclimatic envelope model (DBEM) (Cheung et al. 2009; Tai et al. 2019) and the Fisheries and Marine Ecosystem Model Intercomparison Project (Fish-MIP, www.fishmip.org) (Tittensor et al. 2018a). The DBEM models only include commercially valuable species, and thus exclude many noncommercial invertebrates and benthic fish. The DBEM suggests species turnover in all regions and scenarios, but

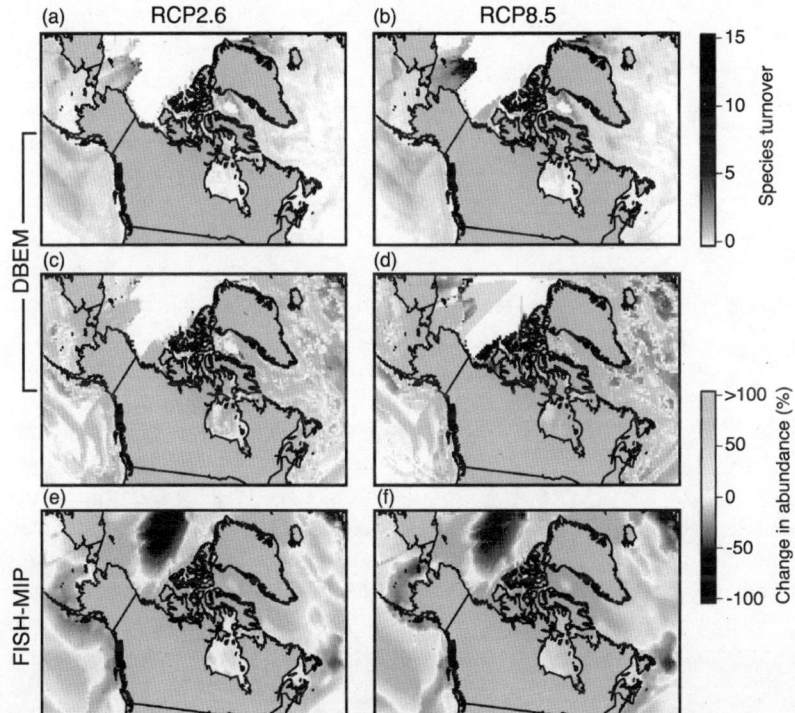

(a) RCP2.6 **(b)** RCP8.5

DBEM

(c) **(d)**

(e) **(f)**

FISH-MIP

Species turnover — 15 / 10 / 5 / 0

Change in abundance (%) — >100 / 50 / 0 / -50 / -100

Figure 4.5 Future climate change scenarios (RCP2.6, left, and RCP8.5, right) of species turnover (a, b) and change in abundance (c, d – DBEM; e, f – Fish-MIP). Species turnover is the sum of invasions and extinctions, and normalized by initial species richness, while change in abundance is the average change in abundance across species in each spatial cell. The DBEM (Cheung et al. 2009; Tai et al. 2019) results represent 939 commercially valuable species (excluding many noncommercial invertebrates and benthic fish). Shown are multi-model averages from simulations with three earth system models (GFDL-ESM2G, IPSL-CM5A-MR, MPI-ESM-MR). Panels (e) and (f) show changes in total higher trophic level biomass from Fish-MIP projections, which used an ensemble of six global marine ecosystem models that were forced with a standardized set of climate change scenarios based on two ESMs, GFDL-ESM2M and IPSL-CM5A-LR (Tittensor et al. 2018a, 2018b). Panels represent annual 2090–99 average, relative to the 1990–99 average.

more so in the Arctic (Figure 4.5a, b). Canada is expected to have increases in catch potential due to the northward shift of many temperate commercial species. The Arctic sees an overall increase, but some areas indicate a larger magnitude of change due to representation of very few species (Figure 4.5c, d). In addition to DBEM, the other five global ecosystem models within the Fish-MIP ensemble are size-class or trophic models, which track the trends in primary production and zooplankton via the flow of energy and production through the food chain, in addition to effects of temperature on metabolism and other factors. Fish-MIP models represent size or trophic groups of commercial and noncommercial animals (including all invertebrates and vertebrates except zooplankton), not individual species, except for DBEM (see discussion in Bryndum-Buchholz et al. 2020). Fish-MIP results hence represent total higher trophic levels and show a decrease in much of the Canadian Pacific and Atlantic (Figure 4.5e, f) with increases in Subarctic and Arctic regions. Biomass and abundance changes are mainly driven by shifts in temperature and primary production, with the latter also depending on nutrient and light availability. Such systematic model intercomparisons and assessments of marine ecosystem changes are essential for understanding the impacts of climate change, including their uncertainty for our oceans and societies (e.g., Lotze et al. 2019). Regional trends and projections from Fish-MIP are summarized in Table 4.1.

Pacific

The Fish-MIP ensemble indicates mean declines of HTL biomass for past and future time periods, with clear differences (about 10%) between RCP2.6 and RCP8.5 emerging in the future (Table 4.1). Around 100 Pacific fish species off Canada are projected to shift their distribution northward, following thermal preferences over the next century (~270–1,630 km for RCP8.5 and ~60 species) (Morley et al. 2018). The northward shift among pelagic species is estimated as ~19 ± 0.5 km and 30 ± 0.6 km per decade for RCP8.5 and RCP2.6, respectively

Table 4.1

Past trends and future projections of mean (± standard deviation, SD) total higher trophic level biomass from the Fish-MIP ensemble for the Canadian exclusive economic zones (EEZs) and the North Atlantic Ocean for RCP2.6 and RCP8.5 (in parentheses)

Region	Hindcast 2010–19 minus 1990–99	Projection 2090–99 minus 1990–99
Canadian Pacific EEZ	−8 ± 8% (−9 ± 8%)	−14.2 ± 14.1% (−21.1 ± 23.3%)
Canadian Arctic EEZ	5.2 ± 7.6% (−6 ± 6.8%)	13 ± 18% (22.5 ± 36.5%)
Canadian Atlantic EEZ	0.1 ± 2.8% (−4.5 ± 5.3%)	−7 ± 3% (−24 ± 9.5%)
North Atlantic Ocean	−2.7 ± 3% (−3.7 ± 3.6%)	−12.4 ± 4.6% (−31.7 ± 14%)

Note: SD of similar magnitude or larger than the mean change indicates a high inter-model variability.
Sources: Bryndum-Buchholz et al. 2019, 2020.

(Cheung et al. 2015). As a result, some species, such as albacore tuna (*Thunnus alalunga*), an important commercial species with a narrow thermal niche, are projected to expand their North Pacific habitat, regardless of scenario (Christian and Holmes 2017), and jointly managed transboundary species such as Pacific halibut (*Hippoglossus stenolepis*) are expected to change their proportion of the catch within Canada and the United States (Palacios-Abrantes, Sumaila, and Cheung 2020). Such change has the potential to shift economic threat points – the minimum fisheries payoffs that countries accept for joint management – and thus challenge the sustainability of shared resources between the countries (Sumaila, Palacios-Abrantes, and Cheung 2020). Increased species invasion along the Pacific coast (Cheung et al. 2015) could bring new fisheries and hence new requirements for management arrangements at the southern and northern US-Canada borders (Pinsky et al. 2018).

In the North and coastal Pacific, oxygen depletion is occurring more rapidly than on the global scale, with loss rates of 0.5–1.0 $\mu mol\ kg^{-1}y^{-1}$ common below the surface mixed layer. The oxygen decline between 250 and 400 m depth along British Columbia's continental shelf affects commercial fish populations by decreasing or compressing available fish habitat (e.g., Okey et al. 2014). On the southern shelf during summer, oxygen-poor waters are upwelled and advected onto the continental shelf, where oxidation of organic material further reduces the oxygen content. Oxygen decline southwest of Vancouver Island (−0.83 $\mu mol\ kg^{-1}y^{-1}$ from 1979 to 2011 [Crawford and Pena 2013]) causes hypoxic waters to enter the tidal passages of Haro Strait, where they are mixed with surface water and enter the deep Georgia Strait. Extrapolation of the long-term trend indicates that parts of Georgia Strait could become episodically hypoxic in the early 2040s (Johannessen, Masson, and Macdonald 2014).

Along the Pacific coast, atmospheric carbon dioxide uptake, in combination with intensified upwelling bringing low-pH water onto the shelves, leads to aragonite saturation levels below the saturation threshold in large portions of the subsurface waters. Haigh and colleagues (2015) summarize ocean acidification changes and relevant risks from OA to Pacific Canadian fisheries and marine ecosystems. They indicate negative OA impacts directly affecting farmed shellfish, especially larval stages, and indirectly affecting finfish through changes to lower trophic levels and habitats (e.g., OA may lead to increased fish-killing algal blooms, affecting the salmon aquaculture industry).

Arctic

The pace of change in the Arctic regions is faster than the global average. In contrast to the Pacific and Atlantic, changes in ocean primary production are suggested to be positive. However, model projections of PP vary significantly depending on whether light increase or nutrient decrease has a more pronounced effect (Vancoppenolle et al. 2013). Sea-ice algae show distinct latitudinal response patterns (Tedesco, Vichi, and Scoccimarro 2019): thinning snow cover causes advancement of algal blooms below latitude 66°N, narrowing growth periods yield small changes from 66°N to 74°N, while shifting ice seasons toward more favourable photoperiods drives increases above 74°N. These responses indicate potentially complex consequences in trophic and phenological cascades in the food web, which may lead to positive and negative impacts on HTL species, many of which are harvested by Inuit peoples (Steiner et al. 2019; see also

UNDERSTANDING COASTAL OCEAN ECOSYSTEM CHANGES WITH THE LOCAL ENVIRONMENTAL OBSERVER (LEO) NETWORK

The LEO Network (leonetwork.org) is an online platform for sharing event-based observations of ecological/ environmental changes through observer-expert collaborations, and for enabling the coordination of local responses to those changes. It is openly accessible and represents a growing community of citizen observers and topic experts, integrating all three lenses identified by the OceanCanada Partnership (local, Indigenous, and scientific knowledge) in collaborative stories of change. Shared stories tend to concentrate along coastlines and rivers, highlighting the importance of aquatic ecosystems to people and the heightened observability of changes in these settings. LEO's origin at the Alaska Native Tribal Health Consortium and the support of its implementation in British Columbia by the First Nations Health Authority exemplifies a contemporary manifestation of a traditional OneHealth approach in which ecosystem health is inherent to the well-being of people and communities.

LEO features curated, quality-assured, and fully attributed first-person observations and news articles that can be queried topically, geographically, and temporally. Each post is linked to observations, articles, and related data and resources, including peer-reviewed science. The content emphasizes the effects of observed changes on the human environment and activities. LEO facilitates solutions to environmental/ ecological problems faced by communities by providing a welcoming space for knowledge sharing and collaboration. LEO stimulates political will for solution making by publicly documenting worrisome changes.

the text box "Country food in Chesterfield Inlet, Nunavut"). Ocean acidification is a key issue in the Arctic Ocean, which is naturally closer to critical acidification thresholds. Subsurface waters in the Beaufort Sea are already routinely below aragonite saturation (favouring $CaCO_3$ dissolution) (e.g., AMAP 2018c).

Projected changes in distribution and abundance of commercially valuable Arctic and Subarctic species suggest a net increase across Canada's Arctic, with a ~70% increase in fisheries catch potential for RCP8.5 (Tai et al. 2019; Figure 4.5c, d). Model projections suggest localized invasions of Subarctic species and range contractions of important Arctic species already at their northern range limit, e.g., Arctic cod (*Boreogadus saida*) (Steiner et al. 2019). The Fish-MIP ensemble indicates modest increases of total HTL biomass for the hindcast and a clear increase in the projections. However, high standard deviations around the ensemble mean indicate high intermodel variability, reflecting a broad range of potential future trajectories. For RCP8.5, decreases are projected along the western coast of Hudson Bay, the northern Beaufort Sea, and parts of the Canadian Arctic Archipelago (Bryndum-Buchholz et al. 2020).

An increase in abundance of fisheries species and ice-related lengthening of fishing seasons will impact the potential development of commercial fishing operations. Current commercial fisheries across Canada's Arctic are minimal, and the fisheries potential is projected at a factor of 20 over the next century (Tai et al. 2019). However, significant uncertainties exist around the impacts of multiple stressors on species, with respect to species interaction and food quality (resident versus incoming species), and ecological and cultural implications (e.g., Tai et al. 2019). Arctic marine resources are highly important for Northern Inuit communities that derive up to 40% of their caloric intake from marine environments, and up to 80% of that from marine mammals alone (Cisneros-Montemayor et al. 2016; Watts et al. 2017). Their dependence on these HTL species is highly contingent on the health of the ecosystem and lower trophic levels as critical links of energy transfer (e.g., Loseto et al. 2018; see also the text box "Country food in Chesterfield Inlet, Nunavut").

Atlantic
Overall, climate change–induced warming in the Canadian Atlantic Ocean is stronger than the average North

Atlantic and global conditions. Chlorophyll a concentrations, representing concentrations of primary producers, have generally been below the 1999–2010 average in Atlantic Canada since 2009 (Bernier, Jamieson, and Moore 2018), and are decreasing faster than the global average (Boyce, Lewis, and Worm 2010). However, the magnitude, duration, and start date of the spring bloom have been highly variable. Within the North Atlantic Ocean and the Canadian Atlantic EEZ, net PP is projected to decline by 2–3% under RCP2.6 and by 8–11% under RCP8.5 by the end of the 21st century relative to 1990–99 (Bryndum-Buchholz et al. 2019, 2020). These net PP decreases are projected to be greater in Atlantic Canada than in other ocean regions (Bopp et al. 2013).

Habitat-forming seaweeds (e.g., kelp and fucoids) and seagrass can be used as sentinel indicator species for changing environmental conditions in coastal habitats (Marbà et al. 2017). So far, large-scale range shifts have not been observed in Atlantic Canada (Merzouk and Johnson 2011), but increasing ocean temperatures have driven declines and species transitions in kelp populations along the Scotian Shelf (Filbee-Dexter, Feehan, and Scheibling 2016) while driving increases in kelp populations in the Gulf of St. Lawrence (Krumhansl et al. 2016). For the dominant eelgrass (*Zostera marina*) population, declines are suggested to be tied to other human stressors (e.g., nutrient pollution, invasive species), while increases around Newfoundland are attributed to warmer temperatures (Bernier, Jamieson, and Moore 2018).

Zostera marina is projected to persist throughout Atlantic Canada by 2100 regardless of emissions scenario, with range contractions occurring only along the US coast (Wilson and Lotze 2019). Similarly, fucoids (*Ascophyllum nodosum, Fucus* spp.) are projected to persist throughout Atlantic Canada, with potential habitat loss along the Bay of Fundy and the Scotian Shelf (RCP8.5) (Wilson, Skinner, and Lotze 2019). Loss of kelp (*Laminaria digitata, Saccharina latissima*) is very likely along the Bay of Fundy, Scotian Shelf, and southern Gulf of St. Lawrence by 2100 under RCP8.5, while invasive green algae (*Codium fragile*) are projected to expand northward (Wilson, Skinner, and Lotze 2019).

Shifts are also reported for marine animals along the northwest Atlantic coast. The zooplankton *Calanus finmarchicus* has been declining since 2009, while smaller, less energy-rich zooplankton species increased in abundance (Bernier, Jamieson, and Moore 2018), affecting the distribution, abundance, and quality of food for higher trophic levels (Worm and Lotze 2016). Several benthic invertebrates and fish species are also shifting, with cold-adapted species moving poleward or toward deeper, colder waters offshore (e.g., Worm and Lotze 2016). Warm-adapted species are moving into Canada's Atlantic waters from the south, creating a turnover in regional species composition. Some species show strong declines or collapses at their southern, warm-water distribution limit, such as Atlantic cod (*Gadus morhua*) in the Gulf of Maine (Pershing et al. 2015). Stock assessments indicate significant decreases in fish recruitment over recent decades, with stocks in the Newfoundland-Labrador and Scotian Shelf region declining by ~7 ± 2% and ~2 ± 1.5% per decade, respectively, compared with global average decreases of 3% per decade relative to the historical maximum; generally, recruitment seems positively related to primary production and negatively to warming waters (Britten, Dowd, and Worm 2016). Groundfish showed the most negative decreases in recruitment, with American plaice (*Hippoglossoides platessoides*) and Atlantic cod among the most strongly affected species (Britten, Dowd, and Worm 2016).

The Fish-MIP ensemble indicates mean past declines of total HTL biomass in the Atlantic EEZ (Table 4.1) at twice the rate of global estimates (−2 ± 1.5%) (Lotze et al. 2019). Projected declines are about three times as high for RCP8.5 compared with RCP2.6, and much higher than the global averages of −5 ± 4% and −18 ± 11% for RCP2.6 and RCP8.5, respectively (Lotze et al. 2019), again with strong regional differences (decreases on the Grand Banks and Scotian Shelf, increases for Baffin Bay and Davis Strait [Bryndum-Buccholz et al. 2020]). Range shifts are projected to occur for many species along the Atlantic coast, with many cold-adapted species moving poleward, or offshore toward deeper, colder water, and warm-adapted species arriving from the south (Cheung et al. 2009; Worm and Lotze 2016). Thus important forage fish, including capelin (*Mallotus villosus*) and sandlance (*Ammodytes hexapterus*) are highly vulnerable to warming, as well as several commercially important fish and invertebrates such as Atlantic cod, pollock (*Pollachius virens*), cusk (*Brosme brosme*), several skate species, and snow crab (*Chionoecetes opilio*). In contrast, American lobster (*Homarus americanus*), Atlantic halibut (*Hippoglossus hippoglossus*), and some squid species are expected to thrive in Atlantic waters (Shackell, Ricard, and Stortini 2014; Stortini et al. 2015). For American lobster, projections show increases in abundance near the Gulf of St. Lawrence and more northern Atlantic regions, while areas around Nova Scotia and south of Newfoundland show decreases. Since it is one of the most lucrative shellfish fisheries, changes in American lobster abundance will have major implications for local communities, and for the local and national economy (Tai et al. 2021).

MOVING FORWARD

The coastal marine and sea-ice ecosystems are among the most productive ecosystems globally, providing sustainable livelihoods and biodiversity. While changes are happening on large regional scales, impacts are felt locally. Hence, response measures are required on large as well as local scales. Scientific research tools and Inuit Traditional Knowledge indicate consistent changes in environmental conditions and ecosystem responses, with IQ highlighting the direct impact of these changes on community well-being. While less visible, changes such as ocean acidification require additional research and communication efforts. The combination of local/Indigenous views, scientific perspectives, and societal impacts are key to understanding and responding to large ecosystem changes.

In addition to concerted, international efforts to reduce carbon emissions impacting environmental changes and species shifts on large scales, specific actions to better manage coastal and marine resources on regional and local scales are required. These include ecosystem-based approaches to fisheries management, spatial planning, effective quotas, protection and management of key marine biodiversity areas, reducing ocean pollution, and working closely with producers and consumers (IPBES 2019). The Intergovernmental Panel on Climate Change (Pörtner et al. 2019) highlights the need for context-specific monitoring and forecasting of changes in the ocean and cryosphere to inform adaptation planning and implementation, and reinforces earlier findings highlighting the benefits of ambitious mitigation and effective adaptation for sustainable development and the escalating

costs and risks of delayed action. Past planning measures have somewhat disregarded local knowledge, and existing management actions have very limited community involvement; hence, there is a limited understanding of the impacts of such changes on communities. The development of community-based and community-led monitoring programs (e.g., Canada's Indigenous Guardians [Government of Canada 2022]) allows for more consistent assessments of environmental changes where they are experienced. Linking these changes with observed changes in species enhances the understanding of potential causes. Newell and colleagues (2019) argue for collaborative frameworks that ensure local authorities prioritize solutions for managing degradation of coastal and marine ecosystems during local and provincial development planning. They highlight effective implementation of participatory approaches in creating resilient communities and coastal cities as sustainable economic hubs, and present the Locally Managed Marine Area (LMMA) approach as a community-scale management mechanism that can be scaled to jurisdictions in Canada. In Canada, significant progress with respect to co-management of marine resources has been made in Indigenous communities with finalized land claims agreements. Joint discussions with and presentations of research projects to Indigenous or local communities (e.g., through regional game council or community meetings, or Indigenous involvement in research grant proposals) are key to building effective collaborations.

New joint management plans are required where fish species move across Canada's Atlantic and Pacific political borders. Such plans will require scientific collaboration to reduce uncertainties in projected species distributions and dynamic management tools, negotiations to resolve changes in economic thread points of already shared stocks, and strong legally binding agreements to solve for potential inclusion of new country members and species in already established treaties.

In addition to the United Nations Sustainable Development Goals (SDGs), which the Government of Canada is committed to supporting, and Canada's 2022–26 Federal Sustainable Development Strategy (Environment and Climate Change Canada 2022), which sets out Canada's sustainable development priorities linked to many SDGs – including SDG 7 (affordable and clean energy), SDG 13 (climate action), SDG 14 (life below water), and SDG 15 (life on land) – other significant changes must be made. These include enhancing human, technical, and financial capacity, and adoption of best fisheries management practices that integrate measures to promote conservation financing and corporate social responsibility. In addition, developing new legal and binding instruments and implementing and enforcing global agreements for responsible fisheries and ocean management remain key to a "healthy oceans and healthy people" vision (Nagabhatla et al 2019; IPBES 2019).

REFERENCES

AMAP (Arctic Monitoring and Assessment Programme). 2018a. *Adaptation Actions for a Changing Arctic: Perspectives from the Bering/Chukchi/Beaufort Region.* Oslo: AMAP.

–. 2018b. *Adaptation Actions for a Changing Arctic: Perspectives from the Davis Strait/Baffin Bay Region.* Oslo: AMAP.

–. 2018c. *Arctic Ocean Acidification.* Tromsø: AMAP.

Bernier, R.Y., R.E. Jamieson, and A.M. Moore. 2018. *State of the Atlantic Ocean Synthesis Report.* Canadian Technical Report of Fisheries and Aquatic Sciences 3167. Moncton: Fisheries and Oceans Canada, Gulf Region.

Bopp, L., L. Resplandy, J.C. Orr, S.C. Doney, J.P. Dunne, M. Gehlen, P. Halloran, et al. 2013. "Multiple Stressors of Ocean Ecosystems in the 21st Century: Projections with CMIP5 Models." *Biogeosciences* 10: 6225–45.

Boyce, D., M. Lewis, and B. Worm. 2010. "Global Phytoplankton Decline over the Past Century." *Nature* 466: 591–96.

Breitburg, D., L.A. Levin, A. Oschlies, M. Grégoire, F.P. Chavez, D.J. Conley, V. Garçon, et al. 2018. "Declining Oxygen in the Global Ocean and Coastal Waters." *Science* 359: eaam7240.

Britten, G.L., M. Dowd, and B. Worm. 2016. "Changing Recruitment Capacity in Global Fish Stocks." *Proceedings of the National Academy of Sciences USA* 113: 134–39.

Bryndum-Buchholz, A., F. Prentice, D.P. Tittensor, J. Blanchard, W.W.L. Cheung, V. Christensen, E.D. Galbraith, O. Maury, and H.K. Lotze. 2020. "Differing Marine Animal Biomass Shifts under 21st Century Climate Change between Canada's Three Oceans." *FACETS* 5 (1): 105–22. https://doi.org/10.1139/facets-2019-0035.

Bryndum-Buchholz, A., D. Tittensor, J.L. Blanchard, W.W.L. Cheung, M. Coll, E.D. Galbraith, S. Jennings, O. Maury, and H.K. Lotze. 2019. "Twenty-First-Century Climate Change Impacts on Marine Animal Biomass and Ecosystem Structure across Ocean Basins." *Global Change Biology* 25: 459–72.

Bush, E., and D.S. Lemmen, eds. 2019. *Canada's Changing Climate Report.* Ottawa: Government of Canada.

CAFF (Conservation of Arctic Flora and Fauna). 2013. *Arctic Biodiversity Assessment: Status and Trends in Arctic Biodiversity.* Akureyri, Iceland: CAFF.

CBC News. 2020. "This B.C. City Had the Rainiest Summer on Record in Over 100 Years." *CBC News,* September 2. https://www.cbc.ca/news/canada/british-columbia/prince-rupert-rain-record-1.5709825.

Cheung, W.W.L., R.D. Brodeur, T.A. Okey, and D. Pauly. 2015. "Projecting Future Changes in Distributions of Pelagic Fish Species of Northeast Pacific Shelf Seas." *Progress in Oceanography* 130: 19–31. https://doi.org/10.1016/j.pocean.2014.09.003.

Cheung, W.W.L., V.W.Y. Lam, J.L. Sarmiento, K. Kearny, R. Watson, and D. Pauly. 2009. "Projecting Global Marine Biodiversity Impacts under Climate Change Scenarios." *Fish and Fisheries* 10: 235–51.

Christian, J.R., and M.G.G. Foreman. 2013. *Climate Trends and Projections for the Pacific Large Aquatic Basin.* Canadian Technical Report of Fisheries and Aquatic Sciences 3032. Sidney, BC: Fisheries and Oceans Canada. https://www.researchgate.net/publication/268513773_Climate_Trends_and_Projections_for_the_Pacific_Large_Aquatic_Basin.

Christian, J.R., and J. Holmes. 2017. "Changes in Albacore Tuna Habitat in the Northeast Pacific Ocean under Anthropogenic Warming." *Fisheries Oceanography* 25 (5): 544–54. https://doi.org/10.1111/fog.12171.

Cisneros-Montemayor, A.M., D. Pauly, L.V. Weatherdon, and Y. Ota. 2016. "A Global Estimate of Seafood Consumption by Coastal Indigenous Peoples." *PLOS One* 11: e0166681.

Crawford, W.C., and M. Angelica Peña. 2013. "Declining Oxygen on the British Columbia Continental Shelf." *Atmosphere-Ocean* 51 (1): 88–103.

DFO (Fisheries and Oceans Canada). 2019. "Canada's Sustainable Fisheries." https://www.dfo-mpo.gc.ca/fisheries-peches/sustainable-durable/fisheries-peches/index-eng.html.

Drost, H., E. Carmack, and A. Farrell. 2016. "Acclimation Potential of Arctic Cod *Boreogadus saida* from the Rapidly

Warming Arctic Ocean." *Journal of Experimental Biology* 219: 3114–25. https://doi.org/10.1242/jeb.140194.

Environment and Climate Change Canada. 2022. *Achieving a Sustainable Future: Federal Sustainable Development Strategy 2022 to 2026.* Ottawa: Environment and Climate Change Canada. https://www.fsds-sfdd.ca/en#/en/goals.

Fennel, K., S. Alin, L. Barbero, W. Evans, T. Bourgeois, S. Cooley, J. Dunne, et al. 2019. "Carbon Cycling in the North American Coastal Ocean: A Synthesis." *Biogeosciences* 16: 1281–1304. https://doi.org/10.5194/bg-16-1281-2019.

Filbee-Dexter, K., C.J. Feehan, and R.E. Scheibling. 2016. "Large-Scale Degradation of a Kelp Ecosystem in an Ocean Warming Hotspot." *Marine Ecology Progress Series* 543: 141–52.

Government of Canada. 2022. "Indigenous Guardians." https://www.canada.ca/en/environment-climate-change/services/environmental-funding/indigenous-guardians.html.

Haigh, R., D. Ianson, C.A. Holt, H.E. Neate, and A.M. Edwards. 2015. "Effects of Ocean Acidification on Temperate Coastal Marine Ecosystems and Fisheries in the Northeast Pacific." *PLOS One* 10 (2): e0117533. https://doi.org/10.1371/journal.pone.0117533.

Harwood, L., T. Smith, J. George, S. Sandstrom, W. Walkusz, and G. Divoky. 2015. "Change in the Beaufort Sea Ecosystem: Diverging Trends in Body Condition and/or Production in Five Marine Vertebrate Species." *Progress in Oceanography* 136: 263–73. https://doi.org/10.1016/j.pocean.2015.05.003.

IPBES (Intergovernmental Science-Policy Platform on Biodiversity and Ecosystem Services). 2018. *Summary for Policymakers of the Regional Assessment Report on Biodiversity and Ecosystem Services for the Americas of the Intergovernmental Science-Policy Platform on Biodiversity and Ecosystem Services.* Bonn: IPBES Secretariat.

–. 2019. *Summary for Policymakers of the Global Assessment Report on Biodiversity and Ecosystem Services of the Intergovernmental Science-Policy Platform on Biodiversity and Ecosystem Services.* Bonn: IPBES Secretariat.

Johannessen, S.C., D. Masson, and R.W. Macdonald. 2014. "Oxygen in the Deep Strait of Georgia, 1951–2009: The Roles of Mixing, Deep-Water Renewal, and Remineralization of Organic Carbon." *Limnology and Oceanography* 59 (1): 211–22.

Kirtman, B., S.B. Power, J.A. Adedoyin, G.J. Boer, R. Bojariu, I. Camilloni, F.J. Doblas-Reyes, et al. 2013. "Near-Term Climate Change: Projections and Predictability." In *Climate Change 2013: The Physical Science Basis. Contribution of Working Group I to the Fifth Assessment Report of the Intergovernmental Panel on Climate Change,* edited by T.F. Stocker, D. Qin, G.-K. Plattner, M. Tignor, S.K. Allen, J. Boschung, A. Nauels, Y. Xia, V. Bex, and P.M. Midgley. Cambridge: Cambridge University Press.

Krumhansl, K.A., D.K. Okamoto, A. Rassweiler, M. Novak, J.J. Bolton, K.C. Cavanaugh, S.D. Connell, et al. 2016. "Global Patterns of Kelp Forest Change over the Past Half-Century." *Proceedings of the National Academy of Sciences* 113 (48): 13785–90.

Kwiatkowski, L., O. Aumont, and L. Bopp. 2019. "Consistent Trophic Amplification of Marine Biomass Declines under Climate Change." *Global Change Biology* 25: 218–29.

Lannuzel, D., L. Tedesco, M. van Leeuwe, K. Campbell, H. Flores, B. Delille, L. Miller, et al. 2020. "The Future of Arctic Sea-Ice Biogeochemistry and Ice-Associated Ecosystems." *Nature Climate Change* 10: 983–92. https://doi.org/10.1038/s41558-020-00940-4.

Levin, L.A. 2018. "Manifestation, Drivers, and Emergence of Open Ocean Deoxygenation." *Annual Review of Marine Science* 10 (1): 229–60. https://doi.org/10.1146/annurev-marine-121916-063359.

Loder, J.W., G. Han, P.S. Galbraith, J. Chassé, and A. van der Baaren, eds. 2013. *Aspects of Climate Change in the Northwest Atlantic Off Canada.* Canadian Technical Report of Fisheries and Aquatic Sciences 3045. Fisheries and Oceans Canada.

Loder, J.W., and A. Van Der Baaren. 2013. *Climate Change Projections for the Northwest Atlantic from Six CMIP5 Models.* Canadian Technical Report of Hydrography and Ocean Sciences 286. Fisheries and Oceans Canada.

Loseto, L.L., C. Hoover, N. Ostertag, D. Whalen, T. Pearce, J. Paulic, J. Iacozza, and S. MacPhee. 2018. "Beluga Whales (*Delphinapterus leucas*), Environmental Change and Marine Protected Areas in the Western Canadian Arctic." *Estuarine, Coastal and Shelf Science* 212: 128–37. https://www.sciencedirect.com/science/article/abs/pii/S027271417311319.

Lotze, H.K., D.P. Tittensor, A. Bryndum-Buchholz, T.D. Eddy, W.W.L. Cheung, E.D. Galbraith, M. Barange, et al. 2019. "Global Ensemble Projections Reveal Trophic Amplification of Ocean Biomass Declines with Climate Change." *Proceedings of the National Academy of Sciences* 116 (26): 12907–12. https://doi.org/10.1073/pnas.1900194116.

Marbà, N., D. Krause-Jensen, B. Olesen, P.B. Christensen, A. Merzouk, J. Rodrigues, S. Wegeberg, and R.T. Wilce. 2017. "Climate Change Stimulates the Growth of the

Intertidal Macroalgae *Ascophyllum nodosum* Near the Northern Distribution Limit." *Ambio* 46 (S1): 119–31. https://doi.org/10.1007/s13280-016-0873-7.

Merzouk, A., and L.E. Johnson. 2011. "Kelp Distribution in the Northwest Atlantic Ocean under a Changing Climate." *Journal of Experimental Marine Biology and Ecology* 400 (1–2): 90–98. https://doi.org/10.1016/j.jembe.2011.02.020.

Morley, J.W., R.L. Selden, R.J. Latour, T.L. Frölicher, R.J. Seagraves, and M.L. Pinsky. 2018. "Projecting Shifts in Thermal Habitat for 686 Species on the North American Continental Shelf." *PLOS One* 13 (5): e0196127. https://doi.org/10.1371/journal.pone.0196127.

Nagabhatla, N., N. Hung, L. Tuyen, V. Cam, J. Dhanraj, N. Thien, and F. Swierczek. 2019. "Ecosystem-Based Approach for Planning Research and Capacity Development for Integrated Coastal Zone Management in Southeast Asia." *APN Science Bulletin* 9 (1). https://doi.org/10.30852/sb.2019.537.

Newell, S.L. 2018. "Social, Cultural, and Ecological Systems' Influence on Community Health and Wellbeing." PhD dissertation, McMaster University.

Newell, S.L., N.C. Doubleday, and Community of Chesterfield Inlet, Nunavut. 2020. "Sharing Country Food: Connecting Health, Food Security and Cultural Continuity in Chesterfield Inlet, Nunavut." *Polar Research* 39. https://doi.org/10.33265/polar.v39.3755.

Newell, S.L., N. Nagabhatla, N.C. Doubleday, and A. Bloecker. 2019. "The Potential for Locally Managed Marine Area (LMMAs) as a Participatory Strategy for Coastal and Marine Ecosystems – The Global Commons." *OIDA International Journal of Sustainable Development* 12 (4): 47–62. https://papers.ssrn.com/sol3/papers.cfm?abstract_id=3439121.

Okey, T.A., H.M. Alidina, V. Lo, and S. Jessen. 2014. "Effects of Climate Change on Canada's Pacific Marine Ecosystems: A Summary of Scientific Knowledge." *Reviews in Fish Biology and Fisheries* 24: 519–59. http://dx.doi.org/10.1007/s11160-014-9342-1.

Palacios-Abrantes, J., U.R. Sumaila, and W. Cheung. 2020. "Challenges to Transboundary Fisheries Management in North America under Climate Change." *Ecology and Society* 25 (4): 41. https://doi.org/10.5751/ES-11743-250441.

Pershing A.J., M.A. Alexander, C.M. Hernandez, L.A. Kerr, A. Le Bris, K.E. Mills, J.A. Nye, et al. 2015. "Slow Adaptation in the Face of Rapid Warming Leads to Collapse of the Gulf of Maine Cod Fishery." *Science* 350 (6262): 809–12.

Pinsky, M.L., G. Reygondeau, R. Caddell, J. Palacios-Abrantes, J. Spijkers, and W.W.L. Cheung. 2018. "Preparing Ocean Governance for Species on the Move." *Science* 360 (6394): 1189–91.

Pörtner, H.-O., and A.P. Farrell. 2008. "Ecology, Physiology and Climate Change." *Science* 322: 690–92. https://doi.org/10.1126/science.1163156.

Pörtner, H.-O., D.C. Roberts, V. Masson-Delmotte, P. Zhai, M. Tignor, E. Poloczanska, K. Mintenbeck, et al., eds. 2019. *IPCC Special Report on the Ocean and Cryosphere in a Changing Climate*. Geneva: Intergovernmental Panel on Climate Change.

Shackell, N.L., D. Ricard, and C. Stortini. 2014. "Thermal Habitat Index of Many Northwest Atlantic Temperate Species Stays Neutral under Warming Projected for 2030 but Changes Radically by 2060." *PLOS One* 9 (3): e90662. https://doi.org/10.1371/journal.pone.0090662.

Steiner, N., J. Christian, K. Six, A. Yamamoto, and M. Yamamoto-Kawai. 2014. "Future Ocean Acidification in the Canada Basin and Surrounding Arctic Ocean from CMIP5 Earth System Models." *Journal of Geophysical Research: Oceans* 119 (1): 332–47. https://doi.org/10.1002/2013JC009069.

Steiner, N., K. Azetsu-Scott, J. Hamilton, K. Hedges, X. Hu, M.Y. Janjua, D. Lavoie, et al. 2015. "Observed Trends and Climate Projections Affecting Marine Ecosystems in the Canadian Arctic." *Environmental Reviews* 23 (2): 191–239. https://doi.org/10.1139/er-2014-0066.

Steiner, N., W.W.L. Cheung, A.M. Cisneros-Montemayor, H. Drost, H. Hayashida, C. Hoover, J. Lam, et al. 2019. "Impacts of the Changing Ocean–Sea Ice System on the Key Forage Fish Arctic Cod (*Boreogadus saida*) and Subsistence Fisheries in the Western Canadian Arctic – Evaluating Linked Climate, Ecosystem and Economic (CEE) Models." *Frontiers in Marine Science* 6. https://doi.org/10.3389/fmars.2019.00179.

Steiner, N., H.E. Drost, and K. Hunter. 2018. *A Physiological Limits Database for Arctic and Subarctic Aquatic Species*. Canadian Technical Report of Fisheries and Aquatic Sciences 3256. Fisheries and Oceans Canada.

Stortini, C.H., N.L. Shackell, P. Tyedmers, and K. Beazley. 2015. "Assessing Marine Species Vulnerability to Projected Warming on the Scotian Shelf, Canada." *ICES Journal of Marine Science* 72: 1731–43.

Sumaila, U.R., J. Palacios-Abrantes, and W.W.L. Cheung. 2020. "Climate Change, Shifting Threat Points, and the Management of Transboundary Fish Stocks." *Ecology and Society* 25 (4): 40. https://doi.org/10.5751/ES-11660-250440.

Swart, N.C., J.C. Fyfe, E. Hawkins, J.E. Kay, and A. Jahn. 2015. "Influence of Internal Variability on Arctic Sea Ice Trends." *Nature Climate Change* 5: 86–89. https://doi.org/10.1038/nclimate2483.

Tai, T.C., P. Calosi, H.J. Gurney-Smith, and W.W.L. Cheung. 2021. "Modelling Ocean Acidification Effects with Life Stage-Specific Responses Alters Spatiotemporal Patterns of Catch and Revenues of American Lobster, *Homarus americanus.*" *Scientific Reports* 11 (1): 23330. https://doi.org/10.1038/s41598-021-02253-8.

Tai, T.C., N. Steiner, C. Hoover, W.W.L. Cheung, and U.R. Sumaila. 2019. "Evaluating Present and Future Potential of Arctic Fisheries in Canada." *Marine Policy* 108: 103637. https://doi.org/10.1016/j.marpol.2019.103637.

Tedesco, L., M. Vichi, and E. Scoccimarro. 2019. "Sea-Ice Algal Phenology in a Warmer Arctic." *Science Advances* 5: eaav4830.

Tittensor, D.P., T.D. Eddy, H.K. Lotze, E.D. Galbraith, W.W.L. Cheung, M. Barange, J.L. Blanchard, et al. 2018a. "A Protocol for the Intercomparison of Marine Fishery and Ecosystem Models: Fish-MIP v1.0." *Geoscientific Model Development* 11: 1421–42.

Tittensor D.P., H.K. Lotze, T.D. Eddy, E.D. Galbraith, W.W.L. Cheung, A. Bryndum-Buchholz, M. Barange, et al. 2018b.

"ISIMIP2a Simulation Data from Fisheries and Marine Ecosystems (Fish-MIP; Global) Sector." GFZ Data Services. http://doi.org/10.5880/PIK.2018.005.

Vancoppenolle, M., L. Bopp, C. Madec, J. Dunne, T. Ilyina, P. Halloran, and N. Steiner. 2013. "Future Arctic Primary Productivity from CMIP5 Simulations: Uncertain Outcome, but Consistent Mechanisms." *Global Biogeochemical Cycles* 27: 605–19. https://doi.org/10.1002/gbc.20055.

Watts, P., K. Koutouki, S. Booth, and S. Blum. 2017. "Inuit Food Security in Canada: Arctic Marine Ethnoecology." *Food Security* 9: 421–40.

Wilson, K.L., and H.K. Lotze. 2019. "Climate Change Projections Reveal Range Shifts of Eelgrass *Zostera marina* in the Northwest Atlantic." *Marine Ecology Progress Series* 620: 47–62

Wilson, K.L., M.A. Skinner, and H.K. Lotze. 2019. "Projected 21st-Century Distribution of Canopy-Forming Seaweeds in the Northwest Atlantic with Climate Change." *Diversity and Distribution* 25: 582–602.

Worm, B., and H.K. Lotze. 2016. "Marine Biodiversity and Climate Change." In *Climate Change: Observed Impacts on Planet Earth,* edited by T.M. Letcher, 195–212. Elsevier.

5

Using Multiscale Scenarios to Investigate the Outlook for Marine Ecosystems and Coastal Communities

Louise Teh and William W.L. Cheung, with Natalie Ban, Russ Jones, Lydia Ross, Nadja S. Steiner, and U. Rashid Sumaila

This chapter uses multiscale scenario modelling as an approach for integrating our present knowledge about ocean changes and social and economic trends to produce potential outlooks for Canadian oceans and coastal communities. The scenarios use fisheries as a lens to investigate future impacts of climate and socio-economic change on Canadian marine ecosystems and society at national and local scales in terms of quantitative indicators. This research is needed because existing scenarios of Canadian oceans lack appropriate frameworks and are too narrow in scope to fully consider the multi-dimensional drivers, social-ecological and policy pathways, and geographical scales necessary for a comprehensive investigation of future ocean sustainability in Canada.

To address these gaps, we describe the development of a framework for integrated, multiscale scenarios that can be used for understanding the sustainability of Canadian coastal communities and marine ecosystems under future environmental and socio-economic change at national, regional, and local scales. We define an "integrated" scenario according to three criteria: (1) it considers the interactions and feedback between social and ecological systems; (2) it addresses the impact from natural and human drivers of change and resulting societal or biological responses; and (3) it combines models of climate, economy, social, and environmental systems (Teh et al. 2017). The ultimate aim of developing multiscale scenarios is to facilitate improved national marine governance under the uncertainty of global change by stimulating dialogue with marine stakeholders about potential ocean futures and pathways toward marine stewardship. This is an advantage over more predictive approaches, which are hindered by future uncertainties in environmental and societal development trajectories.

POLICY CONTEXT FOR CANADIAN SCENARIOS

Canadian ocean management is grounded in a policy framework that encompasses national, regional, and local scales. Canada's Oceans Strategy stipulates the equally important roles federal, provincial, territorial, and local governments have for the management of Canadian oceans. While the federal government has broad responsibilities for the stewardship and management of Canada's oceans and resources, shoreline and seabed areas are the primary responsibility of provincial governments, and municipalities are responsible for land-based activities that affect the marine environment. Canada's national mandate on ocean sustainability covers finer-scale policies for different regions, species, communities, and jurisdictions. For example, the national Oceans Protection Plan (OPP) involves various regional- or local-level initiatives for specific species (e.g., marine mammals, Arctic char), habitats (e.g., estuarine, nearshore), and Indigenous Peoples. Moreover, the federal government's policy on reconciliation inherently involves local-scale interests of Indigenous groups. Consequently, understanding potential futures for Canadian oceans requires a multiscale scenario approach that can inform coherent policies for sustainable ocean management from local to national scales.

Canada's three oceans face increasing threats from a wide range of human pressures, such as overfishing, contaminants, and habitat destruction; these are compounded by global effects from climate change and ocean acidification (Hutchings et al. 2012a, 2012b; Baum and Fuller 2016; AMAP 2018a, 2018b; Oceana 2018; Teh and Sumaila 2020). Improving ocean sustainability is thus critical to stop the declining trend in the state of Canadian marine ecosystems; however, restrictive governance policies in Canada impede efforts to address anthropogenic and environmental pressures (Favaro, Reynolds, and Côté 2012; Bailey et al. 2016). Consequently, there is an urgent need to identify future policy options for strategic ocean planning and sustainable management of Canadian marine living resources (Hutchings et al. 2012a). Scenario analysis can address this uncertainty by exploring alternative futures for Canadian oceans under different pathways of climate change, economic development, and social and policy changes. Such investigations can help inform policy actions for mitigating the effects of global change on Canada's oceans, as well as adaptation efforts at local and national levels.

MULTISCALE SCENARIO FRAMEWORK

A scenario is defined as "a coherent, internally consistent, and plausible description of a possible state of the world" (IPCC 2013). Scenarios represent potential futures of direct and indirect drivers, and policy options. These descriptions are translated into consequences for ecosystems, humans, and their well-being through the use of quantitative or qualitative models (IPBES 2016). Scenarios provide a mechanism for synthesizing data and knowledge to articulate alternative visions about the future, thereby allowing opportunities and risks associated with particular decisions to be identified (Bohensky et al. 2011). As such, they are a useful tool for planning under high uncertainty and for supporting much-needed interventions for sustainable ocean policies.

Figure 5.1 illustrates the conceptual framework we use to investigate scenarios of future environmental and socio-economic development change on Canadian fisheries. The scenarios span a 35-year time period from 2015 to 2050. We use 2015 as the starting point in order to

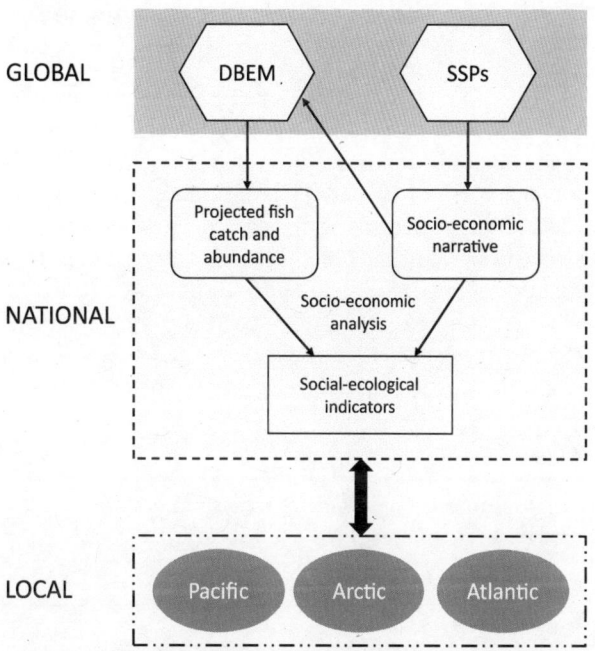

Figure 5.1 Conceptual framework for investigating scenarios of Canadian marine fisheries. Note that the DBEM (dynamic bioclimatic envelope model) is informed by the national-level socio-economic narratives developed for Canada. SSP = Shared Socio-Economic Pathway.

maximize data availability (some data or projections are not available for more recent years). The multiscale scenario development process involves two steps: (1) develop national-scale scenarios by integrating global and national-scale models and data to generate projections of social-ecological outcomes at the national level; and (2) use local-level case studies to add depth and perspective to the projected national outcomes.

Future fisheries catch and abundance are driven by biophysical and socio-economic development trends. In this study, quantitative fisheries projections are derived from a dynamic bioclimatic envelope model (DBEM) (see Appendix 5.2), while future socio-economic narratives are developed based on the global Shared Socio-Economic Pathway (SSP) framework (O'Neill et al. 2013). Thus, projections of fish catch and fish abundance in 2050 are consistent with the particular socio-economic development pathway prevailing in the future.

DEVELOPMENT OF SOCIAL-ECOLOGICAL SCENARIOS

Matrix Framework

The development of national scenarios follows a matrix approach (van Vuuren et al. 2014), in which scenarios are made up of future climate and socio-economic development pathways. Different levels of climate forcing, represented by Representative Concentration Pathways (RCPs) make up one axis, and different socio-economic development storylines (SSPs) make up the second axis. The suite of potential socio-economic and climate pathways forms the two axes of a matrix (Table 5.1). The RCPs and SSPs describe global pathways (e.g., trajectories of global population changes, economic growth, emissions). RCPs present alternative trajectories for atmospheric concentrations of greenhouse gases (van Vuuren et al. 2011). SSPs describe future challenges to mitigation and adaptation in terms of how major socio-economic, demographic, technological, policy, institutional, lifestyle, and other trends may unfold in the future (Table 5.2). Each cell within the matrix thus represents an integrated social-ecological scenario defined by the combination of climate and socio-economic pathways. Plausible scenarios (marked by "×" in Table 5.1) are those that do not have conflicting RCP and SSP pathways. For instance, SSP1 describes a Sustainability pathway, which is not consistent with high RCPs 6.0 and 8.5. The development of national socio-economic development narratives involves the downscaling and translating of global SSPs to a Canadian marine context.

Scenario Narratives

The national narratives were developed in a two-stage process (detailed in Appendix 5.1). As illustrated in Table 5.1, there are many plausible scenarios arising from RCP-SSP combinations. To provide a clear contrast between scenarios, we limit the number of scenarios to two extremes, chosen to reflect futures that are most and least likely to address sustainability goals, plus a middle

Table 5.1

Proposed matrix framework for representing a set of plausible scenarios ("×")

RCP[1]	SSP1[2]	SSP2	SSP3	SSP4	SSP5
RCP2.6	×			×	
RCP4.5	×	×		×	
RCP6.0		×	×		×
RCP8.5			×		×

1 RCP = Representative Concentration Pathway.
2 SSP = Shared Socio-Economic Pathway.
Source: van Vuuren et al. 2014.

Table 5.2

Summary of the five Shared Socio-Economic Pathways

Shared Socio-Economic Pathway (SSP)	Development pathway	Level of challenge to mitigation and adaptation
SSP1: Sustainability	Gradual, persistent shift toward sustainable and environmentally conscious development	Low challenges to mitigation and adaptation
SSP2: Middle of the road	Social, economic, and technology technological trends do not shift much from historical trends	Moderate challenges to mitigation and adaptation
SSP3: Regional rivalry	Weak global institutions for addressing environmental concerns	High challenges to mitigation and adaptation
SSP4: Inequality	Increasing inequality and stratification across and within countries	Low challenges to mitigation, high challenges to adaptation
SSP5: Fossil-fuelled development	Accelerated globalization and rapid development	High challenges to mitigation, low challenges to adaptation

Source: O'Neill et al. 2017.

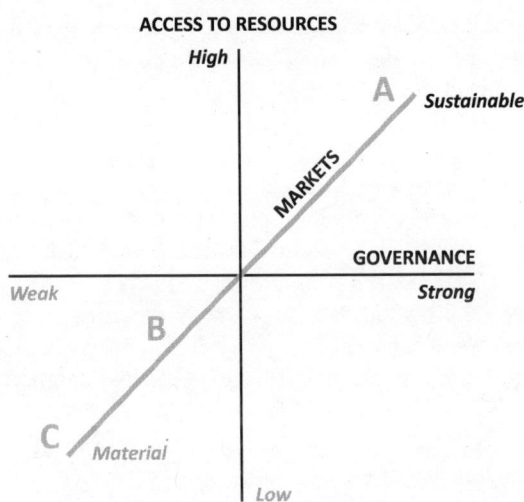

ACCESS TO RESOURCES

Figure 5.2 Framework showing the placement of each scenario's socio-economic pathway (A, B, C) in terms of their position along a continuum of the three main future uncertainties, with the following extremes: Access to Resources – low to high; Markets – material versus sustainable-oriented; Governance – weak to strong.

plausible scenario. Each scenario falls along a continuum of three main future uncertainties: (1) markets, (2) access to resources, and (3) governance (Figure 5.2).

These three scenarios broadly follow the global scenario archetypes outlined by van Vuuren and colleagues (2012): global sustainable development, business as usual, and regional competition (see Table 5.3 below). The plausible RCP-SSP combinations that correspond to each scenario archetype is identified in Table 5.3. Note that Scenario C corresponds to SSP3 with the added emphasis on inequality presented in SSP4, as it is intended to describe a future that presents high challenges to both adaptation and mitigation (Table 5.2). This makes it consistent with an RCP8.5 climate pathway, even though SSP4 by itself would not be consistent with such a pathway (Table 5.1). For each of these three Canadian ocean scenarios, we use the global RCPs as the climate pathways, and translate the global-scale socio-economic development pathways (SSPs) to the national scale in order to create three national-scale ocean narratives for Canada (see Appendix 5.1).

Table 5.3

Summary of scenario archetypes and their application to development of OceanCanada Partnership scenarios

Scenario archetype	Assumption	SSP	Corresponding RCP and scenario narrative in this chapter
Global sustainable development	There is an orientation toward environmental protection, reducing inequality through global cooperation, more efficient technologies, and lifestyle changes. There is a high level of international government cooperation and coordination to address international problems such as poverty alleviation, climate, and nature protection and conservation. This entails regulation of markets at a global scale.	SSP1	RCP2.6, Scenario A
Business as usual	The future will be like a continuation of historical trends. It assumes that historical dynamics will guide the future, i.e., it does not mean that no change will occur.	SSP2	RCP4.5, Scenario B
Regional competition	Regions will focus on national sovereignty and regional identity, resulting in tensions among regions or cultures. Countries are concerned with security and protection, thus there is a priority placed on regional markets, with little attention to common goods. These scenarios are generally pessimistic about global poverty reduction and environmental protection.	SSP3 (with SSP4's depth of inequality)	RCP8.5, Scenario C

Source: Scenario archetypes are based on van Vuuren et al. 2012.

Projected Fish Abundance and Fisheries Catch Potential

Projections of future fish abundance and fisheries catch are simulated from a dynamic bioclimatic envelope model (see detailed method in Appendix 5.2) that has two main drivers: biophysical factors and marine management. Future climate and environmental effects are modelled under two climate change scenarios: RCP2.6 (low emissions) and RCP8.5 (high emissions). The effect of marine management is modelled through assumptions about protected area coverage and fishing mortality rates, which are determined according to each scenario's socio-economic pathway (Figure 5.1, and see Appendix 5.2). Specifically, fishing mortality is used as an input parameter to the DBEM, where it is combined with other environmental and biological variables to project annual fish catch potential and abundance within the Canadian exclusive economic zone (EEZ). Protected area coverage reflects the efficiency of marine management in terms of Canada's ability to fulfill the *Convention on Biological Diversity* target of achieving 10% Marine Protected Area (MPA) coverage by 2020 (Appendix 5.2). To model the potential spillover effects of increased fish stock abundance in protected areas, we use the protected areas identified as Canada's marine conservation targets as the basis for allocating spatial cells within the Canadian EEZ as being closed to fishing.

Economic Analysis

For each scenario, the social-ecological impacts arising from fisheries change and socio-economic development trends in 2050 are quantified through five indicators:

1. Annual landed value of catch = (potential catch × ex-vessel fish price)
2. Net present value (NPV) of projected catch. The NPV of Canadian fisheries in 2050 is equal to the discounted 50-year stream of net benefits from commercial fisheries:

$$\text{NPV} = \sum_{t=0}^{T} \frac{R_{com}}{(1+r)^t}$$

Where:

R_{com} is the net revenue from commercial fisheries, i.e., R_{com} = (landed value – cost of fishing)
T is the end year of the analysis period
r is the discount rate, set at 8%, the rate recommended by the Treasury Board of Canada Secretariat (2007)

Table 5.4

Assumed values for socio-economic variables used to calculate indicators

Assumed levels by 2050		Scenario A	Scenario B	Scenario C	Data source(s)
Canadian population (million people)		46.90	46.90	39.81	Statistics Canada
Fish consumption (kg/person/year)		19.89	22.65	30	FAOSTAT food balance sheets
Fishing cost (% change relative to current)		+15%	+25%	−25%	Current cost based on Teh and Sumaila 2020
Fish price increase by 2050 (%)	High value	+50	–	+50	Assumptions from this study (Appendix 5.3)
	Low value	+25	–	−25	
	Crustaceans	+50	–	−25	
	Molluscs	+50	–	−25	
Fish catch composition in 2050 (% of total projected catch)	High value	30%	20%	10%	Assumptions from this study (Appendix 5.3); current composition based on DFO landings statistics
	Low value	35%	26%	50%	
	Crustaceans	30%	39%	30%	
	Molluscs	5%	16%	10%	

Note: A detailed description for each assumption is provided in Appendix 5.3.

3 Employment (number of fishers). The projected number of fishers in 2050 was estimated based on trends of population change in the number of fishers and the Canadian population. Data on the number of people employed in Canada's fisheries sector were obtained from Fisheries and Oceans Canada (DFO) for the years 2007–17.[1]

4 Fishing income ($/fisher). To calculate individual fishing income, we treat each fisher as an owner-operator, such that income from fishing = (fishing revenue – cost of fishing), where fishing revenue = landed value of potential catch.

5 Food security (kg/person/year) = (catch potential/ Canadian population)

The five indicators are calculated based on assumed future levels of socio-economic variables that vary by scenario, as summarized in Table 5.4. Assumed 2050 values for the different socio-economic variables are derived from a range of published data and/or projections detailed in Appendix 5.3.

Outlook for Canadian Oceans under Context of Global Change

We use national-level scenarios to examine the outlook for Canadian oceans, and use case studies from the Pacific, Atlantic, and Arctic coasts to add local perspectives to the national outlook.

National Outlook

The social-ecological scenarios summarized in Table 5.4 present three potential future scenarios for Canadian fisheries in 2050. Each scenario differs in climate forcing and socio-economic development pathways. Table 5.5 outlines the key uncertainties under each scenario, and the implications for fisheries and marine ecosystems.

National Scenario Outcomes

Projected scenario outcomes in 2050 are represented in terms of socio-economic and ecological indicators. These consist of: (1) quantitative model projections depicting change in future species abundance and fish catch potential under low and high climate scenarios; and (2) socio-economic outcomes for each scenario in 2050 measured by landed value of catch, net present value of Canadian fisheries, employment (number of fishers), fish supply, and fisher revenue.

Under both high and low climate change scenarios, fish catch potential across all three Canadian oceans is expected to decrease in 2050, compared with the present (Table 5.6). The greatest decrease is projected to occur in the Pacific, while the Atlantic is projected to be the least impacted. Conversely, fish species abundance is projected to increase in the Arctic under both climate change scenarios, while it is expected to decrease in the Atlantic and Pacific (Table 5.6). Arctic species abundance appears to do better under a high climate change scenario, whereas projected fish catch potential worsens under RCP8.5.

Under all scenarios, combined climate and socio-economic pathways will result in fewer people fishing and lower total Canadian catches in 2050 relative to a status quo situation with no climate effects. There is overall reduced fish supply per capita compared with the status quo, with the lowest reduction in Scenario C (Table 5.7). Landed value of Canadian fisheries increases substantially under Scenario A, with a lower increase in Scenario B and loss in Scenario C. Revenue to individual fishers is projected to increase slightly under Scenario A, whereas revenue per fisher is reduced by about a third and up to two-thirds under Scenarios B and C, respectively (Table 5.7).

An important point to bear in mind is that increased revenue under Scenario A occurs under more equitable fishing and allocation of fishing rights (see Appendix 5.1). Therefore, individual fishers can retain a lot more of their landed value, as opposed to the current situation in the halibut fishery, for example, where over 70% of landed value goes to the permit holder (Mintz 2018). On the other hand, projected fisheries net present value in 2050 is highest under Scenario C, due mainly to reduced fishing costs, while continuing with the status quo (Scenario B) is estimated to result in NPV loss relative to the present (Figure 5.3).

Regional Scenario Case Studies

National scenario outcomes indicate that the outlook for Canadian fisheries is one where there are fewer fishers, less catch, and reduced fish supply per capita. Marine ecosystems have reduced species abundance, except in the Arctic, where abundance increases. Net economic

Table 5.5

Summary of scenario narratives for three potential pathways and their implications for Canadian fisheries and marine ecosystems

	Scenario		
	A – Sustainability (RCP2.6, SSP1)	*B – Business as usual (RCP4.5, SSP2)*	*C – Decline (RCP8.5, SSP3, and SSP4)*
Key uncertainties			
Governance	Effective cooperation and collaboration across sectors and institutions	Lack of inter- and intra-sectoral cooperation	Weak and fragmented governance, nonparticipatory approach
Access	Increased equity and inclusiveness	Continued marginalization of rural fishing communities	Concentration of power for the elite
Markets	Shift toward sustainability practices, slower economic growth	Slight shift toward environmentally friendly process, moderate economic growth	Deglobalized and insular world, low economic growth
Implications for fisheries and marine ecosystems (qualitative)			
Ocean conditions	Low emissions and ocean acidification (OA)	Increasing emissions and OA	High emissions and OA
Fish stocks and marine habitats	Recovery	Limited recovery	Deterioration
Fishing effort	Decline	Increase	Increase
Fisheries management	Participatory, community-based	Limited community participation	No community input
Coastal community well-being	Improvement	Limited improvement	Increasing marginalization
Seafood demand	Shift to sustainable seafood	Slow and low growth for sustainable seafood	Shift to cheaper species
Implications for fisheries (quantitative inputs to DBEM)[1]			
2050 Marine Protected Area (MPA) coverage	20%	10%	5%
Fishing mortality (% change relative to current)[2]			
High-value fish	−30%	+25%	+30%
Low-value fish	−50%	+25%	+50%
Crustaceans	−50%	+25%	+30%
Molluscs	−50%	+25%	+30%

1 See Appendix 5.2 for detailed description of fisheries implications.
2 Commercial fish catch is divided into different groups of economic value to reflect future socio-economic trends.

Table 5.6

Projected percentage change in species abundance and fish catch potential in 2050 relative to 2015

RCP	Species abundance (% change)	Fish catch potential (% change)
RCP8.5		
Arctic	13.2	−10.8
Pacific	−3.3	−13.9
Atlantic	−3.8	−3.0
RCP2.6		
Arctic	6.7	−8.9
Pacific	−3.1	−11.0
Atlantic	−4.4	−4.9

Notes: Values are averaged over three earth system models (ESMs). The differences in projected values generated by the three ESMs are as follows. Percentages show the range of lowest to highest differences between models, and are reported in sequence for Arctic, Pacific, and Atlantic.

RCP8.5
  Abundance: 1–14%; 2–39%; 0–18%
  Catch: 3–27%; 7–56%; 2–14%
RCP2.6
  Abundance: 5–12%; 13–28%; 2–12%
  Catch: 3–23%; 1–35%; 1–8%

Table 5.7

Estimated employment, fish supply, and change in landed value per fisher in 2050 relative to status quo under each scenario

Indicator	Scenario A – Sustainability	Scenario B – Business as usual	Scenario C – Decline
Employment (number of people)	49,658 (−33%)	57,857 (−21%)	38,615 (−48%)
Fish supply (kg/person)	16.8 (−27%)	17.0 (−26%)	20.0 (−13%)
Landed value/fisher	7%	−37%	−68%

Notes: Status quo = continuation of past trends to 2050 with no climate effect. Figures in parentheses indicate % change.

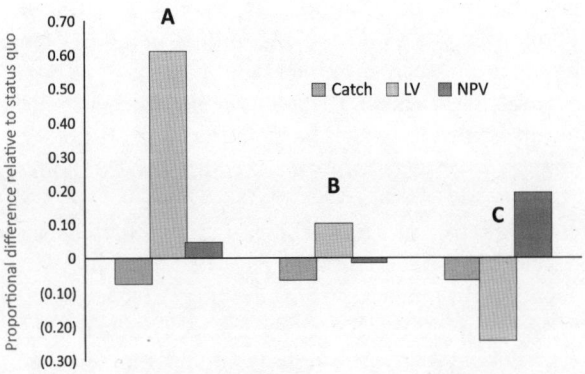

Figure 5.3 Projected change in catch potential, fisheries landed value (LV), and fisheries net present value (NPV) in 2050 relative to 2015 under each scenario.

benefits (i.e., net present value) from Canadian fisheries and revenue from fishing are expected to vary depending on scenario; how individuals, government, and society will deal with these potential issues is a key question to address in order to mitigate adverse social, cultural, economic, and ecological impacts on humans and ecosystems across Canada.

This section presents case studies of prior scenario analyses carried out in Pacific, Atlantic, and Arctic Canada. The case studies, which are detailed in Appendix 5.4 and summarized below, are used to first examine whether

or not the national scenario outlook coincides with local perspectives and priorities. Next, we link local to national scenarios to examine how scenarios at each scale can mutually inform each other in making decisions about the future. This multiscale linkage can help identify potential intervention points and levers for moving toward a preferred oceans future across Canada.

Pacific: Haida Gwaii Marine Plan
Qualitative future scenario analysis was used by the Council of the Haida Nation and the Province of British

Columbia to guide the direction of the Haida Gwaii Marine Plan (MaPP Initiative 2015), one of four local-scale marine plans for northern British Columbia. The future analysis focused on visioning and identified four plausible future pathways: conservation, sustainable technology, local economic growth, and global economy (see details in Appendix 5.4). The governance partners used the scenarios to develop a preferred scenario that was incorporated into the Haida Gwaii Marine Plan in 2015 (Day and Prins 2012; MaPP Initiative 2015).

Atlantic: Port Mouton Bay

The Atlantic case study did not involve a scenario exercise, but rather stimulated future thinking by eliciting local residents' perspectives about climate change impacts and governance in Port Mouton Bay, Nova Scotia. This was done through a workshop and online survey involving 25 community participants and administered by researchers from the Friends of Port Mouton Bay, a local volunteer group formed to protect Port Mouton Bay. This process enables us to determine whether the local outlook is compatible with national scenarios, and identify scenario elements[2] that can form the basis for future scenario development in Port Mouton Bay.

One of the outcomes from engaging Port Mouton Bay residents in future thinking was the realization that governments that are unresponsive to shifts in social and ecological systems are not equipped to adapt to climate change. This shortcoming undermines the ability of communities to protect the social-ecological systems on which they depend, as well as their ability to pursue innovative solutions to ecological changes, such as living shorelines, ecotourism, expanded curricula, and Indigenous reconciliation – this raises a barrier against residents' vision for the future.

Arctic: Impact of Climate Change on Arctic Marine Ecosystem and Subsistence Fisheries

Scenario analysis was undertaken to investigate the ecological and economic impact of future environmental change on Canadian Arctic fisheries. A coupled climate/ecosystem/economic model was applied to assess climate change impacts on Western Canadian Arctic marine ecosystems and consequences to subsistence fisheries (Steiner et al. 2019). The case study focused on Arctic cod

(*Boreogadus saida*), a key forage species in the Inuvialuit Settlement Region. Model simulations estimated a 17% decrease in Arctic cod populations by the end of the century under a high-emission scenario, but suggested increases in abundance of other Arctic and Subarctic forage species (e.g., capelin, sandlance, and herring). Such changes are already being observed by Inuvialuit fishers and in stomach content analysis of harvested species. This case study enables us to assess whether national projection trends coincide with those undertaken at a regional level, and, by doing so, to identify issues that may be overlooked when focusing on just one geographical scale alone.

LINKING NATIONAL AND LOCAL SCENARIOS

Method for Connecting Scenarios of Different Scales

Links between scenarios of different scales can be established upfront, iteratively during the scenario development process, or after scenario development is complete (Appendix 5.5). Two main features of multiscale scenarios are the number of scales at which scenarios are developed, and the strength of connectedness between scales (Biggs et al. 2007). Based on these two features, multiscale scenario exercises can be categorized as: (1) single-scale scenarios; (2) loosely linked scenarios built at two or more scales; and (3) cross-scale scenarios that are tightly coupled (i.e., high level of consistency) across two or more scales (Biggs et al. 2007). Scenarios at different scales can be connected through the development process, or through scenario elements and outcomes (i.e., driving forces, assumptions, scenario logic, boundary conditions, general outcomes) (Zurek and Henrichs 2007). The strength of linkages of scenario elements and/or outcomes across scales can be either equivalent, consistent, coherent, comparable, or complementary (Appendix 5.5). Equivalent, "hard" links provide the strongest linkages between scales, whereas complementarity provides the weakest linkage in elements and outcomes.

Multiscale Linkages for Canadian Scenarios

Building in multiscale linkages enables us to examine local perspectives and concerns about the future in order to understand how they converge with or diverge from the national outlook. It also enables an understanding of local responses to national and global scale drivers, which can help in identifying possible intervention points for communities to move toward their preferred future. Below, we use each of the three regional scenario case studies summarized above as an example to illustrate how national-scale scenarios can be linked to the regional scale through scenario elements, outcomes, and/or the development process.[3]

Pacific: Haida Gwaii

Following Zurek and Henrichs (2007), national and Pacific scenarios can be characterized as being loosely linked, comparable scenarios that share a common focal issue (marine resource and ocean sustainability) but are developed independently at each scale. This enables the scenarios to maintain relevance to users at each scale, and stakeholders to explore issues of concern at each scale, while still allowing for some investigation of cross-scale processes (e.g., climate change). Both the national and Pacific scenarios share similar scenario logic (assumptions about the future) and scenario elements (drivers and uncertainties). The conservation orientation of the preferred scenario chosen by Haida Gwaii stakeholders (see Appendix 5.4) is consistent with national Scenario A's pathway in terms of its sustainability focus. The Haida Gwaii local economy pathway also reflects elements of Scenario C in that both focus on developing insular interests, although at different scales. In this respect, the quantitative outputs from national scenarios can provide Haida Gwaii stakeholders with an idea about the magnitude of change they may face in terms of fisheries and socio-economic impacts under each scenario, thereby enabling them to understand the trade-offs involved in heading toward their preferred future. For instance, fishery managers can use the projected national changes in fish catch potential and fisheries landed value to assess the potential livelihood and food impact of pursuing sustainability or development-oriented pathways.

Atlantic: Port Mouton Bay

While the Port Mouton Bay case study did not involve actual scenario development, the eliciting of local perspectives and future-based thinking it stimulated enabled the identification of scenario elements and local visions that will be applicable to development of future scenarios for the sustainability of Port Mouton Bay. Local scenario elements (i.e., assumptions, driving forces, and their trends) can be set to be either equivalent, coherent, comparable, consistent, or complementary to national-scale scenarios (see Zurek and Henrichs 2007). Due to the rich insights compiled about local perceptions and needs, future work can couple the national-Atlantic scenarios through an independent scenario development process. This process can allow different stakeholder groups' aspirations for the future to be articulated within the boundaries set by the national scenarios, thereby addressing specific stakeholder/policy needs at the local (Port Mouton Bay) scale.

Arctic: Marine Ecosystem and Subsistence Fisheries

There is a fairly strong link between national and Arctic scenarios. According to Zurek and Henrichs's characterization (2007), this national-regional linkage is both consistent and coherent because the scenario logics match, and the main scenario assumptions, driving forces, and their trends are consistent with each other in both scenarios. Moreover, a similar analytical approach is utilized for both the national and Arctic scenarios in terms of models and data because the scenarios' development process involved some of the same OceanCanada team members working on both types of scenarios. The main differentiation point is that the Arctic case study does not incorporate socio-economic development pathways, as its focus is on assessing future climate forcing scenarios.

INFORMING SOLUTIONS: INSIGHTS GAINED FROM MULTISCALE LINKAGES

Connecting scenarios across scales allows for insights that may not have been apparent at only one scale to emerge. These insights can be used to inform the development of locally relevant solutions for addressing priorities that apply across scales. These include climate change adaptation strategies, balancing marine protection with economic development, food security, and integration of traditional, local, and ecological knowledge into assessment of current and future impacts of global change. Specifically:

- The Pacific scenario provides locals' perspective regarding their preferred future. By doing so, it informs us about cultural and community values that did not emerge at the national scale. For instance, participants in the scenario workshop indicated that following a path of industrial development would provide short-term financial benefits but would be detrimental to and misaligned with local values and long-term sustainability.

- The preferred scenario identified qualitatively by Haida Gwaii stakeholders converges on a match with the national sustainability pathway narrative (Scenario A). Quantitative projections generated under Scenario A quantify what the preferred future can potentially provide in terms of indicators (fisheries catch, income). For instance, national fisheries employment is projected to decrease by a third, while landed value per fisher is expected to increase slightly by 7% under Scenario A. Fishers in Haida Gwaii may therefore expect to see a similar decrease in fishing jobs and increase in income if they follow their preferred pathway. This can inform Haida Gwaii stakeholders about the potential costs and benefits, helping them to better prepare for the preferred future that they have qualitatively described.
- The Pacific scenario-planning process identified key stakeholders, institutions, and networks, as well as economic actions needed (e.g., development of low-ecological-impact activities such as ecotourism and small-scale domestic harvesting) for moving Haida Gwaii toward a sustainability pathway. Such locally specific information, which is not apparent at the national scale, is instrumental in helping lead toward national ocean sustainability.
- The variables used for projecting both national and Arctic scenario outcomes are similar. Thus, the local projections help to refine and make nationally projected indicators more relevant to local stakeholders. For instance, while the national scenario projected an overall decrease of 9–11% in fish catch potential for the Arctic, the Arctic case study highlights projected climate change impacts on a culturally and socially crucial fish (Arctic cod) that do not emerge clearly at the national scale. This allows for better understanding of future social, cultural, and economic impacts on Arctic communities, e.g., how future economic development pathways may affect subsistence fisheries and food security.

- The finer-resolution models and data used for Arctic simulations identified gaps and uncertainties that may not have emerged from the more aggregated data used for national-level scenarios. This helps to refine and improve scenario-modelling approach and methods. Importantly, the inclusion of traditional knowledge in the Arctic case study identified further areas where Indigenous and local communities could be directly involved in local environmental and ecological monitoring studies.
- Both the Arctic and Port Mouton Bay scenarios highlighted the importance of community science and integration of traditional knowledge with scientific methods. As these processes are location- and context-specific, they did not emerge in the national scenarios, but are instrumental for collaborative marine resource management.

The Pacific and Atlantic case studies show that local concerns about future impacts of global change, as well as preferred futures, are broadly convergent with those covered by the national scenario socio-economic development pathways. In contrast, the projected increase for Arctic fisheries under future climate change scenarios diverges from the national outlook. To account for these convergent and divergent trends, insights gained from local-scale scenarios can be integrated with national scenarios in several ways. For the Pacific, preferences about protected area coverage can be built into the national scenario model as a spatial input parameter into the DBEM. Likewise, the contribution of local traditional knowledge in the Arctic case study can be incorporated into national scenarios as input parameters into the DBEM (e.g., fishing mortality rate for Arctic cod), or into economic analysis (e.g., landed price for Arctic cod), thereby refining the national-scale projections. Local perceptions about climate change impacts and barriers to governance that emerged from the Port Mouton Bay case study provide a basis for setting scenario drivers, assumptions, logic, and potential outcomes for developing local scenarios. Overall, building in finer-scale resolution data will help improve the applicability of national scenario outcomes across Canada. In addition, the multiscale linkages can help elucidate regionally and locally relevant focal points and priorities for governance. For example, the

Pacific scenario emphasizes stronger co-management relationships, while Port Mouton Bay highlights the urgency of addressing an apparent lack of government responsiveness.

National and local scenarios provide perspectives and quantifiable projections to inform convergent issues that are relevant at both scales. These include balancing marine protection with economic development (Haida Gwaii and Port Mouton Bay), food security (Arctic), and the integration of traditional, local, and ecological knowledge into assessment of current and future impacts of global change. One particular issue that emerged from the Arctic case study, but was not explicit in the national narratives, was the importance of integrating traditional knowledge with scientific methods. Solutions to the issues identified requires delving into the governance of biological, social, and economic systems across geographical scales – a topic that is addressed more thoroughly in Chapter 9.

APPENDIX 5.1

PROCESS FOR DEVELOPING NATIONAL-LEVEL NARRATIVES

The national narratives were developed in a two-stage process and involved a network of researchers from the OceanCanada Partnership (www.oceancanada.org). In the first stage, the main societal trends of the global Shared Socio-Economic Pathways (SSPs) were applied to each national narrative in order to set the context for what type of world Canada would face in terms of global-scale drivers such as international institutions, population change, human and technological development, environmental protection, poverty, social cohesion, and consumer preferences. The resulting narratives were placed along a continuum of three key uncertainties that are particularly relevant to the future sustainability of Canadian oceans: Governance, Access, and Markets (see Figure 5.2). These uncertainties were identified by a group of interdisciplinary ocean researchers during a scenario workshop at the OceanCanada conference held in May 2016 (https://oceancanada.org/conferences/vancouver-2016).

In the second stage, we translated these broad societal narratives into a marine-specific context, as other authors have done for health and demographic research (Ebi 2012; Samir and Lutz 2014). This involved identifying what each narrative implied for Canadian marine fisheries and ecosystems in terms of five social-ecological factors (Canadian population, fish consumption, fishing cost, fish price increase by 2050, and fish catch composition in 2050). This process resulted in three national narratives, detailed below. Each narrative is paired with corresponding Representative Concentration Pathways (RCPs), resulting in three national integrated scenarios (A, B, C) describing contrasting social-ecological futures for Canadian oceans.

Translation of Global SSPs to Canadian Ocean Narratives

Pathway A (Sustainability)

This pathway is based on SSP1, and is characterized by low challenges to mitigation and adaptation. It reflects a world that shifts gradually toward a sustainable path, with a focus on environmentally friendly technologies and low energy demand, leading to low mitigation challenges. There is a value shift toward economic growth that emphasizes human well-being and the increasing effectiveness of global, regional, and national institutions, thus implying low challenges to adaptation.

Access to resources. Global focus on increasing equity leads to increased social cohesion while maintaining social and cultural diversity. This implies that currently marginalized groups, such as Indigenous and/or rural fishing communities, have greater opportunities to participate in mainstream society. Fisheries management systems that are perceived to be unfair and concentrate power in the hands of a few, such as individual transferable quotas (ITQs), are replaced with more inclusive systems that recognize traditional fishing rights and community management systems. Consequently, there is increased marine stewardship, and coastal communities actively participate in managing their coastal resources. The removal of barriers to entering fisheries, and growing demand for sustainable, locally caught seafood favours small-scale operators. Redesigned, more equitable value chains make fishing profitable for these small-scale operators, thus improving the overall economic situation in previously derelict fishing communities.

Governance. Management of environmental resources and the global commons improves over the long term due to effective cooperation and collaboration between national and international organizations and institutions, the private sector, and civil society. Fisheries legislation on protecting vulnerable habitats and species and recognizing traditional fishing rights and territories is strengthened.

Markets. There is a shift toward sustainability practices, and slower economic growth. A better-educated population and higher standards of living lead to a growing demand for local, sustainably caught and produced seafood. Investment in environmentally friendly and efficient technology enables the development of sustainably farmed seafood. Overall, this leads to a decrease in demand for capture fisheries.

Implications for fisheries. Strong governance implies an improvement in the management of Canadian fisheries and ocean environment, including effective enforcement and compliance with regulations protecting marine habitats and species (e.g., *Species at Risk Act*). There is an increase in the number and effectiveness of Marine Protected Areas (MPAs), reduced anthropogenic pressure on the ocean environment (e.g., from pollution and coastal development) and successful co-management arrangements, leading to sustainable use of coastal fisheries. Harmful fisheries subsidies are eliminated, while fuel prices increase due to more stringent climate policies. Both these factors contribute to a decrease in fishing effort. Sustainable aquaculture systems are able to produce sufficient yield to complement capture fisheries. The strengthening of marine awareness among the general public leads to growth in sustainable marine tourism, thereby providing an avenue for private sector investment in protecting the oceans. There is a focus on taking a social and human rights approach to fishing, such that allocation, access, and distribution of fisheries resources is more equitable. This means that quotas are not concentrated in the hands of a few fishing conglomerates, thereby eliminating expensive quota lease fees that fishers have to pay.

Consequently, by 2050 there is recovery of over-exploited and vulnerable fish stocks and damaged marine habitats, leading to improved biodiversity, increased fish biomass, and a healthy/resilient ocean ecosystem. Overall fishing effort and catch decline due to the combination of more effective fisheries regulations, increased marine conservation awareness, decreased consumption of animal products, sustainable aquaculture production, and a shift toward higher-value fisheries (higher-priced fish). The shift toward small-scale producers and sustainable aquaculture systems leads to higher-priced fish, which consumers are willing to pay for because of an overall increase in environmental and social awareness. Higher wages for fishery workers and increased energy costs increases fishing costs, but this is tempered by reduced fuel usage due to improved technology and the recovery of coastal fish stocks, which eliminates the need to travel far offshore to fish. Consequently, while there is an overall increase in fishing cost, it is less than under Pathway B.

Pathway B (Business as Usual)

This pathway is based on SSP2, and is characterized by moderate challenges to mitigation and adaptation. It follows current trajectories, with no marked deviation in terms of social, economic, and technological trends. There is slow progress in achieving sustainable and human development goals. There is emphasis on material growth, with growing consumption of animal products. Equity slowly improves, but inequality and poverty persist for disadvantaged populations, resulting in social stratification and limited social cohesion, as well as presenting food security problems for these groups. Even though fossil fuel dependency slowly decreases, there is growing energy demand, which leads to continuing environmental degradation. While there is moderate environmental awareness and sustainability targets are high on national agendas, implementation is weak due to relatively weak coordination and cooperation among national and international institutions. This presents moderate challenges to mitigation.

Access to resources. The ongoing presence of inequality and poverty among disadvantaged populations means that many rural fishing communities will continue to be marginalized. Insufficient investment in education implies that fishing communities will generally lack the skills necessary to find alternative jobs, even as their access to coastal resources becomes more limited due to the ongoing use of ITQs and concentration of power among fishing conglomerates. The unequal distribution of fisheries benefits means economic disparity for many fishing communities.

Governance. Governance of environmental resources remains relatively weak as there is a lack of progress in implementation of policies, due in part to moderate corruption and limited access to the rule of law. Further, weak coordination and cooperation among national institutions, the private sector, and civil society in addressing environmental issues results in fragmented and generally ineffective management of coastal resources. Lack of inter- and intrasectoral cooperation inhibits participatory and community-based management. It also creates barriers to the creation and development of MPAs, integrated coastal zone management, and resolution of disputes over traditional fishing rights and territories. Consequently, improvement in fisheries management is limited, and anthropogenic pressures on the marine environment persist.

Markets. Economic growth is moderate, and there is a slight but insufficient transformation towards environmentally friendly processes (e.g., aquaculture). However, material-oriented growth and growing consumption of meat products implies that demand for capture fisheries continues to grow. Although there is awareness about sustainability, income and education levels are not high enough to generate a marked change in values, and there is only small growth in the local, sustainable seafood movement, seafood remaining too expensive for the average consumer.

Implications for fisheries. This pathway shows little departure from the current state and management of Canadian fisheries and oceans. Consequently, by 2050 there is limited recovery of overexploited and vulnerable fish stocks and marine habitats, which continue to be threatened by multiple anthropogenic stressors (including aquaculture, which has not transitioned to become fully sustainable). Biodiversity and fish biomass do not improve beyond current levels. Ineffective fisheries governance and rising fish demand lead to continued increases in fishing effort. Fish prices remain relatively stable, but at the same time, economic benefits from fisheries are not equally distributed to the coastal communities that need it the most. Moreover, the lack of

coordination and cooperation between sectors inhibits the growth of community-managed fisheries. This contributes to stagnant catch values, which do not increase due to the relatively low demand for higher-value, sustainably caught local seafood.

Fishing cost. There is a steady increase in fishing costs due to rising energy costs.

Pathway C (Environmental and Social Degradation – Poor Progress to Sustainability)

This pathway combines elements of SSP3 and SSP4 to create a world that faces high challenges to mitigation and adaptation. Global institutions are weak, with limited cooperation and coordination when addressing environmental issues. Instead, policies are increasingly oriented toward national and regional security issues. Countries focus on achieving energy and food security goals, while investment in human and technological development declines. This results in slow economic growth, low technological capacity, material-intensive consumption, escalating inequality, and growing resource intensity and fossil fuel dependency. The strong reliance on fossil fuels, poor energy efficiency, and lack of environmental concern present high challenges to mitigation. Low investment in human capital, combined with weak institutions and leadership, results in high vulnerability and increasing inequality as power becomes concentrated in a small group of business and political elites, even in democratic countries. This degradation of social cohesion brings high challenges to adaptation, and governance systems are unable to address the multiple dimensions of vulnerability.

Access to resources. Power is concentrated in the hands of a few political and business elites, with vulnerable groups receiving little representation in national and global institutions. Social security measures weaken as inequality worsens, and many rural communities get left behind, and therefore face heightened vulnerability to environmental and socio-economic impacts. Further, the consolidation of the fishing industry implies that access to fisheries resources for many small-scale operators becomes limited.

Governance. Lack of cooperation and coordination among global governance institutions leads to weak and fragmented environmental governance regulations that are not effectively implemented or enforced. The focus on national security results in a top-down approach to governance in which society has to support national priorities. This creates entanglement of public and private sector interests, and increases the incidence of corruption. Vulnerable groups receive little representation or access to national and local institutions, culminating in a lack of stewardship for natural resources. Attempts at community and participatory resource management generally fail.

Markets. The world is deglobalized and insular, stalled by a lack of cooperation in human and technological development. Not only does economic growth slow down but trade is restricted, with production of agriculture, fisheries, and energy geared toward national consumption. Financial benefits from natural resource production reach only the affluent in society, who have relatively high material consumption lifestyles. Meanwhile, consumption levels for disadvantaged and vulnerable groups are low as they struggle to maintain living standards. Pursuit of more sustainable lifestyles is low and drops from current levels due to low education levels of the disadvantaged groups, along with the material-oriented consumption of the affluent. The momentum for sustainably caught local seafood that was gaining steam at the beginning of the 2010s gradually wanes, as only the affluent can afford it. Instead, preference shifts to less expensive and ecologically more abundant species that are more affordable and can meet food security demands.

Implications for fisheries. In this world, the status of Canadian oceans and fisheries cannot improve, and instead potentially worsens due to a deterioration in governance, human capital, and environmentally friendly technological advancement. Consequently, management and protection of fish stocks and marine habitats, and sustainable aquaculture, worsen compared to the present; ecosystems and biodiversity gradually decline, leading to a loss in fish populations. There is no increase in the number or effectiveness of MPAs, and lax implementation and enforcement of environmental regulations fail to control anthropogenic pressure on the ocean. As ocean resources deteriorate, livelihoods become increasingly difficult for rural fishing communities, who do not have the economic or human development skills necessary to diversify their livelihoods. This is exacerbated by the dominance of the fishing industry by a few powerful

fishing conglomerates that control the majority of fishing rights and licences, leaving small-scale operators with limited access to fishery resources. Combined with a general decline in social security and equality, these leave fishing communities destitute.

An emphasis on achieving national food security goals drives up capture fisheries production as aquaculture is unable to contribute substantially to supplement fish supply due to lack of technological advancement. Government subsidies to the industrial fishing sector continue due to nationalistic priorities. Much of the increased fishing effort is directed at low-value small fish (small pelagics) due to the continued use of nonselective fishing gears, resulting in deteriorated marine ecosystems. Consequently, fish prices drop substantially, except for large, predatory fish, which are increasingly rare to find. The overall increase in fisheries demand is tempered somewhat by a low population growth rate. However, degraded fish stocks are probably unable to support ongoing levels of fishing pressure, resulting in a collapse of many major fisheries by 2050. The state of the ecosystem necessitates a shift in preference to less expensive fish species (e.g., herring), resulting in low-value fish making up the majority (60%) of the catch, with high-value fish making up only 5% of total catch.

Fishing cost. Low technological innovation keeps fishing labour-intensive, but the dominant fishing companies are able to keep crew wages low. As global fossil fuel demand remains high, fuel prices increase, but subsidies reduce overall fishing costs. The overall effect is a decrease in fishing cost.

APPENDIX 5.2

DYNAMIC BIOCLIMATIC ENVELOPE MODEL

The dynamic bioclimatic envelope model (DBEM) simulates changes in distribution of abundance and maximum catch potential (i.e., maximum potential production from fisheries [Cheung et al. 2010]) of fish and invertebrates over time and space (on a 0.5° latitude × 0.5° longitude grid). These projections are driven by projected changes in oceanographic conditions that are generated using two earth system models (ESMs) (see Cheung et al. 2016 for a detailed description of the DBEM). Projected changes in ocean conditions (e.g., temperature and oxygen concentration) affect habitat suitability for marine species, as well as driving changes to their physiology and life history. The DBEM combines projected changes in oceanographic conditions and resulting changes in habitat suitability and fish species population dynamics to produce projections of changes in species distribution. The projected future distribution of marine species is then combined with projected changes in primary production from earth system models to estimate the annual maximum catch potential of marine species.

Improved fisheries governance and community co-management under Scenario A results in effectively implemented MPAs. MPAs can help restore depleted fish populations and rebuild habitats, thereby strengthening ecosystems' resilience to climate change impacts. In addition, past studies have shown that the spillover effect of MPAs can increase fish catches in adjacent fishing areas; studies from tropical fisheries show that fish catch in adjacent fishing grounds increased by 46–90% (Sala et al. 2013), doubling (Kerwath et al. 2013) and even increasing up to 45 times (Russ et al. 2003). Therefore, to quantitatively portray the effects of governance, we modelled the potential increase/decrease in fishery yields under each pathway.

Protected Area Coverage and Fishing Mortality Assumptions

Scenario A. Canada achieves the *Convention on Biological Diversity* (CBD) target of 10% MPA coverage by 2020, which is increased to 20% MPA coverage by 2050. Due to improved governance, all the MPAs are effective and there is recovery of fish biomass (especially for fish species that are currently assessed as being in the critical zone), as well as spillover effects. Fish stocks also benefit from well-managed fisheries and effective implementation of precautionary regulations. Consequently, there is an overall decline in fishing mortality. The recovery of high-value predatory fish is tempered somewhat by a shift toward high-value fisheries, such that by 2050 the reduction in fishing mortality for this group (–30% relative to present levels) is lower than for low-value fish, crustaceans, and molluscs (–50%).

Scenario B. Canada is able to meet the 10% target, but the MPAs are mostly "paper parks" that are not effectively managed. Biomass recovery is therefore limited as MPAs

fail to protect species and habitats. In addition, there is limited improvement in fisheries management and overfishing continues. Consequently, fishing mortality rates continue to increase across all fish groups (+25% relative to present levels) by 2050.

Scenario C. Canada fails to meet the 10% target and fisheries management progressively deteriorates. Fishing mortality increases and is higher than in Scenario B. Due to the focus on meeting national food security needs and increasing fisheries production, small pelagics and other low-value fish are intensively targeted, such that fishing mortality of this group is higher than others (+50% by 2050 for small pelagics, +30% for all other groups).

Employment

For the past decade (2007–17), the number employed in the primary sector (i.e., fish harvesters/fishers) decreased from 48,239 in 2007 to 45,578 in 2017, while secondary sector workers (fish processors) decreased slightly, from 28,587 to 27,998. Overall, the average percentage change in fishery sector employment for 2007–17 has been almost zero (0.3% for primary, and 0.2% for both primary and secondary). As such, we assume that future fisheries employment will also remain stable (no increase or decrease in fisher population), and that the change in fisher population for 2017–50 will follow the Canadian population growth trend prevailing under each scenario.

APPENDIX 5.3

ASSUMPTIONS AND INPUTS USED FOR DETERMINING FUTURE VALUES OF SOCIO-ECONOMIC VARIABLES

Population

Population estimates were based on demographic model trajectories from Statistics Canada (2015), which provided the Canadian population in 2050 under low-, medium-, and high-growth scenarios (39.81, 46.90, and 54.59 million, respectively). We linearly increased the base year (2015) population of 35,852,000 to the 2050 number for each of the scenarios. Population growth is assumed to be medium for SSP1 and SSP2 in high-income countries (Jiang 2014); therefore, a projected population of 46.9 million (medium growth) is used to represent the Canadian population in 2050 for Pathways A and B. For

SSP3 and SSP4, population growth is expected to be low (Jiang 2014). Pathway C reflects a fragmented and highly unequal future (combined elements of SSP3 and SSP4), in which population growth is expected to be low in high-income countries; hence, we assume a projected low population of 39.81 million in 2050.

Fish Consumption

We obtained data on Canadian apparent fish consumption (kg/person/year) for the period 1961–2013 from Food and Agriculture Organization (FAO) food balance sheets (FAOSTAT) (FAO 2023). The 2013 level of fish consumption of 22.6 kg/person/year was assumed to increase or decrease linearly to 2050, depending on the scenario, as follows:

Pathway A. Fish consumption is expected to decrease. Over the past 50 years, the largest decrease in annual fish consumption rate was 12%; we used this to represent the expected decrease in fish consumption by 2050.

Pathway B. Fish consumption rate is expected to follow historical trends. We extracted annual fish supply per capita data from FAOSTAT and calculated the annual change in fish consumption rate. Data were available for a 50-year period from 1962 to 2013. We applied the average change in fish consumption rate over the past 50 years (0.01) to 2050, and linearly increased current (2015) annual fish consumption to the 2050 level.

Pathway C. Fish consumption is expected to increase. The largest increase in annual fish consumption rate over the past 50 years was 32%, which we used to represent the expected increase in fish consumption by 2050.

Fishing Cost

Current fishing cost of 77% was based on the average fishing costs reported for six fisheries across Pacific and Atlantic Canada (Teh and Sumaila 2020). We then assumed that fishing cost would increase by 15%, 25%, and –25% by 2050 under Pathways A, B, and C, respectively. (See Appendix 5.1 narratives for details.)

Fish Prices

Under Pathway A, an overall decline in fish availability due to reduced catch, combined with a shift toward sustainably caught and certified fisheries, would lead to an increase in ex-vessel fish prices. We assumed an overall increase of 50% by 2050, except for low-value fish, which increases by 25% due to lower demand. Under Pathway B, fish prices would remain stable. Under Pathway C, there would be a general decline in fish prices, driven by the ramping up of fisheries production to satisfy national food security goals; further, the few powerful conglomerates that dominate the fishing industry are able to keep prices low because they can operate at lower costs. As such, fish prices are anticipated to decline by 25% across all fish groups except for high-value fish, for which price is expected to increase by 50% due to their increasing scarcity. We bounded fish price increase for 2015–50 to a maximum of 50% based on a study by Tai and colleagues (2017), which found that global fish ex-vessel prices from 1950 to 2010 generally increased by around 54%.

Fish Catch Composition

The catch composition was assumed to vary under each pathway due to the anticipated difference in fishing practices and ecosystem states prevailing in the future. Following Lam and colleagues (2016), we categorized fish catch into four categories: high value, low value, crustaceans, and molluscs. The 2017 contribution of each group to total catch was assumed to be maintained under Pathway B. We assumed that the recovery of overexploited fish stocks and habitats under Pathway A would be consistent with a higher proportion of predatory fish (i.e., higher-value fish) in the catch, with a corresponding decrease in the other groups. In contrast, degraded marine ecosystems and continued overfishing under Pathway C imply that the proportion of lower–trophic level, low-value fish and invertebrates increases.

Allocation

Current allocation to Indigenous groups was based on First Nations participation in BC commercial fisheries (published in Weatherdon et al. 2016) and Atlantic licences held by the Mi'kmaq and Maliseet First Nations (MMFN) (reported in Thayer Scott 2012). On average, BC First Nations hold 27.5% of BC commercial fisheries licences, although this varies across fisheries, from less than 2% in the geoduck and horse clam fishery to 100% in the Heiltsuk intertidal calm fishery. For the Atlantic region, the MMFN held on average 7.2% of all commercial licences, ranging from 3% in the lobster fishery to 16% for shrimp (Thayer Scott 2012). The regional percentages were then weighted by each region's total fish landings for the period 2000–16 to obtain a national weighted average. The Arctic was not included because the majority of fishing done by Arctic Indigenous groups is subsistence-based.

APPENDIX 5.4
SCENARIO CASE STUDIES

Haida Gwaii Marine Planning Scenarios

A marine advisory committee had supported a draft vision and goals and was reviewing draft objectives and strategies in 2012. The planning team and advisory committee confirmed key topics that needed to be explored in future analysis: climate change, demographics and local economic development, marine energy, fisheries, marine transportation, marine tourism, and shellfish aquaculture. The four alternative futures (Figure 5.4) were fleshed out during a two-day workshop where 23 experts and planning staff reviewed status and trends of key topics, and developed concept maps and storylines (Day and Prins 2012). The concept maps identified internal and external drivers and pressures as well as trade-offs, challenges, risks, and opportunities. The storylines described likely environment, community, economics, and governance outcomes over a 20-year timeframe. Haida Gwaii has a population of about 4,500 people who are primarily dependent on resource extraction such as logging and fishing but are in the early stages of developing as an Aboriginal and ecotourism destination. External drivers and pressures such as climate change and global markets were seen as having a significant impact on the future of Haida Gwaii. Internal drivers included out-migration of youth and its negative effects on community infrastructure such as schools, health care services, and transportation. Potential economic opportunities included shellfish aquaculture, increased local benefits from commercial and recreational fisheries, marine-based tourism, and renewable energy development, such as wind or tidal power. The four scenarios envisioned different levels of marine protection (30%, 20%, 10%, and 5%, respectively).

Key elements of the path as seen 20 years from now are: "Prioritizes culture, healthy intact ecosystems and sustainable communities. Marine use and development are balanced with high environmental protection standards and a comprehensive network of marine protected areas. Marine industries ... generally have low environmental impacts and are consistent with the distinct islands lifestyle. Community growth is based on a diversity of activities that tap into a growing global demand for sustainable seafood and a unique visitor experience" (Day and Prins 2012). The preferred scenario provides a brief one-page description of the intended outcomes of the marine plan that was used to communicate the plan aspirations to the public and the community (see the text box "Haida Gwaii's marine future: A conservation and local economy path," below).

The Haida Gwaii scenarios have been used to help structure Haida Gwaii's marine spatial plan in terms of identifying priorities and actions. Specifically, future scenario analysis was used by the three other Marine Plan Partnership (MaPP) planning regions in northern British Columbia, such that the four plans are aligned in terms of aspirations for marine protection and co-governance. Haida Gwaii priorities are captured in a MaPP regional action framework that identifies joint priorities, such as developing climate change adaptation strategies, implementing ecosystem-based management, advancing collaborative governance between First Nations and the Province of British Columbia, supporting local economic development, and advancing marine protection. MaPP plans are in their fourth year of implementation. Regional and local ecosystem-based monitoring strategies have been developed that focus on performance of marine plans relative to ecosystem, economic, cultural, and governance objectives. Concept maps provide a potential tool for identifying management targets (Francis et al. 2011) as well as periodically assessing changes in internal and external drivers and pressures. The MaPP collaborative governance structure is being adapted to both Marine Protected Area network planning for northern British Columbia and marine shipping with the federal government (Prime Minister of Canada 2018).

LOWER RISK TO MARINE ECOLOGY → **HIGHER RISK TO MARINE ECOLOGY**

OUTCOME 1
Conservation Path

- Ecosystem-based approach to resource and environmental management, with emphasis on precautionary approach, co-management, integration of traditional and local knowledge
- High emphasis on ecological risk (e.g., 30% Marine Protected Areas, high level of requirements for development
- Development of low-ecological-impact activities such as ecotourism, scientific research, marine area management, small-scale domestic harvesting, carbon storage
- Limitation of port and other major infrastructure development
- Focus on local employment targeted to conservation-based opportunities

OUTCOME 2
Sustainable Technology Path

- Ecosystem-based approach with emphasis on structured technical approach decision making, coordination with global environmental protection, and technological innovations
- Moderate-high emphasis on ecological risk (e.g., 20% Marine Protected Areas)
- Expansion of extra-regional markets
- Development of capital-intensive projects with reduced ecological risk, e.g., alternative energy, land-based aquaculture, high-tech fisheries, etc.
- Technological innovation in developing sustainable, low-impact port or infrastructure projects
- Emphasis on knowledge-based jobs; not necessarily local in nature

OUTCOME 3
Local Economic Growth Path

- Emphasis on local benefits and employment
- Expansion of industries requiring low to moderate capital, such as shellfish aquaculture, tourism, recreational and commercial fishing, local processing, etc.
- Marketing, branding, etc., based on local identity and culture
- Low-moderate emphasis on ecological risk (e.g., 10% Marine Protected Areas)
- Protection of local economies from global big business expansion through tax programs, policies, tenures, legislation, and other means; institutional and financial infrastructure to support local planning and management of economic development
- Strong emphasis on local training, capacity building for new and existing opportunities

OUTCOME 4
Global Economy Path

- Economic growth focus with emphasis on national/international economy
- Priority given to access to international markets
- Growth of capital- and profit-intensive industries to meet international demand, e.g., offshore oil and gas, finfish aquaculture, seabed minerals, etc.
- Incentives for large international corporations to participate in local economies
- Growth of local support service industries
- Low emphasis on ecological risk (e.g., 5% Marine Protected Areas)
- Expansion of shipping, transportation, and major port and infrastructure projects
- Emphasis on attracting labour force to area
- External focus/benefits

LOCAL FOCUS/BENEFITS → **EXTERNAL FOCUS/BENEFITS**

Figure 5.4   Four potential scenarios developed by Haida Gwaii stakeholders (adapted from Day and Prins 2012).

HAIDA GWAII'S MARINE FUTURE: A CONSERVATION AND LOCAL ECONOMY PATH

Twenty years from now, Haida Gwaii has followed a path that prioritizes culture, healthy intact ecosystems, and sustainable communities (MaPP Initiative 2015). Marine use and development are balanced by high standards for environmental protection and a comprehensive network of Marine Protected Areas. Marine industries that are supported in and around Haida Gwaii generally have low environmental impacts and are consistent with the distinct islands lifestyle. Community growth is based on diverse activities that tap into a growing global demand for sustainable seafood and a unique visitor experience. Substantial progress in this direction has been made in the realms of environment, economy, community, and governance.

Environment

Haida Gwaii has embraced new conservation efforts by establishing a network of Marine Protected Areas. High environmental standards are required for all developments and activities. New policies and approaches are explored, and fisheries within Haida Gwaii waters are sustainable and include significant levels of local involvement. These actions protect key habitats, marine communities, biodiversity, and culturally significant places, and they buffer against climate change and environmental uncertainties. The result is a resilient and productive marine environment that supports sustainable marine industries.

Economy

Economic development in the marine sector focuses on managed growth of tourism and shellfish aquaculture, slow but steady development of new community fisheries initiatives, and support for new sustainable technology initiatives and research. Haida Gwaii has become known as a premier tourism destination and source of sustainable wild fish and aquaculture products due to concerted efforts in marketing and branding. Economic growth has required sustained investment in human resources and key improvements in infrastructure. These improvements include amenities to support tourism-related service industries, facilities to support fishing, and upgrades in transportation and telecommunication systems to support business development. Focused training efforts prepare island residents and youth for new local opportunities. New partnerships and incentives improve local benefits from recreational and commercial fishing in Haida Gwaii waters. New businesses are successful in part due to strong local leadership, the ability to attract external investment, and sustained government support. Overall, the marine sector provides a greater proportion of local benefits compared with earlier years, and the number of jobs grows at a modest rate, which keeps the islands' population relatively stable.

Community

The Haida's strong cultural attachment to the ocean flourishes while economic opportunities that are a good match with their growing, youthful population are supported. Island residents maintain a high quality of life that results from access to healthy food, fresh air, and the expansive and generally uncrowded inlets and shores. Community cohesion is strong and there is pride in living on Haida Gwaii, particularly with regard to the innovative and progressive management of waters in and around Haida Gwaii. Infrastructure – particularly facilities that support tourism and clean sources of electricity – improves, although it is a struggle to maintain community infrastructure without a large tax base and with a reliance on seasonal jobs.

Governance

Marine and ocean governance systems continue to be refined and improved. The Council of the Haida Nation and provincial and federal governments work together, along with industry sectors, to meet the marine plan objectives, which results in stronger co-management relationships over time. This includes collaborative efforts to manage Marine Protected Areas throughout Haida Gwaii. Stronger relationships between managers, communities, and users generate increased confidence in management decisions. The Haida Nation continues to work with local government to support island initiatives. Governance processes and decisions are sometimes slow but provide a stable base for business development, delivery of services, or advocacy on behalf of island communities and other resource users.

Port Mouton Bay

The Friends of Port Mouton Bay (FPMB), a local volunteer group formed to protect Port Mouton Bay in Queens County, Nova Scotia, held a community workshop in 2018 to identify, record, and categorize community assets.[4] Tangible and intangible values held by the communities included physical and cultural assets, organizations, business and industry, governance and health, and community resilience and well-being. By valuing these assets, residents recognized their implicit vulnerability to climate change from drought, extensive ecosystem changes (ocean warming, acidification, benthic and crustacean population shifts), invasive species, sea-level rise, erosion, and increasing loss of access (fisheries, shoreline).

Other observations of local-scale change included: storm surge effects on beach and dunes from increasingly powerful, more frequent storms; overwash and under-cutting of roads, leaving behind rock debris and sinkholes, along with general subsidence; excessive upland de-forestation causing sedimentation; intensification of Lyme disease–carrying ticks; plastic pollution and up-surge of visitors disrupting sensitive coastal habitats; decreasing herring and mackerel fish for bait fishery; polluted bay from aquaculture, and inadequate coastal and land-use planning.

Participants' feedback revealed an overall sense that the government was unresponsive and ill-suited to tackle climate change. Consequently, it was apparent that community groups needed to respond, but without the financial support that the government could provide if it was committed. To address this lack of government "fit," FPMB used social learning processes to inform decisions that impact the Port Mouton Bay coastal social-ecological system (Armitage and Plummer 2010; Ekstrom and Young 2009; Galaz et al. 2008; Loucks et al. 2017; Wilson 2006). For example, publishing community science based upon local ecological knowledge drew attention to community assets within provincial and municipal regulatory frameworks for policies such as aquaculture moratoriums, nature reserve designations, and land-use planning.

One of the outcomes from engaging Port Mouton Bay residents in future thinking was the realization that governments that are unresponsive to shifts in social and ecological systems are not equipped to adapt to climate change. This shortcoming undermines communities'

ability to protect the social-ecological systems on which they depend, as well as their ability to pursue innovative solutions to ecological changes, such as living shorelines, ecotourism, expanded curriculums, and Indigenous reconciliation – this presents a barrier toward residents' vision for the future. Despite significant work undertaken by FPMB, this largely volunteer-based mechanism for protecting the bay is unsustainable (Loucks et al. 2017). Political will, along with concrete mechanisms for integrating local and ecological knowledge and community science, are needed across multiple scales of management, protection, and monitoring (Galaz et al. 2008). A holistic perspective is also required of electorates, who largely view infrastructure projects and job creation in isolation from the long-term health of the environment. Cohesive action strategies for climate defence and a sustainable future will begin to develop when knowledge-sharing mechanisms meet an openness by government to receiving new information.

Projected Impact of Climate Change on Arctic Marine Ecosystem and Subsistence Fisheries

Scenario analysis was undertaken to investigate the ecological and economic impact of future environmental change on Canadian Arctic fisheries. The case study focused on Arctic cod (*Boreogadus saida*), a key forage species in the Inuvialuit Settlement Region. Model simulations estimated a 17% decrease in Arctic cod populations by the end of the century under a high-emission scenario, but suggested increases in abundance of other Arctic and Subarctic forage species (e.g., capelin, sandlance, and herring). Such changes are already observed by Inuvialuit fishers and in stomach content analysis of harvested species.

To estimate the current and future fisheries potential under high and low climate change scenarios, a coupled climate-ecosystem-economic model was applied to 72 exploited polar and subpolar marine species. Comparisons of the 2004–15 annual reported tonnage and modelled estimates of fisheries potential suggested that annual sustainable fisheries catch potential could be four times greater at 710,000 tonnes than the current catch level of 189,000 tonnes. Landed value potential was estimated to be $779 million, compared with actual fisheries value of $560 million. Under a high climate change scenario, future (2091–2100) fisheries catch potential was projected to increase to 1.27 million tonnes, valued at $1.35 billion, while the future estimate under a low climate change scenario was 833,000 tonnes, valued at $859 million. Ocean acidification may reduce projected increase in catch potential, particularly for invertebrate species.

Simulation results estimated the potential for Canadian Arctic fisheries in Canada; at the same time, it also stressed the importance of considering impacts to ecosystems and human communities in these regions. Proper steps must be taken to ensure the ecological, economic, and social sustainability of Arctic fisheries, and decisions regarding commercial fisheries will need to be precautionary and adaptive in light of the existing uncertainties. The study further highlighted the need to improve the modelling systems to reduce uncertainties, and emphasized that Indigenous and local communities need to be directly involved in these studies to conduct local, unified, and continuous environmental and food chain monitoring to establish baseline information on species distribution, thresholds, and changes over time.

APPENDIX 5.5
METHODS FOR LINKING SCENARIOS ACROSS SCALES
Links between scenarios at different scales can be established upfront, done iteratively throughout the scenario exercise process, or established after the different scenarios have been developed. Upfront linkages are usually established in tightly coupled cross-scale scenarios, and can be done by downscaling global storylines to set the boundary conditions for regional/national/local–scale scenarios. Downscaling refers to the translation of broader-scale scenarios to a finer-scale resolution, and can be done either quantitatively or qualitatively (e.g., downscaling a global storyline to fit the conditions existing at a particular country or region). Upscaling refers to the reverse. Both downscaling and upscaling are needed to incorporate feedbacks and maintain consistency in storylines. Independent scenarios are usually developed within a shared overarching framework and are then linked afterwards. This linkage can be done by categorizing drivers and outcomes in the different scenarios and grouping similar scenarios at different scales (Biggs et al. 2007).

Scenarios at different scales can be connected through the development process, or through scenario elements and outcomes (i.e., driving forces, assumptions, scenario logic, boundary conditions, general outcomes [Zurek and Henrichs 2007]). Linking the development process can be done in several ways, e.g., having the same team of people develop the scenarios at each scale (joint development); running parallel processes in which the same methods are used to build the scenarios (parallel scenario development); using a set of draft scenarios developed at one scale as the starting point for scenario development at another scale, thereby ensuring consistency between the different geographical scales (iterative scenario development); developing and finalizing scenarios at one geographical scale, then, based on these, developing and fleshing out scenarios at another geographical scale while leaving the original scenarios unaltered (consecutive scenario development); developing scenarios at two or more geographical scales fully independent from each other (independent scenario development) (Zurek and Henrichs 2007).

Options to link scenario elements and outcomes include a complete translation of focal questions, assumptions, drivers, and outcomes across the scales, or developing sets of scenarios that address similar broad issues at different scales (Biggs et al. 2007). Loosely linked scenarios that share a common framework or common focal issues (e.g., fisheries sustainability) but are developed independently at each scale (i.e., using scale-specific stakeholder input) tend to be better in terms of maintaining credibility and relevance to users while still allowing for some investigation of cross-scale processes. They also provide more freedom for stakeholders to explore issues of concern at each scale. Loosely linked scenarios can serve as a bridging mechanism for understanding the impact of decisions made at one scale on other scales (Biggs et al. 2007, 10, 16). Compared with tightly linked cross-scale scenarios, this approach requires fewer resources and a more manageable stakeholder engagement process while still allowing for communication of different points of view and perspectives across scales.

The strength of linkages of scenario elements and/or outcomes across scales can range from strong "hard" links in which elements and outcomes are equivalent across scales, through consistency across scales (in which the main scenario assumptions, driving forces, and their trends in the higher-scale scenarios provide boundary conditions for lower-scale scenarios), and coherency across scales (in which scenarios at different scales have matching "logic," i.e., follow the same paradigm or reflect the same scenario archetype), to comparability across scales (in which scenarios are connected by the issue they address, but may be constructed largely independently at different scales). In complementary scenarios, selected information from scenarios at one scale feeds into scenarios at another scale, but logic and assumptions differ across scale. These scenarios have the weakest linkage in elements and outcomes.

NOTES

1 https://www.dfo-mpo.gc.ca/stats/cfs-spc/tab/cfs-spc-tab2-eng.htm#table2-fna. The website has been updated, however, and now shows employment for 2018–21. The older data used in this study is no longer available on the website.
2 Scenario elements are the models, narratives, assumptions, and data that make up the scenario.
3 Note that these actual linkages between regional and national scale have not yet been done, and are part of future work in developing multiscale scenarios for Canadian oceans.
4 An online geographic information system (GIS) asset map can be viewed at https://www.arcgis.com/apps/MapSeries/index.html?appid=4deb908166494097a592c0919aa4b068.

REFERENCES

AMAP (Arctic Monitoring and Assessment Programme). 2018a. *Adaptation Actions for a Changing Arctic: Perspectives from the Baffin Bay/Davis Strait Region.* Oslo: AMAP.
–. 2018b. *Arctic Ocean Acidification.* Tromsø, Norway: AMAP.
Armitage D., and R. Plummer. 2010. *Adaptive Capacity and Environmental Governance.* Berlin and Heidelberg: Springer. https://doi.org/10.1007/978-3-642-12194-4.
Bailey, M., B. Favaro, S.P. Otto, A. Charles, R. Devillers, A. Metaxas, P. Tyedmers, et al. 2016. "Canada at a Crossroad: The Imperative for Realigning Ocean Policy with Ocean Science." *Marine Policy* 63: 53–60. https://doi.org/10.1016/j.marpol.2015.10.002.
Baum, J.K., and S.D. Fuller. 2016. *Canada's Marine Fisheries: Status, Recovery Potential and Pathways to Success.* Toronto: Oceana Canada.
Biggs, R., C. Raudsepp-Hearne, C. Atkinson-Palombo, E. Bohensky, E. Boyd, G. Cundill, H. Fox, et al. 2007. "Linking Futures across Scales: A Dialog on Multiscale Scenarios. *Ecology and Society* 12: 17. http://www.ecologyandsociety.org/vol12/iss1/art17/.
Bohensky, E., J.R.A. Butler, R. Costanza, I. Bohnet, A. Delisle, K. Fabricius, M. Gooch, et al. 2011. "Future Makers or Future Takers? A Scenario Analysis of Climate Change and the Great Barrier Reef." *Global Environmental Change* 21: 876–93. https://doi.org/10.1016/j.gloenvcha.2011.03.009.
Cheung, W.W.L., M.C. Jones, G. Reygondeau, C.A. Stock, V.W.Y. Lam, and T. Froelicher. 2016. "Structural Uncertainty in Projecting Global Fisheries Catches under Climate Change." *Ecological Modelling* 325: 57–66. https://doi.org/10.1016/j.ecolmodel.2015.12.018.
Cheung, W.W.L., V.W.Y. Lam, J.L. Sarmiento, K. Kearney, R. Watson, D. Zeller, and D. Pauly. 2010. "Large-Scale Redistribution of Maximum Fisheries Catch Potential in the Global Ocean under Climate Change." *Global Change Biology* 16: 24–35.
Day, A., and M. Prins. 2012. *Haida Gwaii Marine Planning: Future Scenario Analysis. Report from a Workshop Held July 17 and 18th, 2012.* https://haidamarineplanning.com/resources/.
Ebi, K.L. 2012. "Health in the New Scenarios for Climate Change Research." *International Journal of Environmental Research and Public Health* 11: 30–46.
Ekstrom, J.A., and O.R. Young. 2009. "Evaluating Functional Fit between a Set of Institutions and an Ecosystem." *Ecology and Society* 14 (2): 16.
FAO (Food and Agriculture Organization of the United Nations). 2023. "Food Balances." https://www.fao.org/faostat/en/#data/FBS.
Favaro, B., J. Reynolds, and I.M. Côté. 2012. "Canada's Weakening Aquatic Protection." *Science* 337 (6091): 154. https://doi.org/10.1126/science.1225523.
Francis, T.B., P.S. Levin, and C.J. Harvey. 2011. "The Perils and Promise of Futures Analysis in Marine Ecosystem-Based Management." *Marine Policy* 35 (5): 675–81.
Galaz, V., T. Hahn, P. Olsson, C. Folke, and U. Svedin. 2008. "The Problem of Fit among Biophysical Systems, Environmental and Resource Regimes, and Broader Governance

Systems: Insights and Emerging Challenges." In *Institutions and Environmental Change: Principal Findings, Applications, and Research Findings,* edited by O. Young, L.A. King, and H. Schroeder, 147–86. Cambridge, MA: MIT Press.

Hutchings, J.A., I.M. Côté, J.J. Dodson, I.A. Fleming, S. Jennings, N.J. Mantua, R.M. Peterman, B.E. Riddell, and A.J. Weaver. 2012a. "Climate Change, Fisheries, and Aquaculture: Trends and Consequences for Canadian Marine Biodiversity." *Environmental Reviews* 20 (4): 230–311. https://doi.org/10.1139/a2012-011.

Hutchings, J.A., I.M. Côté, J.J. Dodson, I.A. Fleming, S. Jennings, N.J. Mantua, R.M. Peterman, B.E. Riddell, A.J. Weaver, and D.L. VanderZwaag. 2012b. "Is Canada Fulfilling Its Obligations to Sustain Marine Biodiversity?" *Environmental Reviews* 20 (4): 353–61. https://doi.org/10.1139/er-2012-0049.

IPBES (Intergovernmental Science-Policy Platform on Biodiversity and Ecosystem Services). 2016. *The Methodological Assessment Report on Scenarios and Models of Biodiversity and Ecosystem Services.* Bonn: IPBES Secretariat.

IPCC (Intergovernmental Panel on Climate Change). 2013. "Definition of Terms Used within the DDC Pages." IPCC Data Distribution Centre, http://www.ipcc-data.org/guidelines/pages/definitions.html.

Jiang, L. 2014. "Internal Consistency of Demographic Assumptions in the Shared Socioeconomic Pathways." *Population and Environment* 35: 261–85.

Kerwath, S.E., H. Winker, A. Götz, and C.G. Attwood. 2013. "Marine Protected Area Improves Yield without Disadvantaging Fishers." *Nature Communications* 4: 2347.

Lam, V.W.Y., W.W.L. Cheung, G. Reygondeau, and U.R. Sumaila. 2016. "Projected Change in Global Fisheries Revenue under Climate Change." *Scientific Reports* 6: 32607.

Loucks, L., F. Berkes, D. Armitage, and A. Charles. 2017. "Emergence of Community Science as a Transformative Process in Port Mouton Bay, Canada." In *Governing the Coastal Commons Communities, Resilience and Transformation,* edited by D. Armitage, A. Charles, and F. Berkes, 43–59. Oxfordshire: Routledge.

MaPP Initiative (Marine Plan Partnership Initiative). 2015. *Haida Gwaii Marine Plan.* Prepared by Haida Nation and Province of British Columbia. http://mappocean.org/haida-gwaii/haida-gwaii-marine-plan/.

Mintz, C. 2018. "Seeking an Elusive, Expensive Catch: Quotas." *Globe and Mail,* March 2. https://www.theglobeandmail.com/news/british-columbia/seeking-an-elusive-expensive-catchquotas/article38196750/.

Oceana. 2018. *Fishery Audit 2018: Unlocking Canada's Potential for Abundant Oceans.* https://oceana.ca/en/reports/fishery-audit-2018/. Toronto: Oceana Canada.

O'Neill, B.C., E. Kriegler, K.L. Ebi, E. Kemp-Benedict, K. Riahi, D.S. Rothman, B.J. van Ruijven, et al. 2017. "The Roads Ahead: Narratives for Shared Socioeconomic Pathways Describing World Futures in the 21st Century." *Global Environmental Change* 42: 169–80. https://doi.org/10.1016/j.gloenvcha.2015.01.004.

O'Neill, B.C., E. Kriegler, K. Riahi, K.L. Ebi, S. Hallegatte, T.R. Carter, R. Mathur, and D.P. van Vuuren. 2013. "A New Scenario Framework for Climate Change Research: The Concept of Shared Socioeconomic Pathways." *Climatic Change* 122 (3): 387–400. https://doi.org/10.1007/s10584-013-0905-2.

Prime Minister of Canada. 2018. "Reconciliation Framework Agreement for Bioregional Oceans Management and Protection." Press release, June 21. https://pm.gc.ca/en/news/backgrounders/2018/06/21/reconciliation-framework-agreement-bioregional-oceans-management-and.

Russ, G.R., A.C. Alcala, and A.P. Maypa. 2003. "Spillover from Marine Reserves: The Case of Naso Vlamingii at Apo Island, the Philippines." *Marine Ecology Progress Series* 264: 15–20.

Sala, E., C. Costello, D. Dougherty, G. Heal, K. Kelleher, J.H. Murray, A.A. Rosenberg, and U.R. Sumaila. 2013. "A General Business Model for Marine Reserves." *PLOS One* 8 (4): e58799. https://doi.org/10.1371/journal.pone.0058799.

Samir, K.C., and W. Lutz. 2014. "The Human Core of the Shared Socioeconomic Pathways: Population Scenarios by Age, Sex and Level of Education for All Countries to 2100." *Global Environmental Change* 42: 181–92.

Statistics Canada. 2015. *Population Projections for Canada (2013 to 2063), Provinces and Territories (2013 to 2038).* Catalogue No. 91-520-X. Ottawa: Statistics Canada.

Steiner, N.S., W.W.L. Cheung, A.M. Cisneros-Montemayor, H. Drost, H. Hayashida, C. Hoover, J. Lam, et al. 2019. "Impacts of the Changing Ocean–Sea Ice System on the Key Forage Fish Arctic Cod (*Boreogadus saida*) and Subsistence Fisheries in the Western Canadian Arctic – Evaluating Linked Climate, Ecosystem and Economic (CEE). Models." *Frontiers in Marine Science* 6. https://doi.org/10.3389/fmars.2019.00179.

Tai, T.C., T. Cashion, V.W.Y. Lam, W. Swartz, and U.R. Sumaila. 2017. "Ex-Vessel Fish Price Database:

Disaggregating Prices for Low-Priced Species from Reduction Fisheries." *Frontiers in Marine Science* 4: 363. https://doi.org/10.3389/fmars.2017.00363.

Teh, L.S.L, W.W.L. Cheung, and U.R. Sumaila. 2017. "Scenarios for Investigating the Future of Canada's Oceans and Marine Fisheries under Environmental and Socio-Economic Change." *Regional Environmental Change* 17: 619–33.

Teh, L.S.L., and U.R. Sumaila. 2020. "Assessing Potential Economic Benefits from Rebuilding Depleted Fish Stocks in Canada." *Ocean and Coastal Management* 195: 105289.

Thayer Scott, J. 2012. *An Atlantic Fishing Tale, 1999–2011.* Ottawa: Macdonald-Laurier Institute.

Treasury Board of Canada Secretariat. 2007. "Canadian Cost-Benefit Analysis Guide: Regulatory Proposals." http://www.tbs-sct.gc.ca.

van Vuuren, D.P., J. Edmonds, M. Kainuma, K. Riahi, A. Thomson, K. Hibbard, G. Hurtt, et al. 2011. "The Representative Concentration Pathways: An Overview." *Climatic Change* 109 (1–2): 5–31. https://doi.org/10.1007/s10584-011-0148-z.

van Vuuren, D.P., M.T.J. Kok, B. Girod, P.L. Lucas, and B. de Vries. 2012. "Scenarios in Global Environmental Assessments: Key Characteristics and Lessons for Future Use." *Global Environmental Change* 22: 884–95.

van Vuuren, D.P., E. Kriegler, B.C. O'Neill, K.L. Ebi, K. Riahi, T. Carter, J. Edmonds, S. Hallegatte, S. Mathur, and H. Winkler. 2014. "A New Scenario Framework for Climate Change Research: Scenario Matrix Architecture. *Climatic Change* 122: 373–86.

Weatherdon, L.V., Y. Ota, M.C. Jones, D.A. Close, and W.W.L. Cheung. 2016. "Projected Scenarios for Coastal First Nations' Fisheries Catch Potential under Climate Change: Management Challenges and Opportunities." *PLOS One* 11 (1): e0145285. https://doi.org/10.1371/journal.pone.0145285.

Wilson, J.A. 2006. "Matching Social and Ecological Systems in Complex Ocean Fisheries." *Ecology and Society* 11 (1): 9.

Zurek, M.B., and T. Henrichs. 2007. "Linking Scenarios across Geographical Scales in International Environmental Assessments." *Technological Forecasting and Social Change* 74: 1282–95.

PART 3

Access to Ocean Resources

Lobsters and Livelihoods: Indigenous Rights and Fishery Access

6

Megan Bailey and Anthony Charles

Canada, like many countries, has a history of limiting access to resources. In fisheries, this primarily began with the advent of "extended fisheries jurisdiction" and the creation of the 200-nautical-mile exclusive economic zone (EEZ) under the 1982 *United Nations Convention on the Law of the Sea* (UNCLOS) (ECCC 2022). A large impetus behind UNCLOS was declining fish stocks, particularly cod stocks off the coast of Canada (Scheiber and Carr 1998). Since then, open access, in some cases, and traditional and self-management approaches to limiting access, in other cases, gave way to government restrictions imposing a system of property rights, where users are limited and controlled in the ways in which they access resources. Despite the dominance of this move, issues of access to resources, which exist in a web of constitutional rights and economic privileges (Figure 6.1), continue to be front and centre in the debate over social and economic sustainability across Canada's coastal communities (Bennett et al. 2018).

Pick any fishery across Canada, and access to fisheries resources, and the means of production and sale, will be fraught with conflict (cf. Charles 1992). Fish and fishing opportunities provide food, employment, and income, and support cultural practices and knowledge transfer, across coastal communities in Canada. But access for one group – in the case of a subtractable entity, which fish surely are – means less access for another group. In Canada, several different users seek access to fishing opportunities, and access is granted through provincial, territorial, federal, or Indigenous jurisdiction. Some of these users are rights holders, in that their claims to access come from different forms of Indigenous rights. Other users are commercial fishers who seek and are granted access through licences, most often federal. Still other users may be recreational or sport fish harvesters, where access for those users may or may not come in the form of provincial or territorial licences. In this chapter, we focus on competing access between Indigenous rights holders and commercial fish harvesters, and dig into the constituent parts comprising the ability to use and benefit from fisheries resources. We argue that governing access with this understanding of interconnectedness can help contribute to improved outcomes. We do so by means of a story, which we feel exemplifies the access imperative

The authors thank all of those, Mi'kmaq and non-Mi'kmaq, with whom we have had conversations that helped us understand the topic discussed in this chapter. We ourselves are white settler academics, whose privilege allows us to choose when, where, and how to engage in these conflicts. We recognize that others may not have that privilege. Our hope in writing this chapter is to share but one perspective, that of the access lens, through which the lobster conflict can be viewed. There are many more perspectives that need to be heard.

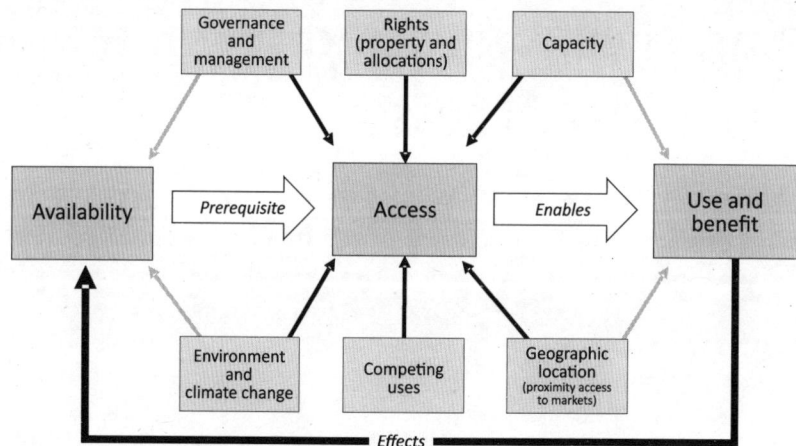

Figure 6.1 An access framework (developed as part of a collaborative workshop and paper, Bennett et al. 2018).

in Canada through the lens of one contentious issue that received heavy media attention in 2020: the Mi'kmaw moderate livelihood fishery in Nova Scotia.

A STORY: NOVA SCOTIA AND THE LOBSTER FISHERY

In September 2020, we, the authors of this chapter, found ourselves in the midst of a storm of controversy.[1]

By way of background, the Mi'kmaq of Atlantic Canada are the main Indigenous Nation in the region, and have inherent rights to practise their traditions and customs, including the right to procure "necessaries" through fishing, hunting, and gathering. Under the Peace and Friendship Treaties signed in the 1700s, codified in Section 35 of the Constitution[2] and reaffirmed by the Supreme Court of Canada in the *Sparrow* decision of 1990 and the *Marshall* decision of 1999,[3] Mi'kmaq have a right to harvest fish for food, social, and ceremonial purposes, and a right to fish for a moderate livelihood. It is the latter right, and its implementation in 2020, in what is referred to as the Mi'kmaw "moderate livelihood fishery," that has led to a months-long horrific conflict that has included violence on and off the water and policing of Mi'kmaw practices instead of policing of settler violence (Forester 2021). What "moderate livelihood" looks like in the context of fisheries remains undefined to this day. Is it a particular income? Is it a certain number of fishing days?

We became involved in that conflict as academics speaking in the media. A key issue related to the sustainability and environmental impact of the moderate livelihood fishery, particularly that of Sipeknekati'k First Nation in St. Mary's Bay (a part of the Bay of Fundy in Nova Scotia), concerns fishing outside the commercial fishing season. Many non-Indigenous commercial fishers expressed concern that the moderate livelihood fishery, then an estimated 250 traps, would damage the lobster stock. We were asked, as academics, if the livelihood fishery at that scale was a concern, and we independently answered that it was not. The level of reported fishing effort would be only 1/1,000 that of the commercial sector inside the lobster fishing area (LFA), in this case LFA 34, and 1/100 that of the commercial sector usually operating in St. Mary's Bay. The impact on the stock, then, would be imperceptible.

We found, however, that what seemed mathematically and scientifically an obvious answer led to many negative responses – weeks of critical messages, calls, and comments. In this section, we attempt to look at that controversy, that experience, through the lens of fishery access. We do so while acknowledging that some of the negative response to the Mi'kmaq fishing was rooted not in concerns over access or fishery sustainability but rather in racism and a denial of the validity of the *Marshall* decision (see Pannozzo 2020). The discussion here focuses not on those aspects but on the fishery-specific issues, specifically access.

Licences and Trap Limits

The Nova Scotian lobster fishery is more or less fully subscribed – that is, there is a cap on licences and on effort, and there is no "extra" unutilized lobster in the ocean. This implies that, although our initial views, expressed in the media, were that the current (late 2020) Mi'kmaw moderate livelihood fishery would not deplete lobster stocks, substantial growth in that fishery may lead to either depletion of the lobster or the need to reduce commercial fishing effort (to adjust for effort increases in the new fishery). From the commercial (largely non-Indigenous) fishers' perspective, both of these options are undesirable, as they have a negative impact on their own livelihoods – either a direct loss of their access (or fishing capability) or an indirect loss as a result of depletion of the lobster.

This reality is well known in the fishery, yet when the Mi'kmaw moderate livelihood fishery began, the Canadian government had no plan to reduce commercial fishing effort accordingly. Some plan is clearly needed, and ideas have been raised, such as to repeat what was done after the *Marshall* decision, when the government bought boats and licences from the commercial sector, for the Mi'kmaq. In that way, individual fishers (and thus effort) were completely removed from the commercial fishery. Another model involves "trap banking" or incremental trap number reductions across the board. For example, Mi'kmaw academic Shelley Denny (2020) published a newspaper article suggesting that the commercial sector could give up 1% of their traps, a marginal loss in catch potential quite small for each fisher but with gains in reconciliation (see, for example, Chapter 2) that could be huge. The commercial sector operates with each boat being allowed to carry a certain number of traps, depending on what LFA they are in. In LFA 34 (the heart of the discussion), it is 350 traps per boat, so those boats would give up 3–4 traps under the proposal.

Conservation and Management Rights

When we made our comments to media about the impact of the Mi'kmaw moderate livelihood fishery, the focus was on the level of fishing, i.e., the number of lobster traps being used. However, the debate moved on to reflect the concerns of many commercial fishers about the Mi'kmaw moderate livelihood fishery as it was introduced, taking place outside the conventional fishing season, and concentrated in a small area of the ocean. These issues are not directly a matter of access, but relate to a connected issue, that of management rights.

The first concern is that fishing outside of the regular season poses a conservation problem by affecting the nature and activity of lobster. The conventional fishing seasons were originally set for biological as well as economic reasons (Wilder 1954). On the latter, the hard-shelled lobsters harvested in those seasons (when the water is colder) are in the best condition marketwise, and thus are worth more; they haul up and ship better, and selling takes place when the markets are strong. Biologically, they are more likely to survive if put back in the water (e.g., if undersized) and they are not reproducing (although they may still be egg bearing) (DFO n.d.; Cook et al. 2020). On the other hand, in warm water, lobsters moult, giving up their hard shell for a period of time, during which they are prone to injury and being crushed in full traps (Dadswell 2020). These soft-shelled lobsters are worth less in a global market, and often go into canned or frozen supply chains instead of the live lobster trade. Notably, however, fisheries for soft-shelled lobster occur in Maine, where fishing is done year-round, so there is a precedent for sustainable year-round harvesting of lobster (Minke-Martin 2020).

Given these aspects, there are valid arguments for not fishing during warmer months of the year, and working within conventional seasons. Indeed, commercial lobster fishers have a long history of developing and implementing conservation measures in the fishery, and have a set of measures they consider proven over time, including the conventional "seasons" for fishing lobster.

If it were just a matter of catch value and revenue, then with the livelihood right being about providing for one's household and family, there is a logic in the conventional season, in having livelihood fishing in a winter fishery over a summer fishery. On the other hand, as we understand it, the Mi'kmaw moderate livelihood fishery is different from a typical commercial fishery in that it is intended to be carried out with smaller boats, in family-oriented activities. Fishing in that way, if done in colder months with worse weather, may be quite unsafe, not to mention undesirable. Adding to that the reality of belligerent behaviour seen in 2020 by some commercial fishers opposing the Mi'kmaw fishery, it may be a difficult and dangerous choice to be out on the water with the commercial sector in the winter.

Accordingly, there are various factors to consider concerning fishing times. But fundamentally, the choice of *when* to fish is up to the Mi'kmaq, through a constitutionally protected right – unless there is a valid, demonstrated conservation concern; there is full consultation; and there is a plan to infringe upon the Indigenous right in the least way possible, after exploring other options for meeting the conservation burden.

A second conservation issue concerns where fishing takes place, as too much fishing concentrated in certain small areas can be damaging. St. Mary's Bay, for example, is a sensitive place that can only take so much activity. To understand the limits of fishing in such locations, it is useful to look at a key index that fisheries scientists use to measure the status of a resource: catch per unit effort (CPUE). In this case, lobster is the unit and the effort invested is a "standard" vessel. While not perfect, the CPUE represents a measure of relative abundance of lobster in a given area. When CPUE falls, it may be a sign that fewer lobsters are available in that particular area, but may or may not signal that the population as a whole may be in trouble. Data for St. Mary's Bay and LFA 34 show that commercial catches have declined in 2017–19

compared with the 2015–16 season. Although the CPUE during that time is at the lower end of the range, these levels are clearly within that range, and seem low only when compared with the highs recorded in 2015–16 (Figure 6.2). Unfortunately, the most recent published assessment for lobster in LFA 34 was from 2013, which at the time indicated that fisheries-dependent indicators (CPUE and landings) were at a record high, but did indicate that there is evidence of increased exploitation in the nearshore (CSAS 2013). A 2019 assessment form LFA 34 has been completed but has yet to be published (personal communication by DFO Lobster Ecology and Assessment Team).

Furthermore, the fall 2020 protests targeted the livelihood fishery as causing a decline, but given the occurrence of a food, social, and ceremonial (FSC) fishery in the same area over a number of years, it seems likely that the lost catches were not due to the fact that the stock was declining, but simply that someone else was catching that fish (Mi'kmaw FSC fishers). The conflict was then not about conservation (although sustainability should be front and centre) but it was precisely about access: who catches the fish (or in this case, the lobster). It is important to note that in the 1990 *Sparrow* decision, the Supreme Court "held that, after conservation and other 'valid legislative objectives,' Aboriginal rights to fish for food, social and ceremonial (FSC) purposes have priority over all other uses of the fishery" (DFO n.d.).

Also crucial to note is that even more than 20 years after the *Marshall* decision, the federal government still does not seem to appreciate that access is not the only ingredient in Indigenous rights. The right to make decisions is also crucial – a matter of management rights (Capistrano and Charles 2012; Charles 2009). How to manage the fishery is considered by each First Nation in developing its fishing plan. What makes the most sense for its community members? What is safe, what is sustainable, and what is profitable? If, given this balance, moderate livelihood fisheries need to operate outside conventional seasons, that can be sustainable as long as the cumulative fishing pressure over the year, by all parts of the fishery, is kept low enough. Yes, one must consider lobster mortalities in planning and managing for a summer fishery, but if the summer is the time to operate a

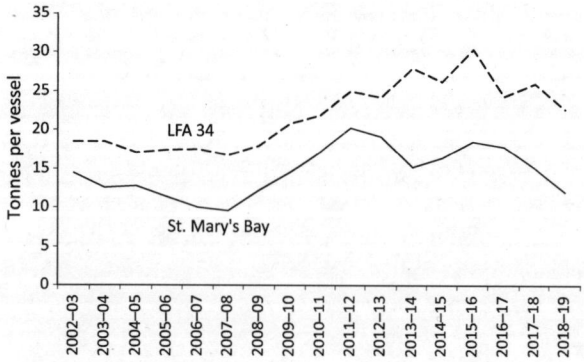

Figure 6.2 Catch per unit effort (CPUE) in St. Mary's Bay and LFA 34 | Data from DFO 2020

small-boat, family-focused lobster fishery, there are ways that can be done.

Livelihoods, Benefits, and Rights

The *Marshall* decision acknowledges the Mi'kmaw treaty right to a moderate livelihood from the fishery. While it indicates that the federal government retains responsibility to manage for conservation and other major goals, it does not limit who among the Mi'kmaq holds the livelihood right. In other words, the government may be able to impose conservation regulations that limit when, where, and how the Mi'kmaq fish for a moderate livelihood, but it cannot remove the right of all Mi'kmaq to that moderate livelihood. This has major implications.

During the 2020 debate, we heard statements and questions raised by commercial fishers and others, such as, "I support treaty rights but why do they need a moderate livelihood fishery if they have commercial communal licences?" and "I support a treaty fishery but where do the benefits go?" A reasonable answer to these questions is, "It's none of your business," because what kind of fisheries the Mi'kmaq operate or how the benefits are used are, indeed, matters for the Mi'kmaq to decide.

Looking broadly at livelihoods, benefits, and rights, it is important to realize that while there is so much discussion of fisheries, and lobster in particular, the livelihood right is not just about fishing but about the right to earn a livelihood from the land and sea and all that the earth may provide. Within the fishery, the Supreme Court decision did not restrict fishing rights exclusively to lobster, and lobster fishing is not the only way to earn a livelihood. In fact, it would be dangerous to follow that particular path: diversification in catch is an important economic and ecosystem imperative (Steneck et al. 2011). Further, the livelihood fishery is only one of several forms of access. There are other ways Mi'kmaq can access fishing. These include: (1) under FSC tags, the inherent right to fish for food, social, and ceremonial purposes; (2) fishing commercial licences as individuals; and (3) fishing the communal commercial licences currently held by some bands.

It is also important to note that recognition of fishing rights, and the acceptance of access, is a necessary but not sufficient condition for Mi'kmaq to benefit from that right. Bodwitch (2017) showed, in looking at the experiences of Maori fishers after a national treaty settlement granted fishing rights to Maori, that the granting of access to a large portion, about 30%, of the commercial fishery could not translate into tangible and adequate benefits for Maori because of consistent legal, social, and economic barriers that continue to this day. Similarly, in Nova Scotia, the catching of lobster under the livelihood fishery did not produce benefits to the Mi'kmaq as great as desired. On the one hand, there was resistance by some to sell bait to Mi'kmaw harvesters, out of protest, forcing harvesters to turn to illegal bait (like female crab) or less desirable bait, leading to smaller yields. On the other hand, current provincial regulations make it illegal to buy lobster outside the commercial season, so some bands were unable to sell their catches. (An illegal market for summer lobster [Cuthbertson 2018] is a predictable outcome, and without supporting legal means of sale, dismantling the illegal market will remain difficult.)

Despite increasing Mi'kmaw participation and self-determination in governing their access to lobster, a key challenge to the future of the lobster fishery is ensuring continued sustainability. While catch is currently sustainable and the stock is considered healthy (Cook et al. 2020), the effort going into it is considered to be at the maximum allowable level. While the initial level of Mi'kmaw moderate livelihood fisheries in 2020 was a small fraction of total annual fishing effort, and thus not a threat to conservation, growth is expected; further, the other two parts of the lobster fishery, the FSC and the conventional commercial fishery, each has its own impacts. Ultimately, conservation means limiting the total fishing pressure from all fisheries together. As moderate livelihood fisheries develop, there will need to be a corresponding decrease in the number of traps used by others, and in the extent of that use, to avoid conservation problems.

That reduction in fishing by non-Indigenous fishers will require new government funding to make the transition. Fairness is an important concept. It is not fair to ask the commercial lobster fishery to bear the sole burden of recognition of the livelihood right. This is a responsibility of all Canadians, in a spirit of reconciliation (see Chapters 2 and 14). The livelihood right comes from the signing

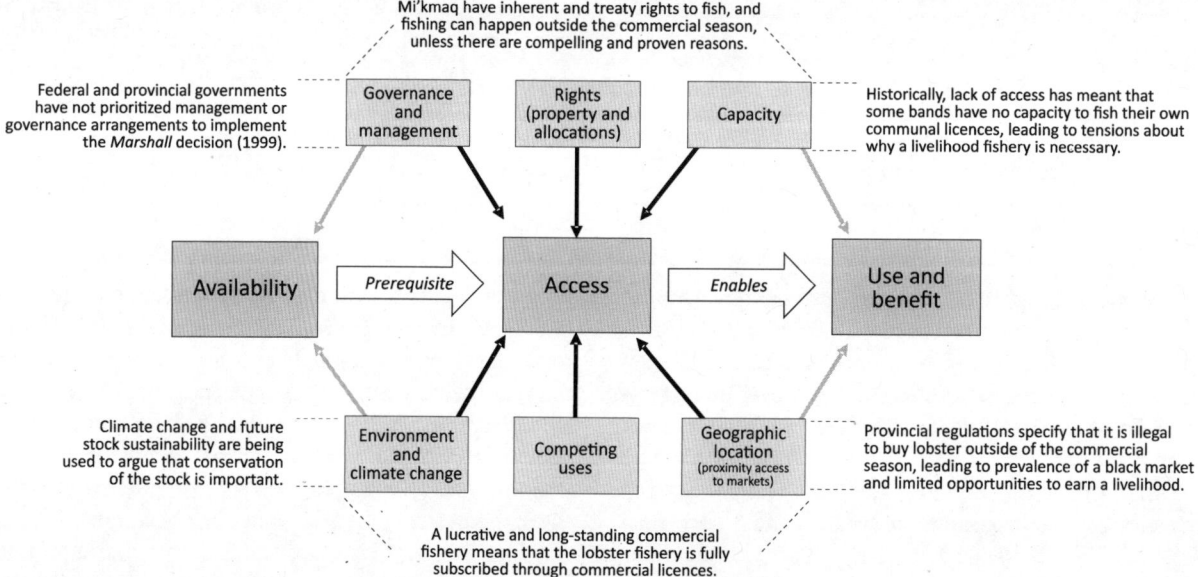

Figure 6.3 A summary of the story of the Mi'kmaw moderate livelihood fishery, as seen through an access lens (adapted from Bennett et al. 2018).

of treaties between the British Crown and the Mi'kmaq, Wolastoqiyik, Abenaki, Penobscot, and Passamaquoddy Peoples. This is Canada's treaty to honour, and fairness dictates that the cost of it should not be shouldered entirely by the commercial lobster fishery sector.

As the Mi'kmaw moderate livelihood fishery develops in coming years, more information will appear on what the livelihood plans look like, the projected scale, and at what times of the year and in what areas they will operate, along with projections on likely impacts to lobster populations when looking at all fisheries together. With greater knowledge, fear and uncertainty in the commercial sector will dissipate.

Who bears the burden of conservation measures will require a deep and uncomfortable conversation for Canadians. What is clear is that treaty rights exist. By supporting them, in all of the imperfect ways they are implemented while trying to get it right, Canadians can honour the foundations what it is to be Canadian. For fisheries researchers, it is perhaps amazing to see that these uncomfortable conversations are playing out on the lobster fishery stage, but that is the reality. Yet we need

to recognize that it goes deeper than lobster, and work on creating the space to support Indigenous rights, while at the same time recognizing that the commercial fishing sector is important to households, communities, and society at large.

ANALYZING ACCESS

The foregoing story speaks to the interrelated aspects of fisheries access and allocation (see also Chapter 7), of inputs and resource availability, of rights and title (see also Chapter 8), and of market access. As a collective, the OceanCanada Partnership defines access as "the ability to use and benefit from a resource or an area," and recognizes that access is influenced by social, legal, political, economic, and environmental factors (the building blocks of the image as first presented in Figure 6.1). Because of the need to both use and benefit from a resource, these influences may enter at one's ability to catch or interact with the resource, or at one's ability to sell or otherwise benefit from that interaction. For example, to catch fish, an abundance of biomass and a right or catch privilege must exist, and one must have a means of harvesting (boat, gear, skills). But in order to benefit from that catch, the catch needs to be edible (free from contaminants, for example), or needs to be sellable. In the case of the lobsters and livelihoods story, the periphery of Figure 6.3, there are multiple barriers that limit the ability of Mi'kmaq to use and benefit from lobster, despite the treaty right that exists.

As many reading this book may be aware, and as we touched on at the beginning of this chapter, what "moderate livelihood" means remains undefined, with misunderstandings even around who has the right to define it and how it can or should be implemented. Inaction by government over the past two decades has led to multiple undesirable consequences, including an illegal market for selling lobster caught outside the commercial season, and now violent behaviour by some in the commercial sector. Confusion about the details of the multiple forms of access that Mi'kmaq do have, for example, in FSC and communal commercial licences, along with a lack of broad understanding of treaty rights, has resulted in the public's questioning of why Mi'kmaq need another form of access under a livelihood right. This is, of course, amplified by the fact that the lobster fishery, while sustainable, is thought to be fully subscribed.

Much has been written about the principles of participation and accountability in good governance (see Molutsi 2001; Chang 2012). In Atlantic Canada, these principles have partially played out in the form of policies (and now regulations) that have engaged the inshore owner-operator lobster fleet in participatory governance processes (see Chapter 7). So, a crucial ingredient in successful conservation and management of lobster has been that those doing the fishing also had a role in setting the rules – a process that in part commits them to following the rules. But when the lobster fishery rules were set, Mi'kmaq were excluded, and instead it was largely non-Indigenous fishers who participated in the rule setting. This cannot be considered a just or sustainable situation. Indeed, it is now changing, as Mi'kmaw fisheries (and First Nations fisheries generally) are developing under their own community-based plans, which include many of the same ingredients of success that have been part of conventional commercial fisheries, but are also utilizing Indigenous wisdom and knowledge. That is reflected in the framing of these fisheries by the Assembly of Nova Scotia Mi'kmaw Chiefs, around the Mi'kmaw approach of *Netukulimk,* described by the Unama'ki Institute of Natural Resources (n.d.) as "achieving adequate standards of community nutrition and economic well-being without jeopardizing the integrity, diversity, or productivity of our environment." Non-Indigenous fishers, with centuries-old knowledge and conservation practices, will certainly identify with this approach.

A FUTURE

What does the future hold? With lobster catches largely at all-time highs recently, and lobster productivity also high (Cook et al. 2020), there is reason to hope that prosperous, equitable, and sustainable lobster harvests in the future should be possible. However, projections regarding potential climate change impacts, like those discussed in Part 2 of this book, suggest that high productivity may not last due to warming and ocean acidification. So, this raises concern and highlights the need for good governance now, to avoid overharvesting in the event of population decline.

Meanwhile, Fisheries and Oceans Canada has declared its rules for moderate livelihood fisheries – specifying, first, that licences would be required to be given by Fisheries and Oceans Canada to Mi'kmaw bands, and second, that fishing and selling can be done only inside the currently specified commercial fishing season. This led to expressions of frustration and disappointment by Mi'kmaw Chiefs and their legal representatives (Withers 2021). From our perspective, the continual imposition of power from Fisheries and Oceans over the behaviour and practices of the Mi'kmaq does not align with the spirit and intent of the treaties, of the Constitution, of the *Marshall* decision, reconciliation, or the *United Nations Declaration on the Rights of Indigenous Peoples* (UNDRIP). Legal pluralism exists in Canada, and essentially means that while Fisheries and Oceans Canada has the authority to manage commercial fisheries, its authority to manage Indigenous fisheries is limited (Fanning and Denny 2020).

It would be far preferable for the government to embrace recent amendments to Canada's *Fisheries Act,* one of the oldest pieces of legislation in Canada (whose inception dates back to 1868, only one year after Confederation). After substantial cuts that weakened the act (see Bailey et al. 2016), the government produced a "modernized" *Fisheries Act* in 2019.[4] The most relevant amendments as related to the livelihood discussion are references to Section 35 of the *Constitution Act, 1982* in Sections 2.3 and 2.4:

2.3 This Act is to be construed as upholding the rights of Indigenous peoples recognized and affirmed by section 35 of the Constitution Act, 1982, and not as abrogating or derogating from them.

2.4 When making a decision under this Act, the Minister shall consider any adverse effects that the decision may have on the rights of the Indigenous peoples of Canada recognized and affirmed by section 35 of the Constitution Act, 1982.

In all of these documents, agreements, and commitments, and in the *Fisheries Act* itself, self-determination and consultation are key. This has yet to be embraced by Fisheries and Oceans Canada, and that should make all Canadians uncomfortable. This does nothing to support Mi'kmaw empowerment in fisheries and livelihood governance, continues to constrain Mi'kmaw access to fisheries resources, and puts the safety of Nova Scotians at risk.

The focus in this chapter is on fisheries, but similar lessons can be drawn in relation to access in a broader context. The ways in which Canadians use and benefit from marine resources and spaces are complex and cannot be understood without understanding historical, legal, economic, environmental, and social contexts. This chapter has shown how using an "access lens" to make sense of the Mi'kmaw livelihood conflict shows the way forward. We remain hopeful that Fisheries and Oceans Canada and the federal government broadly can change course and make room for a Mi'kmaw-governed livelihood fishery.

NOTES

1 This story draws on articles by each author (Bailey 2020; Charles 2020), as well as a talk that Megan Bailey gave at St. Francis Xavier University's Mi'kmaw Livelihood Learning Lodge on November 24, 2020, https://youtu.be/8EQiFr2dP_k.

2 *Constitution Act, 1982,* being Schedule B to the *Canada Act 1982* (UK), 1982, c 11.

3 *R v Sparrow,* [1990] 1 SCR 1075; *R v Marshall,* [1999] 3 SCR 456.

4 *Fisheries Act,* RSC 1995, c F-14.

REFERENCES

Bailey, M. 2020. "Nova Scotia Lobster Dispute: Mi'kmaw Fishery Isn't a Threat to Conservation, Say Scientists." *The Conversation,* October 20. https://theconversation.com/nova-scotia-lobster-dispute-mikmaw-fishery-isnt-a-threat-to-conservation-say-scientists-148396.

Bailey, M., B. Favaro, S. Otto, T. Charles, R. Devillers, A. Metaxas, P. Tyedmers, et al. 2016. "Canada at a Crossroad: The Imperative for Realigning Ocean Policy with Ocean Science." *Marine Policy* 63: 53–60.

Bennett N., M. Kaplan-Hallam, G. Augustine, N. Ban, D. Belhabib, I. Brueckner-Irwin, A. Charles, et al. 2018. "Coastal and Indigenous Community Access to Marine Resources and the Ocean: A Policy Imperative for Canada." *Marine Policy* 87: 186–93.

Bodwitch, H. 2017. "Challenges for New Zealand's Individual Transferable Quota System: Processor Consolidation, Fisher Exclusion, and Māori Quota Rights." *Marine Policy* 80: 88–95. https://doi.org/10.1016/j.marpol.2016.11.030.

Capistrano, R.C., and A. Charles. 2012. "Indigenous Rights and Coastal Fisheries: A Framework of Livelihoods, Rights and Equity." *Ocean and Coastal Management* 69: 200–9.

Chang, Y.-C. 2012. *Ocean Governance: A Way Forward.* Dordrecht, Netherlands: Springer.

Charles, A. 1992. "Fishery Conflicts: A Unified Framework." *Marine Policy* 16 (5): 379–93.

–. 2009. "Rights-Based Fisheries Management: The Role of Use Rights in Managing Access and Harvesting." In *A Fishery Manager's Guidebook,* edited by K.L. Cochrane and S.M. Garcia, 253–82. Oxford: Wiley-Blackwell.

–. 2020. "Moving Forward on Lobster Fishery Means Addressing Access and Conservation." *Policy Options,* October 28.

Cook, A.M., P.B. Hubley, D. Denton, and V. Howse. 2020. "2018 Framework Assessment of American Lobster (*Homarus americanus*)." In *Lobster Fishing Area (LFA) 27–33.* DFO Canadian Science Advisory Secretariat Research Document 2020/017. Ottawa: Fisheries and Oceans Canada.

CSAS (DFO Canadian Science Advisory Secretariat). 2013. "Assessment of Lobster (*Homarus americanus*)." In *Lobster Fishing Area (LFA) 34.* DFO Canadian Science Advisory

Secretariat Research Document 2013/024. Ottawa: Fisheries and Oceans Canada.

Cuthbertson, R. 2018. "Stakeouts and Microchipped Lobster: Inside DFO's Probe of a First Nations Fishery." *CBC News,* October 5.

Dadswell, M. 2020. "Listen to the Science – Lobster Seasons Fine-Tuned to Match Local Ecosystems." *Chronicle Herald,* October 26.

Denny, S. 2020. "Making Room for Mi'kmaw Livelihood Fishery Easier Than You Think." *Chronicle Herald,* October 14.

DFO (Fisheries and Oceans Canada). n.d. "Lobster Fishing Areas 27–38: Integrated Fisheries Management Plan." https://www.dfo-mpo.gc.ca/fisheries-peches/ifmp-gmp/maritimes/2019/inshore-lobster-eng.html#toc2.

ECCC (Environment and Climate Change Canada). 2022. "Law of the Sea: United Nations Convention." https://www.canada.ca/en/environment-climate-change/corporate/international-affairs/partnerships-organizations/law-sea-united-nations-convention.html.

Fanning, L., and S. Denny. 2020. "Conflict over Mi'kmaw Lobster Fishery Reveals Confusion over Who Makes the Rules." *The Conversation,* October 29. https://theconversation.com/conflict-over-mikmaw-lobster-fishery-reveals-confusion-over-who-makes-the-rules-148978.

Forester, B. 2021. "DFO, RCMP Knew Violence Was Coming but Did Nothing to Protect Mi'kmaw Lobster Harvesters: Documents." *APTN News,* February 10.

Minke-Martin, V. 2020. "Mi'kmaw Fishery Dispute Is Not about Conservation, Scientists Say." *Hakai Magazine,* October 9.

Molutsi, P. 2001. "Tracking Progress in Democracy and Governance around the World: Lessons and Methods." United Nations Online Network in Public Administration and Finance. https://www.eldis.org/document/A42699.

Pannozzo, L. 2020. "In Search of Common Ground: An Interview with Arthur Bull about the Lobster Fishery Crisis in St. Mary's Bay." *Halifax Examiner,* November 1.

Scheiber, H.N., and C.J. Carr. 1998. "From Extended Jurisdiction to Privatization: International Law, Biology, and Economics in the Marine Fisheries Debates (1937–1976)." *Berkeley Journal of International Law* 10: 15.

Steneck, R.S., T.P. Hughes, J.E. Cinner, W.N. Adger, S.N. Arnold, F. Berkes, S.A. Boudreau, et al. 2011. "Creation of a Gilded Trap by the High Economic Value of the Maine Lobster Fishery." *Conservation Biology* 25: 5. https://doi.org/10.1111/j.1523-1739.2011.01717.x.

Unama'ki Institute of Natural Resources. n.d. "Netukulimk." https://www.uinr.ca/programs/netukulimk/.

Wilder, D.G. 1954. *The Lobster Fishery of the Southern Gulf of St. Lawrence.* General Series Circular No. 24. St. Andrews, NB: Atlantic Biological Station.

Withers, P. 2021. "Nova Scotia Mi'kmaq Unified in Opposition to Moderate Livelihood Rules." *CBC News,* March 5. https://www.cbc.ca/news/canada/nova-scotia/nova-scotia-mikmaq-opposition-moderate-livelihood-rules-1.5937465.

7

The Impact of Quotas on Canada's Fisheries and the Response of Fish Harvesters

Evelyn Pinkerton, Marc Allain, Danielle Edwards, Phillip Saunders, and Charlotte Whitney

Individual quotas (IQs) are a share of the Total Allowable Catch (TAC) allocated to an individual operating unit such as a country, a vessel, a company, or an individual fisher, depending on the allocation system. IQs are used as an output control for limiting the total catch of individual fishers, allocating catch, and spreading catch over time. When IQs are permanently allocated to the individual and transferable via the market as individual transferable quotas (ITQs), with little or no restrictions on who buys or leases them, they have been found to create serious distributional inequities by shifting benefits from fishers and fishing communities into the hands of investors, processors, and/or offshore entities (FOPO 2019; CIFHF 2015). This chapter analyzes the development and implementation of IQs and ITQs in Canada in the context of fisheries management policy. It explores how the public interest and management objectives could be better served by alternative approaches.

Canada initiated modern fish management in the mid-1970s by creating the Department of Fisheries and Oceans (DFO), also known as Fisheries and Oceans Canada. The federal government was expanding its role in fisheries science and management in preparation for extended fisheries jurisdiction (EFJ) in 1977, and the coming into force in 1994 of the 1982 *United Nations Convention on the Law of the Sea* (UNCLOS). The government's plan to address the challenges and opportunities of EFJ were spelled out in the 1976 *Policy for Canada's Commercial Fisheries* (Fisheries and Marine Service 1976), a remarkable public policy document, given what was to transpire in subsequent decades.

The policy attributed the main problems in Canada's commercial fisheries to the "tragedy of the commons" (Hardin 1968), i.e., overcapacity stemming from the race for fish under open access to common pool resources, but explicitly rejected private property rights and market mechanisms for determining fisheries access. Instead, it declared that the state responsibility for allocating access could not be delegated, and proposed government intervention in fisheries development "in the best interest of Canadian society" (Fisheries and Marine Service 1976, 53) and of "the people who depend on the fishing industry" (5). While the policy called for a major transformation of the fishery from seasonal, small-boat fleets making day trips toward a class of larger, intermediate mobile vessels covering more of the fishing grounds and supplying plants for longer periods, it also expressed a strong concern about fishers and their communities during the transition. It recognized that "abrupt action" could destroy the livelihoods of people and erode the economic base of communities (56). Government decisions concerning the allocation of fisheries access would therefore involve trade-offs between economic efficiency objectives and the interests of fishers and their communities in favour of the latter.

This fisher- and fishing community–centric vision for Canada's fisheries was short-lived. A series of major crises – the bankruptcy of Atlantic Canada's vertically integrated offshore groundfish fleets in the early 1980s, conservation and over-capacity crises in the owner-operator groundfish fleets on the Scotian Shelf and in British Columbia in the mid to late 1980s, and the fiscal crisis of the federal government in the 1990s – conspired to shift government policy 180°, opening the door to radical experiments with neoliberal rationalization in DFO's Pacific and Maritimes-Scotia-Fundy regions (Barnett, Messenger, and Wiber 2017). By the mid-1990s, DFO had abandoned any pretence of social and economic objectives for the fishery beyond economic efficiency, and refocused its mandate on resource conservation, regardless of social, economic, or distributional equity considerations (Demont 1998).

While the government's initial socio-economic policy objectives were not officially revoked, it was left up to the country's owner-operator fleets to defend them. Over almost four decades, the leaders of these fleets resisted the tide of neoliberal economic thinking that swept over DFO, protecting policies in Atlantic Canada such as fleet separation, which prohibited processors from controlling licences in the inshore fishery, and proposing new ones to keep fishing licences in the hands of working fishers.

Here, we provide a description of BC fleet rationalization through the introduction of IQs and ITQs in more than a dozen BC fisheries between 1979 and 2006, and review the impact of ITQs on BC fleets and fishing-dependent communities. We then present a contrasting history of fisheries in Atlantic Canada, where the owner-operator fishery has thrived under a policy framework that prohibits vertical integration of fishing and fish-processing operations in

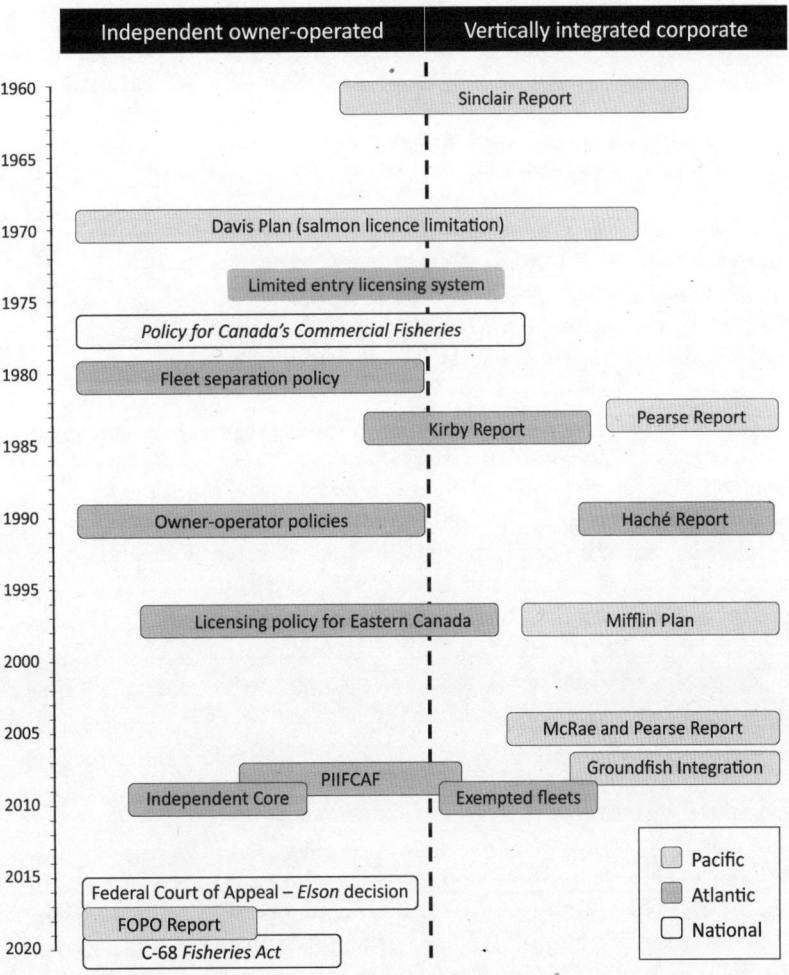

Figure 7.1 Timeline of select Canadian fisheries policy, legislation, management decisions, and DFO-commissioned reports, organized by the extent to which they promoted and adhered to independent owner-operated or vertically integrated corporate fisheries approaches. The regional application of each is denoted by colour: national in scope (grey), Atlantic (blue), and Pacific (green).

the inshore fishery. We then briefly discuss how two recent events – a decision of the Federal Court of Appeal upholding the Minister of Fisheries and Oceans' authority to adopt policies in support of social and economic objectives for its fisheries, and changes to the *Fisheries Act* (Bill C-68) that enshrine this authority in legislation[1] – allow Canada to consider new directions for its fisheries policies. We end with a discussion of scenarios for transition back to owner-operator in Pacific ITQ fisheries (see Figure 7.1).

BRITISH COLUMBIA AS A LABORATORY FOR DEVELOPING ITQS

The Justification for Fisheries Privatization

From the mid-1970s onward, policies based on the assumptions of the "tragedy of the commons" (Hardin 1968) formed the overarching paradigm through which DFO and neoliberal economists chose to understand fishery dynamics. Hardin imagined state intervention or privatization as the only options to avoid resource depletion, and advocated for privatization. The prevailing assumption was that fisheries were predominantly open-access and that fishers, pursuing their self-interest, would individually overinvest in boats and gear in a race to catch fish, leading to overcapacity and the dissipation of profit.

While some form of access limitation is widely acknowledged to be necessary, access limitation has been unnecessarily conflated with resource privatization, with devastating consequences for fishing-dependent communities. The limited-entry program advocated by economist Sol Sinclair and implemented by Minister Jack Davis in 1969, which not only limited licences but also made licences transferable via an unregulated market, was an abject failure because it increased both access and fishing effort (Pinkerton 2015). This failure opened the door to consideration of ITQs as a better mechanism for limiting access and effort.

Starting with the 1978 Powell River Symposium, economists focused on the efficiency of allocating a fixed amount or percentage (quota) of the TAC of a species to individual fishers, who would thus not have to "race" to take their allocation. They maintained that if individual quotas were transferable via the market, they would end up in the hands of the most efficient fishers (those who could operate at lowest cost), obviating the need for state allocations. In so doing, they ignored previous self-regulation efforts, such as "layup system" in the BC halibut fishery. The "race for fish" had been eliminated under the layup system by requiring fishermen to stop fishing for about a week after delivery, which staggered and spread out fishing effort. For many years in British Columbia, a temporary monthly allocation system was also employed successfully for rockfish management before the system was abolished in favour of ITQs in 2006. A similar option in western Newfoundland currently allocates time periods for nontransferable individual allocations of halibut to working fishers, in keeping with the distributional objectives of the Food, Fish and Allied Workers Union of Newfoundland and Labrador (Pinkerton et al. 2018).

In contrast, ITQs were premised on the presumed behaviour and motives of fishers as disembedded, self-interested, individual rational actors who would not cooperate for the greater good. But this presumption ignored the evidence that fishers' organizations had indeed cooperated to solve the race for fish in the BC and Newfoundland halibut fisheries. In letting the market dictate quota allocation, ITQ policy ignored these solutions and how benefits could be equitably distributed and support existing fishing-dependent communities.

Impact of Unlimited Transferability and Unequal Distribution

Unlike nontransferable individual allocations (IQs), ITQs transform the permit/quota into a tradable commodity like stocks in a stock market. Neoliberal economists Donald McRae and Peter Pearse (2004) argued against any limits on transferability based on the assumption that since ITQs would automatically be bought by the most economically efficient fishers, this would produce the most efficient use of capital. This argument ignored the reality of unequal access to capital in the fishing fleet, and the fact that processors and investors with far greater access to capital could (and did) buy up ITQs and licences (Pinkerton and Edwards 2009). ITQs and licences increased in value by 600% over ten years in British Columbia's halibut fishery (Pinkerton 2013), and owner-operator quota ownership fell from 90% in 1991 to 15% in 2016. Owner-operators who have entered the fishery within the last 15 years collectively own less than 1% of the quota, while investors (whose only role in the fishery is to collect lease fees that often exceed 80% of the landed catch value) own 42% of the quota (Edwards and Pinkerton 2019b).

The impact of quota leasing on working fishers is crushing. The average lessee fisher in the halibut fishery, catching about $580,000 worth of halibut, receives between

$70,000 and $120,000, out of which all of the fishing expenses, including crew wages, vessel maintenance, fuel, bait, and monitoring fees, must be paid. The remainder, as much as $510,000, or 88% of the landed value, is paid in lease fees to the quota owner. Under these leasing and ownership conditions, owner-operators are not financially viable, struggle to cover operating costs, and fail to cover long-term costs such as vessel upkeep and replacement, with the result that the fishery is no longer self-sustaining as an owner-operator fishery (Edwards and Pinkerton 2020). Furthermore, in the BC halibut fishery, processors control quota not only through ownership – with processor ownership more than doubling in the halibut fishery since IQ implementation, to just under 10% in 2016 – but also through quota brokering, through which processors control more than half of halibut quota leasing (Edwards and Pinkerton 2019a). This has meant that processors are able to secure catch from fishers not through offering the best prices but through control of quota.

As ITQ ownership has become concentrated in urban and urban-adjacent regions of British Columbia, ITQs have contributed to community and intergenerational disenfranchisement and the loss of access to adjacent fishery resources (Edwards et al. 2006), as fishers' numbers overall also decreased by 70% from 1980 to 2016 (Stocks 2017). For halibut in 2002, 96% of the catch was in the rural North/Central Coast and West Coast of Vancouver Island regions, but less than 10% of the quota was owned in these regions – a catch/ownership pattern also observed in other ITQ fisheries. As ITQs increased in value in the absence of ownership restrictions, rural and Indigenous fishers could not compete because of lower incomes and lower property values, which significantly reduced their access to capital, leading to the loss of access in communities that historically depended on these fisheries. As discussed below, regulating *who* can own ITQs and IQs, and *limiting their transferability*, are key to preventing the situation that developed in British Columbia.

ATLANTIC CANADA AS THE LABORATORY FOR OWNER-OPERATOR AND FLEET SEPARATION

The introduction of ITQs in Atlantic fisheries followed a pattern similar to the processes in Canada's Pacific fisheries, with government using market-driven measures to address serious conservation and rent dissipation

problems stemming from management's inability to constrain growth in fishing effort, which its own policies had largely fuelled. However, in most fisheries in the Atlantic an alternative fisheries management model emerged, restricting fishing licences to working fishers and prohibiting the vertical integration of fishing and fish-processing operations. This owner-operator model has permitted fleet-managed allocation processes in some fisheries, based on distributional equity considerations that prevent the accumulation and concentration of quota that occurs under fully tradable quasi-property rights regimes that separate access to fishery resources from the communities that depend on them.

Over more than four decades, the struggle to develop, enhance, and protect the owner-operator fishery from vertical integration led to new government policies and a series of legal victories, culminating in the *Elson* decision of the Federal Court of Appeal[2] and significant amendments to Canada's *Fisheries Act*. During this time, the owner-operator fleets not only prospered but have emerged as the dominant socio-economic fleet in Canada's Atlantic fisheries.

The Promise and Prescriptions of the Early Years

Before extended fisheries jurisdiction, Canada was taking only 20% of the northwest Atlantic's total allowable groundfish catch, with a fishing fleet composed of three distinct fleet sectors: 160 large offshore trawlers over 100 feet in length owned by 12 vertically integrated fish companies; 500 midshore vessels mostly under 65 feet (small draggers, longliners, large gillnetters); and 10,000–15,000 small inshore vessels, fishing day trips near their home port communities and involved in two or more fisheries (Fisheries and Marine Service 1976, 23).

The government plan to maximize the benefits from Atlantic groundfish called for a "fundamental restructuring" of the fishery. The small-boat inshore fleet was identified as a focus for capacity reductions given its "low productivity" (Fisheries and Marine Service 1976, 56, 64). In this restructuring, however, government policy clearly favoured fishers and their communities, ensuring that they received the primary benefits from EFJ. Minister Roméo LeBlanc acted on this policy commitment in 1979 by adopting the fleet separation policy, which prohibited vertical integration in the inshore fisheries by banning fish-processing companies from acquiring licences in fisheries prosecuted by vessels under 65 feet. This policy extended protections from corporate concentration to the Atlantic fishery, protections that had been in place for more than a decade in the Pacific fishery through a policy that limited corporate ownership of licences to 12% by monitoring and moral suasion.

Corporate Financial Crisis and the Introduction of Enterprise Allocations and ITQs

A mere six years after the 1976 policy was announced, its vision for the Atlantic fishery was in shambles. Exogenous factors such as high interest rates and fuel costs had pushed the offshore, vertically integrated fish companies into bankruptcy, and a government task force (Kirby 1983) proposed a sharp reversal of the 1976 policy, recommending that offshore licences become divisible and transferable, i.e., sold or traded through the granting of Enterprise Allocations (EAs), a form of ITQ. EAs eliminated the race for fish by providing the two large offshore corporate entities that emerged from the debt crisis with their own share of the TAC, which they could harvest at moments best suited to their business operations. While EAs were applied only to the offshore vertically integrated corporate sector, they were widely touted as proof of concept of the efficiency of ITQs and market mechanisms for rationalizing fisheries facing serious overcapacity problems (DFO 1994).

The next decade in the Atlantic fisheries culminated in the collapse of the northern cod fishery and the first groundfish moratorium in 1992, highlighting the state's inability to control both destructive fishing practices (dumping and highgrading are encouraged by EAs and ITQs) and growth in fishing capacity and effort in all fleet sectors. In fisheries managed under TACs only, problems and conflicts sharpened as signs of declining abundance led to TAC reductions in many fisheries. During this period, ITQs spread from the corporate offshore fleet to the owner-operator inshore and midshore fleets as DFO attempted to control overcapacity through the allocation of TAC first by fleet sector (e.g., mobile and fixed gear in groundfish) and then allowing/encouraging fleet sectors to move to IQs and ITQs.

This process of ITQ introduction in the Atlantic owner-operator fleets was complex, spread over time and space,

and poorly documented. Generally, however, ITQs took hold initially and most extensively in the midshore specialized fleet sectors along the Scotian Shelf – the area and fleets identified in the 1976 policy as having the most promise for developing a more industrial and productive fishery. Initial quota allocations were provided free of charge based on prior fishing histories, which resulted in the transfer of a valuable public resource to private control. Although all of these fleets were originally owner-operator and under fleet separation when they entered the ITQ system, the majority of them have since been formally exempted from the fleet separation and owner-operator policies because of their de facto corporate control through the logic of the ITQ process (described above for BC halibut) and the use of legal mechanisms to strip control from owner-operators, as discussed below.

The Emergence of the Owner-Operator Inshore Fleet as the Dominant Fleet Sector

The great irony of this recent Atlantic fishery history is that the fleet targeted for elimination in 1976 because of its low incomes and productivity – the small-boat, inshore fleet of multi-species vessels – emerged by the mid-1990s as the economic engine of the Atlantic fishery. As groundfish landings disappeared, the value of fishery landings shifted massively to valuable shellfish fisheries (e.g., lobster and crab), which this fleet controlled almost exclusively.

This inshore fleet sector – now known officially as the Independent Core sector – still represents approximately 10,000 licence holders and remains the dominant employer in the Atlantic fishery. Moreover, it has increased its dominance in landed value. By 2017, over 70% of the Atlantic fishery's landed value was in lobster and crab. When the inshore sector's majority participation in the shrimp fishery, and major positions in virtually all other fisheries because of its multi-species licensing system, is considered, the inshore, owner-operator fleet is by far the dominant economic performer in the Atlantic fishery.

The Origins of the Owner-Operator Policy as a Complement to Fleet Separation

The two policies that directly support the inshore fishery – the fleet separation and owner-operator policies – did not develop simultaneously. Fleet separation was a government initiative to stimulate dockside competition among processors for the products of the inshore fishery (Gough 2000), while the owner-operator policy was a unique contribution of the Atlantic inshore fleets. It emerged initially as four separate policies, as DFO officials in the four distinct DFO administrative regions of the Atlantic (Quebec, Gulf, Scotia-Fundy, and Newfoundland regions) followed through on the 1976 policy commitment to consult fishermen's organizations in their respective regions on how to implement the policy, particularly around licensing.

Inshore fishermen's organizations – the Regroupement des pêcheurs professionnels du sud de la Gaspésie (RPPSG) in the Quebec region; the Maritime Fishermen's Union (MFU) and the Eastern Fishermen's Federation (EFF) in the Gulf and Maritimes Scotia-Fundy regions; and the Fish, Food and Allied Workers Union (FFAW-Unifor) in Newfoundland – used these opportunities to fashion a licensing system for the inshore sector built around the traditional fishing strategies of small vessels fishing a range of resources easily accessible from their communities during the fishing seasons. The main features of these owner-operator policies were multi-species licence portfolios issued to individuals who had to be on board and actively directing fishing operations on vessels they owned. These features were consistent across regions and contained surprisingly detailed processes for vetting of licence transfers to new entrants by fishermen at the port level, to ensure that fishing access and its attendant employment and income benefits remained attached to local communities.

By 1989, all four DFO administrative regions in the Atlantic had fully developed owner-operator policies to accompany fleet separation. A subsequent reform of DFO's licensing policy in 1995–96 merged these into an Atlantic-wide policy that became known as the Core Licensing Policy. By the mid-1990s, the inshore fleet in Canada's Atlantic fisheries had clear policy protections prohibiting vertical integration and stipulating that the holders of inshore fishing licences had to be actively participating in their fishing activities. Their fleets also had either exclusive or majority access to the region's most valuable fisheries. Concerns about ITQs expanding into their fisheries were also dealt with through the 1995 Montreal Round Table, a joint DFO and fishing industry process,

which outlined conditions under which ITQ programs could be introduced in the future: they could not confer or take away access to the fishery; they must be supported by a clear majority of licence holders in the fishery in question; they must include restrictions on the transferability of licences to prevent undue accumulation of quota; and they must include an intervener process for any group of fishers who believed they would be adversely affected by the introduction of ITQs (Standing Senate Committee on Fisheries 1998).

The reaction of corporate interests to these restrictions was to use private contractual arrangements to exploit the definitional imprecision in the policies to assert effective control over the fishing licences they were denied under policy. The first iteration of these efforts was through the use of "trust agreements" whereby the *beneficial interest* in the inshore fishing licence (including the ability to direct its use and to request replacement licences) was transferred to a third party not entitled to hold one (DFO 2003). These agreements effectively defeated the purpose of both the fleet separation and owner-operator policies and permitted the kind of accumulation and commodification of licences and any attached quota that occurred via the unrestricted ITQs in the Pacific.

Protecting the Independence of the Inshore Fleet in Canada's Atlantic Fisheries

The growth and spread of these contractual arrangements and the lack of DFO enforcement of its own policies created apprehension among owner-operator fleets across the Atlantic. These arrangements were perceived as an existential threat to their fleets, especially in light of the effects of corporate concentration on owner-operators in British Columbia since the introduction of ITQs in Pacific fisheries. During the 2000–04 Atlantic Fishery Policy Review (AFPR), the Canadian Council of Professional Fish Harvesters (CCPFH), a national organization of owner-operator fleets, drew attention to the growing corporate control over their fisheries and were successful in getting the federal government to adopt a remedy – the Policy for Preserving the Independence of the Inshore Fleet in Canada's Atlantic Fisheries (PIIFCAF) – in 2007.

In response to the use of trust agreements to circumvent DFO licensing policies, and to preclude the utilization of other contractual arrangements that resulted in third-party control over a licence, PIIFCAF required all Atlantic inshore fleet Core Licence holders to declare whether or not they were party to what were now defined as Controlling Agreements (CA). These were broadly defined to encompass any private contractual arrangements (including trust agreements) that ceded certain forms of control over the licence to a third party, particularly those that influenced a licence holder's decision on requests to issue a replacement licence, commonly referred to as a "licence transfer." Under PIIFCAF, licence holders unfettered by CAs were redesignated as Independent Core harvesters, while all others were given until 2014 to extricate themselves from the agreements or risk nonrenewal of their licences. Significantly, PIIFCAF also provided an exemption mechanism for fleets operating under ITQ programs. As elsewhere, those owner-operator fleets along the Scotian Shelf (operating under ITQs since the late 1980s and 1990s) had come under corporate control through CAs and were allowed to exempt themselves from the new policy.

The period leading up to the 2014 deadline for PIIFCAF compliance was feverish in terms of Atlantic fisheries policy advocacy. The holders of CAs, particularly processors who controlled valuable inshore lobster licences in Nova Scotia and inshore crab licences in Newfoundland, not only lobbied to have their agreements grandfathered but continued acquiring additional Independent Core licences through agreements deemed PIIFCAF-compliant by DFO. Alarmed by these developments and a sudden DFO consultation on management policy launched in early 2012 and perceived as a prelude to the elimination of PIIFCAF and the owner-operator and fleet separation policies, the Atlantic's Independent Core fleets created a new national organization that their BC counterparts immediately joined: the Canadian Independent Fish Harvesters' Federation (CIFHF), dedicated exclusively to policy development and advocacy for Canada's community-based owner-operator fisheries. Within months, the CIFHF generated motions in support of the owner-operator and fleet separation policies from four provincial legislatures (New Brunswick, Nova Scotia, Quebec, and Prince Edward Island) and the Federation of Canadian Municipalities. This broad demonstration of support and unity of purpose led the Minister of Fisheries and Oceans to enforce the 2014 PIIFCAF deadline.

Advocates for the owner-operator fishery faced an additional challenge, however. Shortly after the adoption of PIIFCAF, senior policy officials within DFO began to raise doubts privately about the Minister's constitutional authority to adopt management policies unrelated to the conservation of fisheries resources, suggesting that PIIFCAF could be *ultra vires* because it attempted to regulate private contracts, a provincial responsibility under the *Constitution Act, 1867* (personal communication by senior DFO officials and ministerial staff, 2010–17). Since the owner-operator and fleet separation policies and PIIFCAF were all directed at management's social and economic objectives, these appeared to be at risk. The CIFHF felt obliged to commission an independent legal opinion on the scope of the Minister's authority presented to Minister Shea in early 2015, an intervention thought to have influenced her subsequent decision to fully enforce PIIFCAF (personal communication by DFO ministerial staff, 2016). DFO's doubts regarding the constitutional validity of PIIFCAF emboldened those violating the policy, however. A crab licence issued to Kirby Elson, an inshore fisher from Labrador, declared to be under a CA with two Newfoundland processing firms, remained noncompliant when Elson failed to end the CA by the 2014 PIIFCAF deadline, apparently as a test case of the policy's legal validity. In early 2016, Elson, having exhausted his rights to administrative appeals of the decisions of two successive Ministers (Gail Shea and Hunter Tootoo) not to grant him an exemption to the policy, sought judicial review of the Ministers' decisions in the Federal Court.

The 2017 Federal Court (FC) ruling in *Elson v Canada (Attorney General),* subsequently upheld by the Federal Court of Appeal (FCA) in 2019, is a significant decision in Canadian fisheries law.[3] It unequivocally confirmed and clarified the broad scope of the Minister's authority to manage fisheries for purposes beyond conservation and to "carry out social, cultural or economic goals or policies" (para. 51, FC). The courts rejected Elson's argument that the CAs were simply matters of contract law and thus within provincial jurisdiction, holding that the federal fisheries power (which overrides provincial jurisdiction) extended to the broader social and economic concerns underlying PIIFCAF (para. 22, FCA). The final word in the matter came later in 2019, when the Supreme

Court of Canada denied Elson's request for leave to appeal the FCA decision. Weeks earlier, owner-operator fleets and fishery-dependent coastal communities had received further guarantees that their access to the fishery could be protected when the House of Commons adopted Bill C-68, which amended the *Fisheries Act* to enshrine in legislation the purposes that may be pursued by the Minister, including "social, economic and cultural factors in the management of fisheries" (Section 2.5[g]). Furthermore, in a clear reference to PIIFCAF, these purposes include "the preservation or promotion of the independence of licence holders in commercial inshore fisheries" (Section 2.5[h]). A further amendment supports PIIFCAF's interdiction of CAs, authorizing the Minister to promulgate regulations prohibiting "the transfer of the use and control of rights and privileges" associated with licences (Section 4[g.01]). In addition to these 2019 amendments to the *Fisheries Act,* the DFO acquiesced to a long-standing request from owner-operator fleets to bind itself to implement the fleet separation and owner-operator policies and PIIFCAF through regulatory reform by adding key elements of these policies to the *Atlantic Fishery Regulations* and *Maritime Provinces Fishery Regulations.* These changes came into force in December 2020 and April 2021.

THE NEW CONTEXT: VISIONING A RETURN TO THE BEGINNING

With the Federal Court and Federal Court of Appeal decisions in *Elson* and the 2019 changes to the *Fisheries Act,* Canadian fisheries policy has come full circle. After nearly four decades of experiments with the neoliberal prescriptions of ITQs that marginalized the interests of fishers and fishing communities, it is possible to imagine a return to fisher- and fishing community–centric fisheries policies as envisioned in the 1976 *Policy for Canada's Commercial Fisheries.* Further indication of such change came in the May 2019 report of the House of Commons Standing Committee on Fisheries and Oceans (FOPO 2019). This all-party consensus report recommended that DFO develop "a new policy framework for Pacific fisheries" (41), including the establishment of "an independent commission to transition the West Coast fishery to a 'made-in-BC' owner-operator model" (43). Below, we propose some scenarios from the theory and

real-world practice of fisheries management that could be considered for this kind of transitioning.

Scenario 1: The Need to Put an End to "Business as Usual"

Based on the evidence presented above, it is clear that the introduction of ITQs has had adverse effects on the socio-economic health and well-being of owner-operator fishers and fishery-dependent coastal communities, despite the fact that the economic health and well-being of both fishers and fishing communities remain among Canada's key fisheries management objectives. Continuing with ITQs could, therefore, reasonably be expected to continue undermining key government objectives. Fishing effort continues to be concentrated on fewer, larger corporate vessels, with a net increase in overall fishing capacity in the BC groundfish fisheries. Since effort concentrating onto larger vessels requires large, repeated catches to cover costs, pressure on the resource can be expected to increase alongside reliance on state subsidies (Sumaila et al. 2019). It is difficult to see how any of the above could be considered to be in the public interest. Based on the poor performance of ITQ fisheries against management objectives (Pinkerton 2015; FOPO 2019; Edwards and Pinkerton 2020), one would expect the government to eliminate this option from any future consideration and begin seriously considering alternatives.

Scenario 2: Continuing with ITQs but with Significant Restrictions

It could be argued that with significant restrictions and ongoing evaluation against objectives such as fleet stability and viability, equitable distribution of access, stability of adjacent communities, ITQ-managed fisheries could improve their socio-economic outcomes. Under this scenario, ITQs would remain, but mechanisms to increase ownership in fishing-adjacent rural areas (such as restricting licence transfers to owner-operators or to residents or institutions in a geographic area) would be implemented. When transitioning an ITQ fishery with few restrictions to one with restrictions, industry participants testifying before the House of Commons Standing Committee on Fisheries and Oceans suggested that new restrictions could be phased in over five to seven years, as was done in the case of Controlling Agreements under

PIIFCAF; the committee noted this suggestion in its final report (FOPO 2019, 43). If significant restrictions were implemented and/or maintained, many negative outcomes associated with ITQ fisheries management could be ameliorated (but likely not eliminated). Regular monitoring and evaluation of continuing ITQ fisheries would be needed to ensure that restrictions are maintained and benefits and outcomes are as intended.

Scenario 3: Transitioning ITQ Fisheries – Starting with a Multi-Stakeholder Process

Fortunately, multiple tools are available to governments to retroactively address socio-economic and conservation objectives for fisheries. A method proven to achieve positive outcomes involves multi-stakeholder, consensus-based approaches (Innes and Booher 2018), as recommended in the FOPO (2019, 41) report. Such a process has three general steps: objective identification, consultation and decision making, and evaluation. Given the issues identified earlier with ITQ fisheries and efforts to undermine restrictions, this process would be ongoing and iterative, to re-evaluate and adjust the system as needed to meet objectives over time.

Below, we briefly review what might be considered variants of the licence bank option for transitioning to owner-operator/fleet separation in the under 65-feet fleet on the West Coast without undue pain to fishermen who own ITQs or to processors. Licence/quota banks are government or nongovernmental organizations that own and manage licences and/or quotas for the benefit of their members. We also consider alternative management options that have proven successful on the east coast.

Northern Native Fishing Corporation

The first variant is the licence bank of 252 salmon gillnet licences owned since 1982 by the Northern Native Fishing Corporation (NNFC), comprising three tribal councils in northern British Columbia (Haida, Tsimshian, Nisga'a). The NNFC leases out these licences to its members at affordable prices on an annual basis, retaining licence ownership and control of allocation policies.

Alaska Community Development Quota Model

The second variant is the Alaska Community Develop-

ment Quota (CDQ) model, in operation since 1992. In western Alaska, six regional place-based nonprofit organizations (Community Development Quota groups) manage approximately 7–10% of most groundfish ITQs. The CDQ groups lease their share of the quota to offshore trawlers (who were allocated 90–93% of the quota), eventually accumulating enough lease fees from these trawlers to purchase their own quota and/or licences for inshore fisheries, which are then leased to their traditional inshore fishers (Pinkerton and Langdon 2019).

Northern Shrimp

Foley, Mather, and Neis (2015) document a third variant of this multi-stakeholder licence/quota bank whereby the federal government in the late 1990s and early 2000s created "special allocations to communities" from increases in northern shrimp Total Allowable Catch. Under this program, the existing fishers' Fogo Island Cooperative and isolated coastal communities in Newfoundland's northern peninsula, grouped under a new nonprofit organization, received a new allocation of quota that was fished by the inshore fleet and processed in local plants, contributing significantly to the local economy.

Atlantic Halibut Sustainability Plan

Since 2013, the Atlantic Halibut Sustainability Plan for the fixed-gear fleet in Division 4R off western Newfoundland and southern Labrador has been temporarily assigning equal individual nontransferable allocations to qualifying fishers, allowing them to choose time periods for taking their allocation, thus spreading out effort so there is no race for fish. This approach has improved conservation results, delivered strong economic returns, distributed benefits widely and equally to active multi-species owner-operator fishers, and allowed an even flow into the market. The legitimacy and effectiveness of this system, which cost government nothing but meeting time, would not have been achievable without its owner-operator foundation (Pinkerton et al. 2018).

CONCLUSIONS

Since the mid-1970s, Canada has had a broad sweep of management objectives for its fisheries, starting with conservation of the resource and extending to the profitability of fishing enterprises for fishers and the socio-economic well-being of fishing communities. For the better part of four decades, however, under the influence of neoliberal ideology, the objectives related to the interests of fishers and fishing communities were sacrificed in favour of investors and corporations and their profits through the introduction of ITQs. Canadian fisheries policy from the early 1980s to the 2000s introduced quasi-property rights schemes and market-driven mechanisms to determine access to fishing licences and quotas that have proven disastrous for the distribution of fisheries benefits to fishing communities and those who fish, effectively undermining the country's socio-economic objectives for the fishery. In Atlantic Canada, a large inshore fleet, with effective leadership and representation and support from provincial governments and fishery-dependent coastal communities, effectively fought against corporate control of their fisheries. Fishers' organizations were able to resist the neoliberal prescriptions advocated by corporate interests through creative mechanisms for socially determined, collective means to allocate fisheries access to maximize returns to fishers and coastal communities. The fisher-focused fishery model that they were able to put in place ensures that recent windfalls in abundance and values of the Atlantic fishery's products are equitably distributed and contribute to thriving local economies. In comparison, when a corporate agenda took hold, implementing ITQ fisheries with few to no restrictions in many of British Columbia's fisheries, the result was loss of fleets, concentration of wealth in very few hands, and an impoverished employment base in coastal communities. Through the Pacific halibut ITQ example, we have shown that completely transferable, market-driven ITQs lead to eventual accumulation, commodification, and dispossession of small-scale, inshore fishers in fishing-dependent communities. Given the failure of neoliberalism to deliver on the full spectrum of Canada's objectives for its fisheries, now is the time for developing alternative approaches that build on proven working models, such as those that we have identified, which place restrictions on ownership, do not commodify access privileges, and establish democratic ownership/control over access rights through fishers' organizations, cooperatives, community licence/quota banks, or comparable arrangements.

NOTES

1 *Fisheries Act*, RSC 1985, c F-14.
2 *Elson v Canada (Attorney General)*, 2017 FC 459 (CanLII); aff'd 2019 FCA 27 (CanLII); leave to appeal to SCC dismissed, 2019 CanLII 67973.
3 Ibid.

REFERENCES

Barnett, A.J., R.A. Messenger, and M.G. Wiber. 2017. "Enacting and Contesting Neoliberalism in Fisheries: The Tragedy of Commodifying Lobster Access Rights in Southwest Nova Scotia." *Marine Policy* 80: 60–68.

CIFHF (Canadian Independent Fish Harvesters' Federation). 2015. *Proceedings, Fisheries Management Policy Workshops*.

Demont, J. 1998. "Turning the Tide." *Maclean's*, June 20.

DFO (Fisheries and Oceans Canada). 1994. *Experience with Individual Quota and Enterprise Allocation (IQ/EA) Management in Canadian Fisheries, 1972–94*. Halifax: DFO. https://waves-vagues.dfo-mpo.gc.ca/Library/40597301.pdf.

–. 2003. "Preserving the Independence of the Inshore Fleet in Canada's Atlantic Fisheries." Discussion document.

Edwards, D., and E. Pinkerton. 2019a. "The Hidden Role of Processors in an Individual Transferable Quota Fishery." *Ecology and Society* 24 (3): 36. https://www.ecologyandsociety.org/vol24/iss3/art36.

–. 2019b. "Rise of the Investor Class in the British Columbia Pacific Halibut Fishery." *Marine Policy* 109: 1–9.

–. 2020. "Priced Out of Ownership: Quota Leasing Impacts on the Financial Performance of Owner-Operators." *Marine Policy* 111: 1–10.

Edwards, D., A. Scholz, E. Tamm, and C. Steinbeck. 2006. "The Catch-22 of Licensing Policy: Socio-Economic Impacts in British Columbia's Commercial Ocean Fisheries." In *North American Association of Fisheries Economists Forum Proceedings*, edited by U.R. Sumaila and A.D. Marsden, 65–76. Fisheries Centre Research Reports 14 (1). Vancouver: Fisheries Centre, University of British Columbia.

Fisheries and Marine Service, Department of the Environment. 1976. *Policy for Canada's Commercial Fisheries*. Ottawa: Fisheries and Marine Service, Department of the Environment.

Foley, P., C. Mather, and B. Neis. 2015. "Governing Enclosure for Coastal Communities: Social Embeddedness in a Canadian Shrimp Fishery." *Marine Policy* 61: 390–400.

FOPO (House of Commons Standing Committee on Fisheries and Oceans). 2019. *West Coast Fisheries: Sharing Risks and Benefits.* Report of the Standing Committee on Fisheries and Oceans. Ottawa: House of Commons.

Gordon, H.S. 1954. "The Economic Theory of a Common Property Resource: The Fishery." *Journal of Political Economy* 62: 124–42.

Gough, J. 2000. *Managing Canada's Fisheries: From Early Years to the Year 2000.* Montreal and Kingston/Sillery, QC: McGill-Queen's University Press/Septentrion.

Hardin, G. 1968. "The Tragedy of the Commons." *Science* 162 (3859): 1243–48. https://www.science.org/doi/10.1126/science.162.3859.1243.

Innes, J.E., and D.E. Booher. 2018. *Planning with Complexity.* London: Routledge.

Kirby, M.J. 1983. *Navigating Troubled Waters: A New Policy for the Atlantic Fisheries.* Ottawa: Canadian Government Publishing Centre.

McRae, D.M., and P.H. Pearse. 2004. *Treaties and Transition: Towards a Sustainable Fishery on Canada's Pacific Coast.* Vancouver: Hemlock Printers.

Pinkerton, E. 2013. "Alternatives to ITQs in Equity-Efficiency-Effectiveness Trade-Offs: How the Lay-Up System Spread Effort in the BC Halibut Fishery." *Marine Policy* 42: 5–13.

–. "The Role of Moral Economy in Two British Columbia Fisheries: Confronting Neoliberal Policies." *Marine Policy* 61: 410–19.

Pinkerton, E., M. Allain, D. Decker, and K. Carew. 2018. "Atlantic and Pacific Halibut Co-Management Initiatives by Canadian Fishermen's Organizations." *Fish and Fisheries* 19 (6): 984–95. https://doi.org/10.1111/faf.12306.

Pinkerton, E., and D. Edwards. 2009. "The Elephant in the Room: The Hidden Costs of Quota Leasing." *Marine Policy* 33: 707–13.

Pinkerton, E., and S. Langdon. 2019. "Indigenizing and Co-Managing Local Fisheries: The Evolution of the Alaska Community Development Quota Program in the Norton Sound Region." In *The Rights of Indigenous Peoples in Marine Areas,* edited by Stephen Allen, Nigel Bankes, and Eyvind Ravna, 375–400. London: Hart Publishing.

Sinclair, S. 1960. *License Limitation – British Columbia: A Method of Economic Fisheries Management.* Ottawa: Fisheries and Oceans Canada.

Standing Senate Committee on Fisheries and Oceans. 1998. *Privatization and Quota Licensing in Canada's Fisheries.* Report of the Standing Senate Committee on Fisheries and Oceans. Ottawa: Senate of Canada.

Stocks, A. 2017. *The State of Coastal Communities in British Columbia.* Vancouver: T. Buck Suzuki Environmental Foundation. https://d3n8a8pro7vhmx.cloudfront.net/tbuck suzuki/pages/158/attachments/original/1576611259/The_State_of_Coastal_Communities_in_British_Columbia_2017.pdf?1576611259.

Sumaila, U.R., N. Ebrahim, A. Schuhbauer, D. Skerritt, Y. Li, H.S. Kim, T.G. Mallory, V.W.L. Lam, and D. Pauly. 2019. "Updated Estimates and Analysis of Global Fisheries Subsidies." *Marine Policy* 109: 103695. https://doi.org/10.1016/j.marpol.2019.103695.

8

The Role of Fisheries Co-Management in Addressing Access and Allocation Inequities in Eastern Inuit Nunangat

Carie Hoover, Jason Akearok, Tommy Palliser, Amber Giles, Mark Basterfield,
Aaron Dale, Melina Kourantidou, Ashlee Cunsolo, Megan Bailey, and Jamie Snook

We would like to dedicate this chapter to Jorgen H. Bolt (1961–2020). Jorgen was from the community of Kugluktuk, Nunavut, and was appointed to the Nunavut Wildlife Management Board (NWMB) in 2017, serving until he passed away in 2020. During his time on the board, he was heavily involved in commercial fisheries issues and decisions. During the 2018 revisions to Nunavut's Allocation Policy for Commercial Marine Fisheries, he attended stakeholder meetings to ensure consistency and account-ability with regard to allocation issues and decisions. In addition, he made significant contributions to the first meeting of the three co-management boards (Nunavut, Nunavik, Nunatsiavut) to discuss commercial fisheries and the reports that serve as the recommendations for this chapter. Besides his role on the NWMB, Jorgen was known for being a master hunter, trapper, and guide. He will be remembered by his colleagues for sharing hunting stories over a meal or his open invitations to join him on hunting trips. His warm, inviting, sharing nature is greatly missed. This profound loss to his family, friends, and community and to wildlife management is far-reaching. We hope to continue sharing knowledge of wildlife management and fisheries in his honour.

Inuit have been asserting, negotiating, finalizing, and implementing modern-day land claims agreements (LCAs) since the early 1970s (see also Chapters 2 and 14). Together, these agreements make up Inuit Nunangat (Inuktitut for "homeland"), a term that includes water and ice in addition to land (Inuit Tapiriit Kanatami 2020). The agreements have brought significant changes in Inuit governance and sovereignty, and greater involvement of local resource users and Inuit community members in decision making compared with the initial settlement periods.

In the eastern portion of Inuit Nunangat – Nunavut, Nunavik, and Nunatsiavut – there is commercial fisheries access at both the local (community) and territorial (offshore) levels. The main commercial fisheries[1] are focused on Greenland halibut (also referred to as turbot in Canada, *Reinhardtius hippoglossoides*), northern shrimp (*Pandalus borealis*), striped shrimp (*Pandalus montagui*), and snow crab (*Chionoecetes opilio*). Under each of the LCAs, territorial co-management boards have been created: the Nunavut Wildlife Management Board (NWMB), the Nunavik Marine Region Wildlife

Board (NMRWB), and the Torngat Joint Fisheries Board (TJFB) in Nunatsiavut. Each co-management board has a responsibility to ensure implementation of respective LCAs and to provide advice on commercial fisheries management in alignment with the goals and aspirations of the LCAs. These three boards are part of a network of co-management boards in Canada, and have a track record of learning from, and incorporating, different knowledge systems as a result of navigating through the complex interests and positions of diverse stakeholders and rights holders in their regions (Snook, Cunsolo, and Dale 2018; Snook, Cunsolo, and Morris 2018).

While co-management boards hold various responsibilities for managing and allocating access in commercial fisheries, overall authority and responsibility for resource conservation and management rests with the federal Minister of Fisheries and Oceans (DFO 2018). Federal policies such as the New Access Framework of Fisheries and Oceans Canada (DFO) (DFO 2008) describe how new or additional access to commercial fisheries should be implemented. This includes recognition of treaty rights, and conservation and equity goals, designed to ensure management that does not exacerbate or create interregional access disparities, with access being additionally evaluated based on adjacency (those closest to the resource), historical dependence, and economic viability (DFO 2004). Further development of a Fisheries and Oceans Canada and Canadian Coast Guard Reconciliation Strategy in 2019 highlighted long-term co-management objectives so that "Indigenous groups' role in management and decision-making is well defined, implemented, accepted by all parties" and "Indigenous groups effectively manage their own fisheries in their territories" (DFO 2019b). While promising, these federal policies have not succeeded in ensuring either equitable access to adjacent commercial fisheries or consistent application of co-management decision making.

In this chapter, we discuss the process of fisheries allocation decision making in Nunavut, Nunavik, and Nunatsiavut, keying in on federal allocations, territorial sub-allocations, and better allocation decision-making processes. Our objective is to highlight the robustness of co-management board recommendations and decisions over their territorial and adjacent commercial fisheries

resources, and to identify inequities in access to commercial fisheries and ways to move toward self-determination of Inuit commercial fisheries. We begin by reviewing the rise of co-management in Inuit Nunangat, and continue with descriptive analyses of fisheries and allocations in each of the regions of interest in this chapter. Lastly, we offer policy recommendations (see also Chapter 14, with which these align) for moving toward more equitable access to adjacent fisheries in Inuit Nunangat.

TAKING STOCK: THE RISE OF CO-MANAGEMENT IN INUIT NUNANGAT

The implementation of LCAs created a system of co-management boards and processes across Inuit Nunangat to support resource management, including fisheries, for socially desirable outcomes. While there is no one common definition of co-management, it may be conceptualized as a "shared space" where the different organizational structures, roles, knowledge systems, processes, and responsibilities in the management of fisheries come together so that actors can reach consensus on management decisions (Berkes 2009; Snook, Cunsolo, and Dale 2018). Outside of this shared space, the signatories to co-management agreements retain their respective roles, responsibilities, and powers. This definition takes account of the spirit and intent of LCAs and may be juxtaposed against DFO's definition, where co-management is described as "the sharing of responsibility and accountability for results between Fisheries and Oceans Canada and resource users, and will *eventually* also encompass the sharing of authority for fisheries management" (emphasis added) (DFO 2004).

Land Claims, Fisheries Policies, Access, and Decision Making

In Nunavut and Nunavik, the NWMB and NMRWB make access (Total Allowable Catch [TAC] for areas outside land claims jurisdiction or Total Allowable Harvest [TAH] for areas within land claims jurisdiction) and sub-allocation decisions on commercial fisheries stocks inside the settlement areas (Nunavut Settlement Area [NSA] and Nunavik Marine Region [NMR], respectively), while the TJFB makes recommendations for commercial stocks inside and adjacent to the Labrador Inuit Settlement Area (LISA)

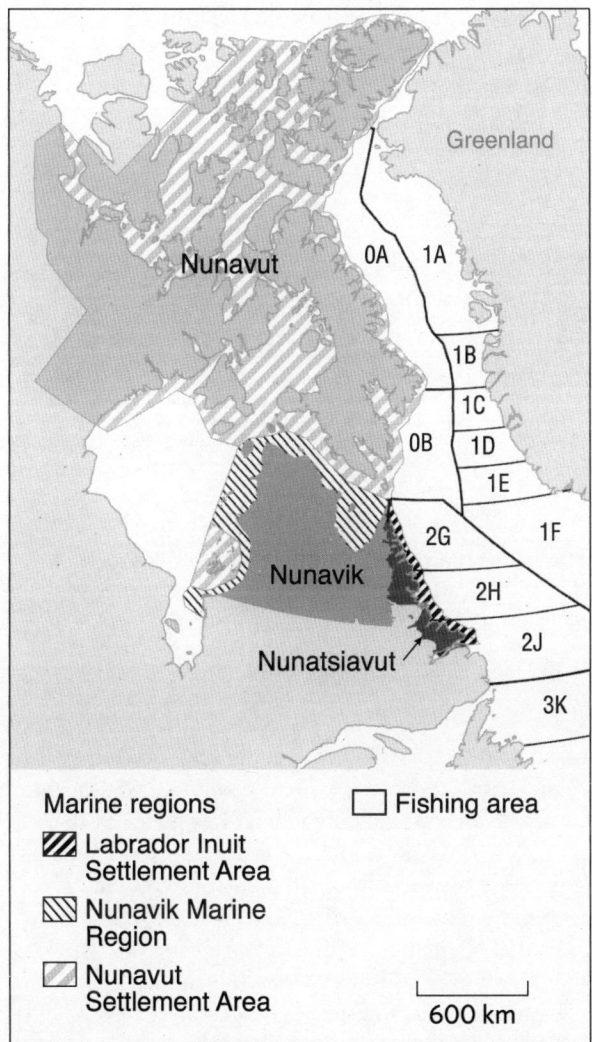

Figure 8.1 Territories of Nunavut, Nunavik, and Nunatsiavut, including marine areas under jurisdiction of each land claims agreement: the Nunavut Settlement Area (NSA), the Nunavik Marine Region (NMR), and the Labrador Inuit Settlement Area (LISA), respectively. Northwest Atlantic Fisheries Organization (NAFO) marine fishing areas are presented for reference.

(*Nunavut Agreement* [NA], Section 5; *Nunavik Inuit Land Claims Agreement* [NILCA], Section 5; and *Labrador Inuit Land Claims Agreement* [LILCA], Section 13.11) (Figure 8.1). In the offshore areas outside the settlement areas but still adjacent to the territories, all three boards provide advice to the Minister (NA, Section 15; NILCA, Section 5.5; LILCA, Section 13.11). Once each co-management board identifies who has access to this sub-allocation, each board then makes its recommendations to DFO. In the past, the Minister has varied (i.e., modified or changed) many decisions relating to recommendations and decisions inside and outside the settlement areas.

REGIONAL SNAPSHOTS

Nunavut

Within Nunavut, the NWMB is the main instrument of wildlife management and the main regulator of access to wildlife in the Nunavut Settlement Area (NA, Section 5.2.33), with a mission of "conserving wildlife through the application of Inuit Qaujimajatuqangit (IQ) and scientific knowledge." IQ is a term that captures Inuit experience in the world – world view – and includes the "principles, beliefs, and skills which have evolved as a result of that experience" (Tagalik 2010). The NWMB, an independent administrative tribunal, is part of the overall structure of public government within Nunavut. It relies upon government (local, territorial, and federal), Inuit, and other stakeholders for advice and technical assistance. Its members base their decisions on their own independent and impartial findings of facts, consideration of issues raised, and any applicable law. Under the *Nunavut Agreement*, the NWMB's wildlife management (including commercial fisheries) decisions inside the NSA are subject to the Minister's final say (NA, Section 5.3.22). Outside of the NSA, the NWMB provides advice on the TAC (NA, Section 15) and on sub-allocations of the quota for the Minister's consideration.

Access to commercial fisheries through the NWMB depends on multiple factors. DFO allocates to the NWMB, which then sub-allocates (Figure 8.2A) in tonnes once every five years (NWMB 2019). During the call for commercial fisheries applications, the NWMB determines the amount to sub-allocate to each of the Nunavut-owned fishing enterprises, of which there are currently four. In making these sub-allocation decisions, the NWMB enlists the expertise and advice of a Fisheries Advisory Committee (FAC) to review applications and

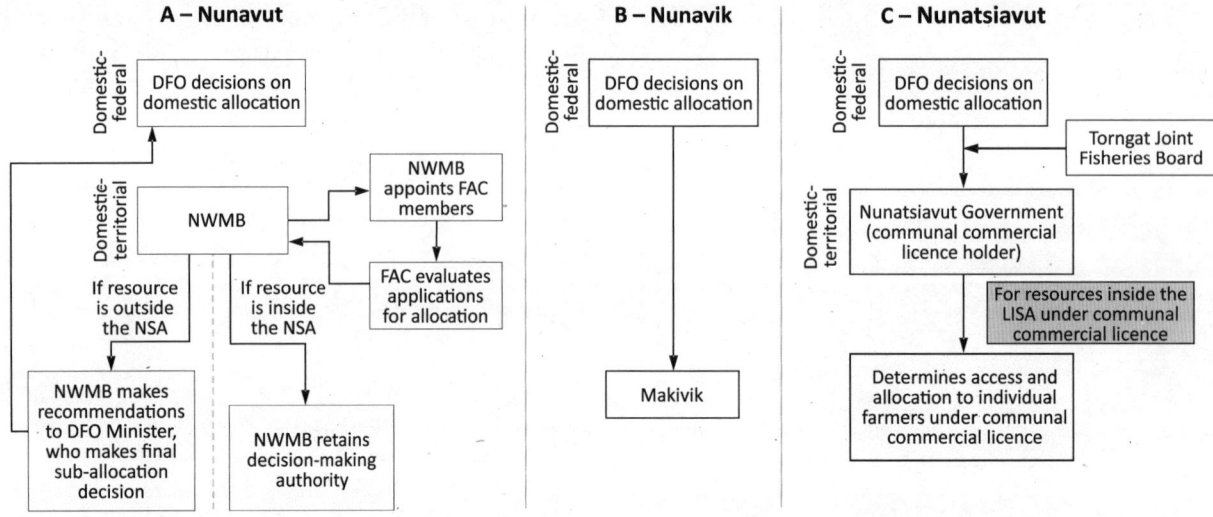

Figure 8.2 The role of the federal government (DFO), co-management boards, and advisory committees in territorial decision making regarding fisheries allocations. See each relevant section for a description of decision making and policies associated with fisheries allocations. FAC = Fisheries Advisory Committee; LISA = Labrador Inuit Settlement Area; NSA = Nunavut Settlement Area; NWMB = Nunavut Wildlife Management Board.

identify adherence to fishery policy procedures (NWMB 2019). Once these recommendations are made, the NWMB may recommend or advise that DFO increase, maintain, decrease, or, in extreme cases, terminate sub-allocation quotas to each of the Nunavut fishing companies. Within the NSA, the NWMB makes decisions, and the Minister has 60 days to respond to such decisions, after which, if there is no response, the decision is deemed accepted. Outside the NSA, the NWMB provides advice that the Minister shall consider and there is no prescribed timeline. Whether inside or outside the NSA, the Minister has the final say.

Under the NWMB's allocation policy (2019), access to the fisheries is focused on Nunavut Inuit-owned fishing enterprises (primarily owned by hunters' and trappers' organizations (HTOs), regional wildlife organizations (RWOs), hamlets, and Designated Inuit Organizations (DIOs). Two enterprises own and operate multiple vessels in the shrimp and halibut fisheries, and the other two are royalty-based. The NWMB's allocation policy (NWMB 2019), which was first developed in 2007, with revisions in 2012 and most recently in 2019, prioritizes quota access

to Nunavut Inuit fishing enterprises that can demonstrate their contribution to the socio-economic well-being of Inuit in Nunavut, including direct benefits back to communities (dividends, jobs, or other non-cash benefits). The policy also recognizes the need for people to work together to achieve a sustainable and prosperous fishery (see the IQ principle of *piliriqatigiingniq* [working together for the common good]: Tagalik 2010).

Nunavut access to fisheries ranges from 20.78% of the TAC for northern shrimp in Davis Strait West to 100% for Greenland halibut in 0A (Figure 8.3b). Since 2002, all 0A and 0B (outside the NSA but adjacent to Nunavut) increases in the Greenland halibut quota have gone to Nunavut for sub-allocation, thereby increasing Nunavut's access. This is interpreted as a sign of support for fishery development on the part of the Government of Canada, along with recognition of the economic importance of commercial fisheries to the Nunavut economy (DFO 2019d). The landed value of these two commercial fisheries is significant for the region, totalling over Cdn$150.1 million (Table 8.1), with Greenland halibut contributing close to Cdn$90 million, or roughly 60% of landed value

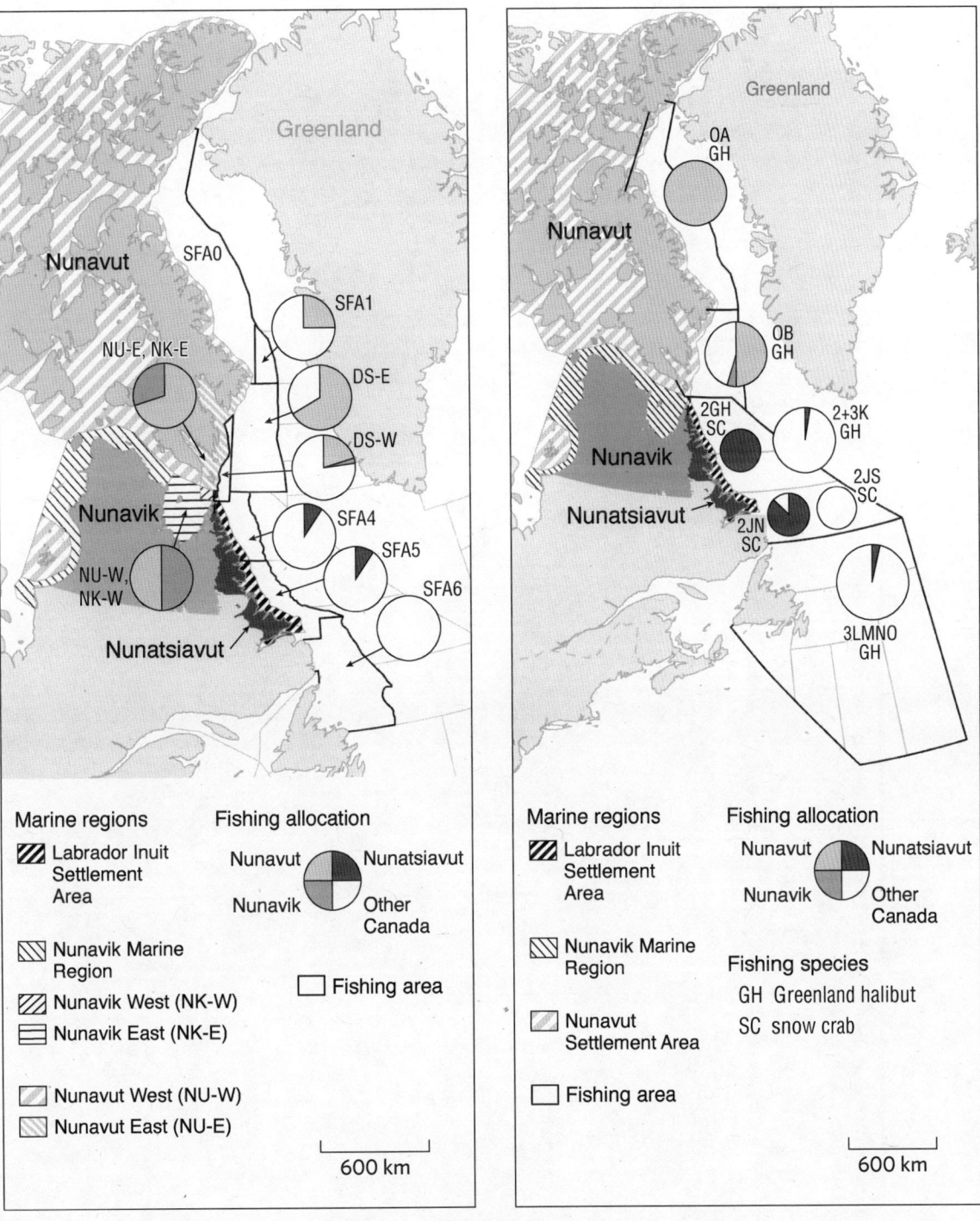

Figure 8.3 Inuit regions in the Canadian Eastern Arctic, and NAFO divisions showing the relative access to each of the Inuit territories. Each NAFO fishing area or shrimp fishing area (SFA) is shown with the representative allocations to each of the territories and the remainder of Canada, as noted in Table 8.1. Greenland halibut (GH) and snow crab (SC) are shown together with some allocations covering multiple fishing areas. SFA allocations are calculated using the targeted species (not the bycatch species). Other Canadian allocations represent allocations outside of the three territories, as noted in this chapter.

Table 8.1

Access in 2019 to adjacent areas as defined by the allocation under direct control of co-management boards or land claims agreements directly presented in tonnes and percent of landed value

Species by area		Canadian TAC tonnes (t)	Nunavut		Nunavik		Nunatsiavut		Other Canadian	
			Allocation: NWMB (t)	Landed value ($ million)	Allocation: NILCA (t)	Landed value ($ million)	Allocation: NG (t)	Landed value ($ million)	Allocation (t)	Landed value ($ million)
Shrimp[1]										
SFA1-N		14,875 (B)	3,722 (B) (25.1%)	$20.02					11,153 (74.9%)	$60.00
Eastern — Davis Strait East (DS-E)		2,406 (B-exploratory)	1,604 (B) (66.7%)	$8.63					802 (33.3%)	$4.31
Eastern — Davis Strait West (DS-W)		5,217 (B)	1,084 (B) (20.7%)	$5.83	120 (2.3%)	$0.65	0	$0	4,013 (76.9%)	$21.59
Eastern — Nunavut/ Nunavik East (NU/NK-E)		430 (M) 217 (B*)	301 (M) (70.0%) 210 (B) (80.2%)	$1.72	129 (M) (30.0%) 53 (B) (19.8%)	$0.59				
Western — Nunavut/ Nunavik West (NU/NK-W)		11,975 (M) 3,163 (B)	5,987.5 (M) (50%) 1,581.5 (B) (50%)	$24.08	5,987.5 (M) (50%) 1,581.5 (B) (50%)	$24.08				
SFA4 2019/2020		10,845 (B)					935 (10%)	$5.03	8,841 (90%)	$47.56
SFA5 2019/2020		22,100 (B)					2,188 (9.9%)	$11.77	19,912 (90.1%)	$107.13
SFA6		8,960 (B)					0	$0	8,960 (100%)	$48.20

Assessment Zone

Greenland halibut (turbot)[2]	TAC	Allocation	Value	Allocation	Value	Allocation	Value	Allocation	Value
0A	9,592.5	9,592.5 (100%)	$62.09					0	$0
0B	8,592.5	4,283.25 (49.84%)	$27.73	449.25 (4.94%)	$2.91			3,860 (45.2%)	$24.99
2 + 3K	4,273					140 (3.38%)	$0.91	4,129 (96.62%)	$26.71
3LMNO[3]	1,833.9					62 (3.38%)	$0.40	1,771.9 (96.62%)	$11.47
Snow crab[4]									
2GHJ 2GH	100					100 (100%)	$0.97		
2J North	310					270 (87.1%)*	$2.61	40 (12.9%)	$0.39
2J South	1,455					0	$0	1,455 (100%)	$14.08
Total values			$150.10		$28.23		$21.69		$366.43

Notes: Economic benefits are estimated from the landed value for each territory by species and fishing area. Refer to Figure 8.3 for a map of each area. Allocations presented here are those over which the territory has control, i.e., it is the territory's decision or recommendation (Nunavut Wildlife Management Board [NWMB], Nunavik Inuit Land Claims Agreement [NILCA] directly, Nunatsiavut Government [NG]) as a percentage of the total Canadian TAC (Total Allowable Catch). Shaded cells indicate that a fishing area is not adjacent to the territory. Enterprises operating within the territories may also access commercial fisheries resources through alternate DFO means (e.g., enterprise allocations, competitive fishery), but as these are under DFO control and outside the territorial jurisdictions, they are captured in the "Other Canadian allocation" column. Exploratory fisheries are not included in this table.

1 For northern shrimp, the 2015 inshore landed value ($3,870/tonne) and offshore landed value ($5,380/tonne) (DFO 2018) were used, along with a landed value ($3,000/tonne) for striped shrimp (personal communication, Brian Burke Nunavut Fisheries Association, 2020) and 2019 quota values (DFO 2019c, 2019f, 2019g). Shrimp fishing areas include northern shrimp, Pandalus borealis (B), and striped shrimp, Pandalus montagui (M). Asterisk indicates bycatch species allowance.

2 The landed value of Greenland halibut ($6,473/tonne) was taken from the average of 2013–17 values (DFO 2019d), and using 2019 quota values (DFO 2019d, 2020).

3 Food and Agriculture Organization (FAO) area 3LMNO is not considered adjacent to Nunatsiavut but was included in the analysis as the NG has access to allocations in this area.

4 Landed value of snow crab in 2017 ($4,390/tonne) and 2019 quota values (DFO 2019a, 2019e). Asterisk indicates bycatch species allowance.

in 2019. Fishing, hunting, and trapping were valued at Cdn$2.9 million (2012 chained dollars[2]), accounting for 0.09% of the 2019 GDP for Nunavut (Department of Executive and Intergovernmental Affairs 2020). However, based on the landed value of $150 million (excluding any other direct, indirect, and induced economic impacts), it is likely that fisheries contribute upward of 4.75% of the Nunavut GDP.

Despite progress, Nunavut ownership and access within commercial fisheries have not developed without challenges in operationalizing co-management. At the last call for commercial fisheries applications in 2015, the NWMB, with advice from the FAC, made recommendations to the Minister. At the time, only one enterprise had access to shrimp quota inside the NSA, and the NWMB made its initial decision for this to continue. However, the Minister allowed a second enterprise access to this resource (Ducharme 2016). This varied ministerial decision used criteria not included in the NWMB's allocation policy to reduce an existing Nunavut allocation holder's quota (i.e., not harvesting all quota allocated, a criterion not outlined in Nunavut's Allocation Policy). A new enterprise for which the NWMB had not granted access or allocation (see NWMB 2019, Section 10) was rewarded with access to the fishing area. While the NWMB reconsidered and made a final decision in line with the Minister, it had noted that "repeated requests from [a fishing enterprise] for fishing opportunities in the area [NSA], should not be a factor in [the NWMB's] final decision" (Ducharme 2016).

Here the Minister applied criteria outside of the NWMB's policy in the decision. After discussion, and in the spirit of co-management, the NWMB did move forward to enact the Minister's decision, but this has resulted in a precedent for bypassing established decision-making processes. The vice president of one of the fishing enterprises observed that "this creates uncertainty and means there are no clear rules in the allocation process. Banks like as much certainty as possible. This will make it much more difficult for all Inuit fisheries enterprises to get financing to buy vessels and take control of their companies" (Ducharme 2016).

The NWMB decision regarding resources *inside* the NSA includes the use of time, energy, and expertise during a formal decision-making process. The Minister varied the decision, resulting in a loss of faith by industry in the NWMB's ability to govern within its own land claims region. There were also immediate economic implications: at the time (2016), the gross value of shrimp inside the NSA was estimated at over Cdn$33 million (Ducharme 2016); allowing a second entrant with 30% access resulted in a financial loss for the existing quota holder. The impact of this one ministerial decision was a driving factor in the NWMB making changes to its 2019 allocation policy, and resulted in long-term negative impacts in the relationship between industry and the NWMB.

Nunavik

In 2006, the *Nunavik Inuit Land Claims Agreement* was signed by Makivik Corporation, an Indigenous political organization representing the Inuit of Nunavik since 1987, and the Government of Canada, and it came into force in 2008.[3] The NILCA addresses the ownership and use of the Nunavik Marine Region, including James Bay, Hudson Bay, Hudson Strait, and Ungava Bay and surrounding lands (Figure 8.1) (Indigenous and Northern Affairs Canada 2017). Article 5 of the NILCA established the NMRWB as the main regulator of access to wildlife, with primary responsibility for wildlife management in the NMR. The NMRWB is involved in allocations within the land claim area only, including regulating levels of total allowable take, creating nonquota regulations, and funding and directing research (NILCA, Sections 5.2.3 and 5.2.4). In fulfilling its mandate, the NMRWB strives to ensure that IQ and western science are given equal weight in all decisions taken. It is responsible for ensuring that wildlife populations in Nunavik are healthy and are harvested sustainably, while ensuring that the Inuit of Nunavik retain their harvesting rights, with their concerns, principles, and knowledge reflected in the wildlife management regime. To strike this balance, the NMRWB depends on the active participation of Nunavimmiut in the management process. It is thus crucial that the NMRWB work closely in collaboration with the Regional Nunavimmi Umajulirijiit Katujjiqatigiinninga (RNUK) and Local Nunavimmi Umajulirijiit Katujjiqatigiinningit (LNUK), who represent Nunavik Inuit resource users and bring to the NMRWB the wildlife knowledge they have acquired from their elders and that obtained through personal experience.

The NMRWB advises DFO on Total Allowable Harvest levels inside both the NMR and the NSA.[4] As shared stocks exist in shrimp fishing areas (Nunavut and Nunavik East and West; see Figures 8.1 and 8.3), the NMRWB and NWMB regularly consult with each other and make unified recommendations to the Minister on the TAH in these areas. Within the NILCA, Makivik and its subsidiaries known as Makivik Designated Organizations (MDOs) are explicitly given allocations to fisheries adjacent to the NMR (Figure 8.2B). For example, regarding 0B Greenland halibut access, Makivik (or MDOs) is noted to receive 2.54% of the TAC established by the Minister, if equal to or less than 5,500 tonnes, and 10% of the TAC above 5,500 tonnes (NILCA, Section 5.4.8). In addition, Sections 5.4.12 and 5.4.16 of the NILCA outline shrimp increases, noting that 7% of TAC increases in Southern Davis Strait (Davis Strait East [DS-E] and West [DS-W] in Figure 8.3a) and 8.8% of increases in Northern Davis Strait (SFA0 in Figure 8.3a) are to be provided to Makivik. Within the NMR, access ranges from 19.8% (Nunavik East) to 50% (Nunavik West) of the total quota, noting these are both within the NMR, and are shared stocks with Nunavut. Outside of the NMR, although access levels are directed by the LCA, they remain low, ranging from 2.3% (shrimp in Davis Strait West) to 4.94% (Greenland halibut in 0B).

The benefits from commercial fisheries in terms of landed values for 2019 are estimated to be Cdn$28.23 million, with most of these benefits arising from the Nunavik West shrimp fishing area, where the landed value of Cdn$25.32 million contributes 90% of total landed value for the territory (Table 8.1). Fishing operations are royalty-based, meaning that Makivik does not fish its own quotas directly. Profits from commercial fisheries are combined with other ventures (construction, airlines, north warning defence systems, fur tanning, fuel providers) under Makivik, then distributed and/or invested for the benefit of Nunavik Inuit (Makivik Corporation 2019b).[5] Makivik and its subsidiaries employed 134 individuals in 2019, and have contributed over Cdn$150 million between 2002 and 2019 to over 56 programs or entities benefiting education, promoting welfare, and protecting and preserving the Inuit way of life (Makivik Corporation 2019a). In the 2018–19 fiscal year, over Cdn$4.2 million was distributed to the region, with the majority (Cdn$3.5 million) allocated to community and regional donations (primarily for local and regional initiatives) and the remainder for supporting elders' activities, Nunavik festivals and recreational activities, education, and churches (Makivik Corporation 2019a). While fisheries are only a portion of the cumulative benefits generated by Makivik, they contribute to many social programs aimed at improving the lives of Nunavimmi.

Nunatsiavut

In 2005, after almost four decades of negotiations, the *Labrador Inuit Land Claims Agreement* between the Labrador Inuit Association, the Government of Canada, and the Government of Newfoundland and Labrador was signed, resulting in contemporary self-governance in Nunatsiavut.[6] The LILCA established the Torngat Joint Fisheries Board in Chapter 13, which outlines its role in the conservation of fish and the management of commercial fisheries. The TJFB is the primary body making recommendations to the Minister regarding conservation of fish stocks, species, and habitats, as well as fisheries management (LILCA, Section 12.9), TACs, and allocations within the Labrador Inuit Settlement Area (LILCA, Sections 13.11.1–13.11.2). The TJFB hosts regular workshops involving regional rights holders and stakeholders to discuss improvements in fisheries management and means of achieving equity in access to adjacent fisheries, to promote a self-sustaining industry and maximize economic and social value for the region.

The introduction of co-management has brought issues of inequitable access to fisheries for Nunatsiavummiut to the surface, and provides quality information and analysis for consideration in future recommendations by the TJFB (Kourantidou et al. 2021; Mills et al. 2020). Commercial allocations to Nunatsiavut are made to the Nunatsiavut Government (NG) from DFO, and include Greenland halibut, northern shrimp, and snow crab, and are issued as a communal commercial licence.[7] Allocations to the NG are determined on an annual basis, in line with DFO policies for licensing (Figure 8.2C). Based on the Commercial Fishery Designation Policy of the Nunatsiavut Government (2012), access to commercial fishing opportunities is reviewed annually and granted to those meeting specific designation criteria, including beneficiary status under the LILCA, having or planning to obtain a Professional Fish Harvester (Level II) Certification, and three or more years of participation in the commercial fishery. After eligibility criteria are met, preference is given to those who invest in the fishery, are able to fish the quota themselves, create economic stimulus for beneficiaries and Nunatsiavut, and have a recent history of harvesting and compliance with the Nunatsiavut Government (2012). Recipients of allocation are required to pay an access fee to the NG ($0.02 per pound on landed value), contributing to the Nunatsiavut Commercial Fisheries Fund, which aims to support further commercial fisheries development within the territory.

Access to northern shrimp in waters adjacent to Nunatsiavut currently ranges from 0% for shrimp in Davis Strait West and SFA6 to about 10% in SFA4 and SFA5. Access to Greenland halibut is 3.38% in NAFO divisions 2+3K and 3LMNO, while snow crab access ranges from 0% in 2J South to 100% in 2GH.[8] The landed value for all fisheries has been estimated at Cdn$21.69 million for 2019, with shrimp in SFA4 and SFA5 contributing Cdn$16.80 million, or 77% of the total landed value. Nunatsiavut has zero access to many of its adjacent fisheries (Davis Strait West and SFA6 shrimp and 2J South snow crab; see Table 8.1). In addition to the landed value, additional benefits to the region come from the employment of fishers and processing facility workers, and from contributions to the Nunatsiavut Commercial Fisheries Fund. In 2017, for example, the northern shrimp fishery in SFA4 and SFA5 employed 54 fishers in the harvesting sector, and contributed roughly $65,000 to the fund (TJFB, unpublished data). The communal allocation for snow crab in area 2HJ had an estimated value of Cdn$3.7 million in 2016 and employed 9 beneficiaries in the harvesting sector and 77 in the processing sector (snow crab and turbot).

A VISION FOR THE (NEAR) FUTURE

In early 2019, the three co-management boards met and developed insights on how they would like commercial fisheries co-management to function in the future, both within and between different regions.[9] This meeting marked the first time the three boards had met to collectively discuss ongoing issues and offer insights on how co-management should progress in an era of reconciliation. While the Minister retains absolute discretion under the *Fisheries Act,*[10] improvements in the co-management relationships between each of the boards and the federal government (DFO) could be achieved through the following four recommendations (Snook et al. 2019a, 2019b): (1) acknowledging the principle of adjacency in allocation decisions to remove inequities; (2) building confidence and trust in co-management processes, recommendations, and decisions; (3) responding to substance with substance (i.e., providing in-depth and thorough responses to co-management advice); and (4) honouring the spirit and intent of the LCAs.

Recommendation: Increased Inuit Access to and Benefits from Adjacent Fisheries

Access for the three territories combined ranges from 100% for shrimp in the western assessment zone (50% Nunavut, 50% Nunavik) and in the Nunavut/Nunavik East portion of the eastern assessment zone (70% Nunavut, 30% Nunavik), to 0% of the most southern adjacent stocks such as shrimp in SFA6. Conversely, for non-Inuit interests (i.e., access acquired directly from DFO outside of the LCAs), access is 100% in many of the southern areas, and access in all other offshore areas (outside of land claims) ranges from 33.3% for shrimp in Davis Strait East (DS-E in Figure 8.3a) to 96.6% for Greenland halibut in 2 + 3K (see Figure 8.3b). Only one offshore fishing area, 0A for Greenland halibut, provides 100% access to Inuit.

DFO's own New Access Framework (DFO 2008) identified adjacency as one of three key criteria (along with historic dependence and economic viability): "Priority access should be granted to those closest to the fishery ... [o]n the implicit assumption that access based on adjacency will promote values of local stewardship and local economic development."

During the 2019 meeting of co-management boards, this position that Inuit should have priority access to their adjacent resources was echoed: "If Nunavut, Nunavik and Labrador Inuit were to have access to their adjacent share similar to other Canadian jurisdictions, the potential socio-economic benefits for Inuit and the respective jurisdictions would be significant" (Snook et al. 2019b).

The result of current inequities is made clear when we consider the total landed value for the Inuit-adjacent fisheries: Cdn$566.45 million for 2019 (Table 8.1), with only 35.3%, or Cdn$200.02 million, being under control of the three co-management boards or LCAs (Cdn$150.10, $28.23, and $21.69 million for Nunavut, Nunavik, and Nunatsiavut, respectively).[11] The 64.7% of cumulative commercial fisheries resources allocated through DFO processes equates to a landed value of Cdn$366.43 million (Table 8.1) of Inuit-adjacent resources allocated to non–Inuit-controlled interests (as represented by the "Other Canada" category in Figure 8.3a, b). This is clearly inequitable.

While progress has been made, opportunities to increase access to adjacent resources remain limited. Such access increases have historically occurred only when there is an increase in the stock and subsequent TAC. As Nunavut has had access to 100% of Greenland halibut 0A since 2005, subsequent increases in TAC have gone to Nunavut, which is adjacent to this fishing area (Figure 8.3b). In 0B, an area primarily adjacent to Nunavut but also offshore to Nunavik and Nunatsiavut, a 2017 increase in turbot TAC resulted in Inuit receiving 100% of the increase, with a 90/10 split between Nunavut and Nunavik. The Nunavut Fisheries Association (NFA), a group representing commercial fishing interests in Nunavut, has long advocated for increased adjacent access (Rogers 2018) and noted their success when DFO recognized Inuit adjacency, specifically that NFA had "fought long and hard to get to the point where we are getting the majority of increase" (Brown 2019). DFO's integrated fisheries management plans (IFMPs) generally promote, as part of their long-term objectives, fair access and equitable sharing of resources to stakeholders and Indigenous groups (e.g., see shrimp IFMP [DFO 2018]). To facilitate this, the three co-management boards note that additional increases in TAC should go to Inuit first, to decrease disparity in access to adjacent commercial fisheries (Snook et al. 2019b). Furthermore, when TAC decreases do occur, they should be taken on first by non-Inuit allocation holders to reduce inequities in fisheries access and benefits.

Recommendation: Trust in Co-Management Institutions

The *United Nations Declaration on the Rights of Indigenous Peoples* (UNDRIP) (2007) recognizes that Indigenous Peoples have the right to participate in decision making in matters that affect their rights, in accordance with their own procedures. Articles 19 and 40 state that states shall consult and cooperate in good faith with Indigenous Peoples before adopting and implementing legislative measures that may affect them; further, Indigenous Peoples must have access to prompt decisions through just and fair procedures for the resolution of conflicts and disputes with states or other parties. In 2018, the Canadian Department of Justice issued principles outlining how the Government of Canada could commit to reconciliation through renewed relationships with Indigenous Peoples. As noted under DFO's reconciliation strategy, a long-term objective of strengthened Indigenous-Crown relationships should be achieved through "Indigenous groups' role in management and decision-making" being "well-defined, implemented, accepted by all parties" (DFO 2019b). Furthermore, self-reliance, determination, and community control in fisheries management is a key guiding principle under the Nunavut Fisheries Strategy (Government of Nunavut and Nunavut Tunngavik Inc. 2005). Despite these stated articles, principles, and long-term objectives, decision making across Inuit Nunangat still progresses largely without trust in the co-management boards.

Decisions and/or recommendations made by the co-management boards regarding setting TAC levels or sub-allocations within the territories (the Fisheries Advisory Committee for the NWMB) involve multiple sources of input: legal advice, scientific advice, stakeholder workshops, and the consideration of Inuit world view (or IQ) when making recommendations. For example, the IQ principle of *piliriqatigiingniq,* meaning to work together for a common purpose, operates through a consensus approach to making allocation decisions/recommendations in the case of the NWMB, and through collaborative approaches between NMRWB and NWMB in making joint unified decisions to the Minister regarding shared shrimp stocks (i.e., setting TAH levels and percentage of TAH for each territory).

Despite growth and progress in co-management, however, ministerial discretion continues and is not limited to commercial fisheries. In Nunavik, Makivik Corporation (representing Northern Quebec Inuit) challenged a decision made by the Minister of Environment regarding the polar bear TAT (Total Allowable Take, equivalent to TAC in fisheries) in Southern Hudson Bay.

In 2012, the NMRWB was asked to establish management decisions for polar bears, including setting a TAT. In July 2015, the NMRWB submitted its decisions based on the latest scientific information as well as on Inuit Traditional Knowledge (ITK) collected from within the region (NMRWB 2018). The Minister rejected this initial decision, and a second decision was submitted by the NMRWB in December 2015, establishing a TAT of 28 bears per year. In October 2016, the Minister(s) varied this decision, and lowered the TAT on polar bears to 23 (Makivik Corporation 2017). In accordance with the NILCA, the NMRWB's final decision had taken into account extensive community interviews and validation of Inuit Traditional Knowledge on polar bear migration, feeding, body condition, mating and denning, habitat, abundance, and human–polar bear interactions (NMRWB 2018), as well as scientific information. Makivik's legal challenge was based on its claim that the Minister's decision to vary the board decision relied solely on scientific information and ignored Inuit Traditional Knowledge, without explaining why – and this violated the NILCA. Ultimately, the Minister's decision was upheld by federal courts, although poor communication in the decision making was noted by the judge (Forrest 2019). This most recent example – involving the Department of Environment and Climate Change – suggests that federal departments have established a pattern of disagreeing with local boards, thus highlighting the need for more respect for the spirit and intent of co-management processes. The inclusion of Inuit Traditional Knowledge in policy-making under the LCAs is a central aspect of self-determination across Inuit Nunangat (Kourantidou, Hoover, and Bailey 2020). Inuit Traditional Knowledge is not simply a body of information but a world view that is an important, and often dominant, "way of knowing" in Inuit communities (Karetak, Tester, and Tagalik 2017).

Similarly, in Nunavut, the Minister of Environment has varied recommendations of the Nunavut Impact Review Board (NIRB) regarding land-based renewable resource extractions within Nunavut on two occasions.[12] In both cases (the Sabina Gold and Silver Corporation's Back River project and Baffinland Iron Mines Corporation's Production Increase proposal), the Minister approved activities that led to increased resource extraction, against the recommendation of the NIRB. In the first case, the Minister did not accept the NIRB's rejection of the Sabina gold mine proposal and allowed it to continue further in the environmental review process. In the second case, the Minister approved an increased production rate for Baffinland Iron Mines against the NIRB recommendations. While the NIRB is subject to public consultation and discourse as set out in the *Nunavut Agreement*, the Minister carried out directed consultations and assessments that were not witnessed by the concerned parties, and then used that information when making resource extraction decisions on lands within the LCA (Arngna'naaq et al. 2020).

In the case of Nunatsiavut, where 100% of snow crab TAC in 2HJ north is under control of Nunatsiavut interests (the Nunatsiavut Government and Torngat Fish Producers Co-op[13]), there is more control over shaping long-term strategies. Starting in 2015, 15% of the snow crab quota was left unfished over concerns of low recruitment and a potentially decreasing stock (Snook, Cunsolo, and Morris 2018). This conservationist approach was implemented due to stock uncertainties, continued for over five years, and was made possible by the cooperation of the designates and their commitment to long-term resource sustainability (Foley et al. 2017). The fact that there were no outside (non-Inuit) competing interests in the fishery may have played a role in leaving quota unfished: co-management boards not only are mandated under the LCAs to effectively manage their resources but have also demonstrated the ability to provide extra safeguards to the long-term stability of stocks.

Moving toward self-determination for each territory means managing their resources and being empowered to do this effectively. These examples of ministerial divergence in the co-management process, including the Minister's varying of shrimp allocations inside the NSA noted earlier, are not aligned with either international or domestic goals for reconciliation. Specifically noteworthy is DFO's exclusion of IQ in decision making even though it is considered especially important in managing resources in accordance with Inuit world views, as it allows Inuit to participate in decision making in accordance with their own procedures (UNDRIP 2007). To realign the co-management process in Inuit Nunangat with these federal and international targets, respect for co-management as an institution is paramount.

Lack of inclusion in decision making and acceptance of Inuit Traditional Knowledge or IQ hinders self-determination and undermines trust between resource users and federal management authorities. This lack of inclusiveness does not align with DFO's long-term Reconciliation Strategy objective of "Strengthened Indigenous-Crown Relationships," whereby Indigenous groups' role in management and decision making is well-defined, implemented, and accepted by all parties (DFO 2019b).

We now offer two key approaches that will achieve a more balanced co-management process: (1) responding to substance with substance, and (2) honouring the spirit and intent of the land claims agreements (Snook et al. 2019a).

Recommendation: Responding to Substance with Substance

Recommendations from co-management boards are the result of carefully designed processes that include specific policies about how boards reach these recommendations/decisions. These policies are found in LCAs, and emphasize the inclusion of Inuit Traditional Knowledge (e.g., IQ, Traditional Ecological Knowledge, or other forms), fisheries experts, legal advisers, and/or scientific advisers. Despite the effort and expertise that goes into the recommendations of the boards to the Minister, many ministerial decisions fail to integrate board recommendations. Such decisions are often made without the necessary justification, and this failure to respond to substance with substance "prevents an understanding of Ministerial decisions, limits shared learning opportunities, and disengenders an authentic sense of co-management" (Snook et al. 2019a). This could be remedied by providing in-depth and thorough ministerial responses and explanations, beyond the current decision with no indication of how co-management advice or decisions were considered. Furthermore, wider recognition of Inuit Traditional Knowledge and IQ principles in decision making and ongoing management of resources is needed, with explicit acknowledgement of its value.

Recommendation: Honouring the Spirit and Intent of the LCAs

The 2019 workshop with the three co-management boards stated that "one of the main goals of the land claim agreements was to encourage Inuit self-reliance through cultural and social well-being, livelihoods, and protected rights to hunt and fish" (Snook et al. 2019b).

That said, the co-management processes continue to grow and mature even after these LCAs were signed and should be viewed as a step for Inuit to regain control of their resources. LCAs provide "an opportunity for change but that opportunity must be seized" (Foley et al. 2017). Rather than relying on the literal legal interpretation of each LCA, acting in the "spirit and intent" would refocus the interactions from unidirectional to integrative decision making, and would strengthen decision making for the co-management boards while supporting self-determination of the Inuit communities they represent. "The land claim agreements may best be thought of as opportunities to encourage dialogue that reflects the intent and spirit of the claim, new possibilities that reflect contemporary and changing contexts, and ultimately points to a shared objective that ensures healthy Inuit coastal communities as meaningful co-participants in the co-management system" (Snook et al. 2019b). This could be achieved through more regular, open communication in co-management, in addition to the other recommendations presented above.

MOVING FORWARD: NEW OPPORTUNITIES

There are always opportunities for change. A new Arctic region has been established by DFO, as announced in October 2018. During the 2019 Eastern Arctic co-management workshop, there was a sense of hope that this signalled more opportunities for enhancing fisheries co-management in the region, with more space for dialogue and collaboration. Indeed, this may be the ideal time for a recommitment to co-management: as the co-management boards continue to change over time, this will be reflected in commercial fisheries decision making and increased engagement by the Minister. In line with these expectations, a participant at the 2019 workshop summed up the spirit of the workshop as follows:

I hope that the future includes a lot more trust on behalf of the mature co-management network that is there and people will look to these boards in the future to see what their advice and decisions are going to be, and everyone can have confidence that these decisions went through good process and whatever they end up being they went through a process that everyone agreed to and trust and we can move forward together. (Snook et al. 2019b)

These boards that have been building co-management experience are well positioned to play a lead governance role in their respective territories and across Inuit Nunangat and the Arctic (Snook et al. 2019a). With increasing discussion and pressure for a more open and inclusive co-management approach, there are expectations for renewed faith in the established co-management systems.

ACKNOWLEDGMENTS

In addition to OceanCanada, the authors would like to acknowledge the financial contributions from the Social Sciences and Humanities Research Council of Canada for funding the 2019 co-management workshop mentioned in the text. There were also numerous significant contributions by Paul Irngaut, Peter Rose, Neil Greig, Todd Broomfield, Colin Webb, Robert Moshenko, Allen Gorden, Daniel Shewchuk, David Kritterdlik, Jorgen Bolt, John Mercer, Chesley Anderson, Rick Comerford, Roland Andrews, and Keith Watts at this 2019 workshop.

C. Hoover would like to thank ArcticNet for funding and Duncan Burnside, the Nunatsiavut Government, and DFO for assistance with Figures 8.1 and 8.3.

NOTES

1 We use the definition of the Organisation for Economic Co-operation and Development to describe commercial fishing as "harvesting of fish, either in whole or in part, for sale, barter or trade" (OECD 1998). For the purposes of this chapter, commercial fisheries are limited to marine species, including Greenland halibut, striped shrimp, northern shrimp, and snow crab for consistent comparison across regions.
2 Chained dollars is a method of adjusting real dollar amounts for inflation over time, to prevent bias when comparing figures from different years (Newfoundland and Labrador Canada 2023).
3 *Nunavik Inuit Land Claims Agreement Act,* SC 2008, c 2.
4 TAH refers to harvest levels, or catches in the case of fisheries, within a land claims area. It is similar to the TAC set by DFO in offshore areas, but takes into account the basic needs level of Inuit to harvest the resource in nearby communities.
5 In addition to commercial fisheries, business subsidiaries and joint ventures of Makivik include: airlines (Canadian North and Air Inuit), fuel delivery (Halutik Enterprises), construction (Kautaq Construction), tanning and taxidermy (Nunavik Furs), a mapping consulting firm (Nunavik Geomatics), marine shipping (NEAS), Pan Arctic Inuit Logistics (PAIL), renewable energy (Tarquti Energy Corporation), and a commercial fisheries venture (Unaaq Fisheries). Unaaq Fisheries shares a shrimp fishing licence obtained directly through DFO processes, and therefore is not included in Table 8.1.
6 *Labrador Inuit Land Claims Agreement Act,* SC 2005, c 27.
7 *Aboriginal Communal Fishing Licences Regulations,* SOR/93-332, https://laws-lois.justice.gc.ca/PDF/SOR -93-332.pdf.
8 While the access to 2GH is 100% and 2J North is 87.1%, these are sub-areas of the larger 2GHJ fishing area. Cumulatively, Nunatsiavut has access to 21.9% of all crab resources in 2GHJ.
9 For a short video about the meeting, see https://youtu. be/2m5OUP49Tbo.
10 *Fisheries Act,* RSC 1985, c F-14.
11 Note that there are cases where Inuit obtain licences directly from DFO, or allocations to the Innu Nation, Nunatu-Kavut Community Council, and Imakpik (DFO 2018).
12 The NIRB is a co-management board whose mandate is to protect and promote the well-being of the environment and Nunavummiut through the impact assessment process. It operates in a similar manner to the NWMB in terms of providing advice to the Minister of Environment for land-use issues (NA, Section 12.2.2).
13 Note that the Torngat Fish Producers Co-op is not under the jurisdiction of the Nunatsiavut Government, but does represent Nunatsiavut interests, among others.

REFERENCES

Arngna'naaq, K., H. Bourassa, D. Couturier, K. Kaluraq, and K. Panchyshyn. 2020. *Realizing Indigenous Law in Co-Management.* Toronto: Gordon Foundation.

Berkes, F. 2009. "Evolution of Co-Management: Role of Knowledge Generation Bridging Organizations and Social Learning." *Journal of Environmental Management* 90 (5): 1692–1702. http://dx.doi.org/10.1016/j.jenvman.2008. 12.001.

Brown, B. 2019. "Nunavut Fishers Get More Turbot Quota from Ottawa." *Nunatsiaq News,* January 29. https://nunatsiaq.

com/stories/article/nunavut-fishers-get-more-turbot
-quota-from-ottawa/.

Department of Executive and Intergovernmental Affairs.
2020. "Economic Data: Nunavut Real GDP by Industry
2012 to 2019." https://www.gov.nu.ca/executive-and-inter
governmental-affairs/information/economic-data.

DFO (Fisheries and Oceans Canada). 2004 (updated in
2008). *Atlantic Fisheries Policy Review: A Policy Framework for the Management of Fisheries on Canada's Atlantic Coast.* http://www.dfo-mpo.gc.ca/fm-gp/policies
-politiques/afpr-rppa/framework-cadre-eng.htm#a621
AdoptingaMoreInclusiveApproachtoPolicyPlanning.

–. 2008. "New Access Framework." http://www.dfo-mpo.
gc.ca/fm-gp/policies-politiques/access-acces-eng.htm.

–. 2018. "Northern Shrimp and Striped Shrimp – Shrimp
Fishing Areas 0, 1, 4–7 , the Eastern and Western Assessment Zones and North Atlantic Fisheries Organization
(NAFO) Division 3M." https://www.dfo-mpo.gc.ca/
fisheries-peches/ifmp-gmp/shrimp-crevette/shrimp
-crevette-2018-002-eng.html.

–. 2019a. "2019 Snow Crab Fishery, Newfoundland and
Labrador." Fisheries management decision. https://www.
dfo-mpo.gc.ca/fisheries-peches/decisions/fm-2019-gp/
atl-11-eng.html.

–. 2019b. "DFO–Coast Guard Reconciliation Strategy."
September. https://waves-vagues.dfo-mpo.gc.ca/library
-bibliotheque/40947208.pdf.

–. 2019c. "Eastern and Western Assessment Zones: Northern
Shrimp Management Units Davis Strait East, Davis Strait
West, Nunavut East, Nunavik East, Nunavut West and
Nunavik West – 2018/19." https://www.dfo-mpo.gc.ca/
fisheries-peches/decisions/fm-2019-gp/atl-03-eng.html.

–. 2019d. *Integrated Fishery Management Plan: Greenland
Halibut (Reinhardtius hippoglosssoides) Northwest Atlantic
Fisheries Organization Subarea 0.* Winnipeg: Fisheries and
Oceans Canada. https://www.dfo-mpo.gc.ca/fisheries
-peches/ifmp-gmp/groundfish-poisson-fond/2019/halibut
-fletan-eng.htm.

–. 2019e. "Integrated Fisheries Management Plan: Snow
Crab Newfoundland and Labrador Region." https://www.
dfo-mpo.gc.ca/fisheries-peches/ifmp-gmp/snow-crab
-neige/2019/index-eng.html.

–. 2019f. "Management Decision for 2019/20 – Northern
Shrimp Fishing Areas (SFAs) 4, 5, 6." https://www.dfo-mpo.
gc.ca/fisheries-peches/decisions/fm-2019-gp/atl-17-eng.html.

–. 2019g. "Northern Shrimp – Shrimp Fishing Areas 0, 1 and
7." https://www.dfo-mpo.gc.ca/fisheries-peches/decisions/
fm-2019-gp/atl-04-eng.html.

–. 2020. *Groundfish Newfoundland and Labrador Region
NAFO Subarea 2 + Divisions 3KLMNO.* https://www.dfo
-mpo.gc.ca/fisheries-peches/ifmp-gmp/groundfish
-poisson-fond/2019/groundfish-poisson-fond-2_3klmno
-eng.htm#toc6.

Ducharme, S. 2016. "QC Gets 30 Per Cent of Nunavut
Shrimp after Ottawa Leans on NWMB." *Nunatsiaq News,*
July 22. https://nunatsiaq.com/stories/article/65674qc_
gets_30_per_cent_of_nunavut_shrimp_quota_after_
ottawa_leans_on_nwmb/.

Foley, P., C. Mather, R. Morris, and J. Snook. 2017. *Shrimp
Allocation Policies and Regional Development under Conditions of Environmental Change: Insights for Nunatsia-
vutimmuit.* Report prepared for the Leslie Harris Centre
of Regional Policy and Development, Memorial University
of Newfoundland. https://research.library.mun.ca/
12788/1/FOLEY_ARF_15_16.pdf.

Forrest, M. 2019. "Federal Court Sides with Ottawa in
Decision to Reduce Nunavik Polar Bear Hunt." *National
Post,* November 19. https://nationalpost.com/news/
politics/federal-court-sides-with-ottawa-in-decision
-to-reduce-nunavik-polar-bear-hunt.

Government of Canada. 2018. Department of Justice
*Principles Respecting the Government of Canada's Rela-
tionship with Indigenous Peoples.* http://www.justice.gc.
ca/eng/csj-sjc/principles-principes.html.

Government of Nunavut and Nunavut Tunngavik Inc. 2005.
Nunavut Fisheries Strategy. https://www.tunngavik.com/
documents/publications/2005-03-00-Nunavut-Fisheries
-Strategy-English.pdf.

Indigenous and Northern Affairs Canada. 2017. *Nunavik Inuit
Land Claims Agreement: Implementation Report Fiscal Years
from 2011–2012 to 2014–2015.* Indigenous and Northern
Affairs Canada. https://www.rcaanc-cirnac.gc.ca/DAM/
DAM-CIRNAC-RCAANC/DAM-TAG/STAGING/texte
-text/2011-2012To2014-2015ImplementationReportOf
TheNunavikInuitLandClaimsAgreement_152967940
6743_eng.pdf.

Inuit Tapiriit Kanatami. 2018. "Fisheries and Oceans Canada, the Canadian Coast Guard and Inuit Tapiriit Kanatami
Announce New Arctic Region." Press release, October 24.
https://www.itk.ca/dfo-and-canadian-coast-guard-and-itk
-announce-new-arctic-region/.

–. 2020. "Inuit Nunangat Map." https://www.itk.ca/inuit
-nunangat-map/.

Karetak, J., F. Tester, and S. Tagalik. 2017. *Inuit Qaujimajatuqangit: What Inuit Have Always Known to Be True.*
Black Point, NS: Fernwood Publishing.

Kourantidou, M., P. Hoagland, A. Dale, and M. Bailey. 2021. "Equitable Allocations in Northern Fisheries: Bridging the Divide for Labrador Inuit." *Frontiers in Marine Science* 8: 590213. https://doi.org/10.3389/fmars.2021.590213.

Kourantidou, M., C. Hoover, and M. Bailey. 2020. "Conceptualizing Indicators as Boundary Objects in Integrating Inuit Knowledge and Western Science for Marine Resource Management." *Arctic Science* 6: 279–306. https://doi.org/10.1139/as-2019-0013.

Makivik Corporation. 2017. "Nunavik Inuit File Judicial Reviews Challenging Nunavut and Canadian Decisions on Polar Bear Quotas in Hudson's Bay." Press release, January 19. https://www.makivvik.ca/les-inuits-du-nunavik-contestent-devant-les-tribunaux-les-decisions-du-nunavut-et-du-canada-concernant-les-quotas-de-chasse-lours-blanc-dans-la-baie-dhudson/.

–. 2019a. *Makivik Corporation Annual Report October 1, 2018 to September 30, 2019.* https://www.yumpu.com/en/document/read/63488924/2018-2019.

–. 2019b. "Our Mandate." https://www.makivik.org/.

Mills, M., A. Dale, J. Chen, A. Cunsolo, P. Foley, C. Mather, A. Saunders, and J. Snook. 2020. *Taking Stock of Beginnings and Endings: Commercial Fisheries in Nunatsiavut.* Happy Valley–Goose Bay, NL: Torngat Wildlife, Plants and Fisheries Secretariat.

Newfoundland and Labrador Canada. 2023. "Finance: Glossary of Terms." https://www.gov.nl.ca/fin/economics/mnglossary/.

NMRWB (Nunavik Marine Region Wildlife Board). 2018. *Nunavik Inuit Knowledge and Observations of Polar Bears: Southern Hudson Bay Subpopulation.* Project conducted and report prepared for the NMRWB by M. Basterfield, K. Breton-Honeyman, C. Furgal, J. Rae, and M. O'Connor. https://nmrwb.ca/wp-content/uploads/2017/05/NMRWB-Nunavik-Inuit-knowledge-and-Observations-of-polar-bears-SHB-subpopulation.pdf.

Nunatsiavut Government. 2012. *Nunatsiavut Government's Commercial Fishery Designation Policy.* Department of Lands and Natural Resources.

NWMB (Nunavut Wildlife Management Board). 2019. *2019 Allocation Policy for Commercial Marine Fisheries.* https://www.nwmb.com/en/component/fileman/file/2019%20NWMB%20Allocation%20Policy%20for%20Commercial%20Marine%20Fisheries_ENG.pdf?routed=1&container=fileman-attachments.

OECD (Organisation for Economic Co-operation and Development). 1998. "Commercial Fishing." *Review of Fisheries in OECD Countries: Glossary, February 1998.* https://stats.oecd.org/glossary/detail.asp?ID=2990.

Rogers, S. 2018. "Northern Fishing Groups Want Help Getting Fair Share of Quotas." *Nunatsiaq News,* February 5. https://nunatsiaq.com/stories/article/65674northern_fisheries_say_they_need_help_to_catch_up_to_southern_counterp/.

Snook, J., J. Akearok, T. Palliser, with A. Cunsolo, C. Hoover, and M. Bailey. 2019a. "Enhancing Fisheries Co-Management in the Eastern Arctic." *Northern Public Affairs* 6 (2): 70–74.

Snook, J., J. Akearok, T. Palliser, A. Cunsolo, C. Hoover, M. Bailey, M. Basterfield, A. Dale, and A. Giles. 2019b. *"The Opportunity for Inuit in the Commercial Fishery Is Pretty Significant": Enhancing Fisheries Co-Management in the Eastern Arctic.* Report prepared for the Social Sciences and Humanities Research Council.

Snook, J., A. Cunsolo, and A. Dale. 2018. "Co-Management Led Research and Sharing Space on the Pathway to Inuit Self-Determination in Research." *Northern Public Affairs* 6, 1 (2018): 52–56.

Snook, J., A. Cunsolo, and R. Morris. 2018. "A Half Century in the Making: Governing Commercial Fisheries through Indigenous Marine Co-Management and the Torngat Joint Fisheries Board." In *Arctic Marine Resource Governance and Development,* edited by N. Vestergaard, B.A. Kaiser, L.M. Fernandez, and J.N. Larsen, 53–73. Cham, Switzerland: Springer. https://doi.org/10.1007/978-3-319-67365-3_4.

Tagalik, S. 2010. *Inuit Qaujimajatuqangit: The Role of Indigenous Knowledge in Supporting Wellness in Inuit Communities in Nunavut.* National Collaborating Centre for Aboriginal Health. https://www.ccnsa-nccah.ca/docs/health/FS-InuitQaujimajatuqangitWellnessNunavut-Tagalik-EN.pdf.

UNDRIP (*United Nations Declaration on the Rights of Indigenous Peoples*). 2007. https://www.un.org/development/desa/indigenouspeoples/declaration-on-the-rights-of-indigenous-peoples.html.

PART 4

Ocean Governance

9

Transforming the Governance of Canada's Oceans and Coasts from the Bottom Up

Derek Armitage, Nathan J. Bennett, Rachelle Beveridge, Anthony Charles, Nancy Doubleday, Inka Milewski, Sarah Newell, Lydia Ross, Ruth E. Smith, and the Community of Chesterfield Inlet

From offshore fishing banks to rugged shores, pristine beaches to biodiversity-rich estuaries, Canada's dynamic oceans and coasts help to define this country. However, our oceans and the coastal communities they support are exposed to diverse and multiple unprecedented changes (see Chapters 3 and 4). Coastal communities in many parts of Canada are under significant stress (Ommer 2007). Collapsed resource stocks (Charles 2001), changing climatic conditions (Cheung, Watson, and Pauly 2013), unsustainable economic development activities (Ban, Alidina, and Ardron 2010), and the concentration of access rights away from small-scale or community-based enterprises (Ecotrust Canada 2004, 2009, 2018; Silver and Stoll 2019) are just a few of the drivers of change that undermine the health and well-being of coastal ecosystems and communities (Stocks 2016; Dolan et al. 2005; Ommer 2007). Canada's success as an ocean nation requires that we do better.

Communities and their role in shaping our coastal and ocean spaces have been the focus of diverse projects and research initiatives internationally and across Canada's three oceans (Charles et al. 2020; Berkes 2015). These initiatives have led to valuable insights about priority issues and concerns confronting coastal communities and their interactions with marine systems. Key issues include the influences of climate change; the degradation of coastal resources, livelihoods, and economic security; outmigration of youth; health challenges; access to resources; and a sense that coastal communities are too often disconnected from governance processes (Charles 2012; Ommer

2007; Stephenson et al. 2018). While past research has focused on vulnerability of, and declining conditions in, coastal communities, attention is increasingly shifting to the strategies through which communities are proactively fostering change and deliberately seeking *transformations* toward sustainability (Armitage, Charles, and Berkes 2017; Bennett et al. 2019; see also Chapter 11).

We define transformation here as deliberate efforts to encourage fundamental change when existing conditions (economic, social, and ecological) are untenable (see Walker et al. 2004). Actions to deliberately transform ocean governance can be taken by governments and communities. Here, we focus on efforts by communities and community-supporting actors to transform ocean governance from the bottom up. Such transformations typically involve a shift in how communities govern themselves and their interactions with the ecosystems upon which they depend. Governance includes the structures and processes through which communities, societies, and organizations make decisions about our ocean and coastal resources. Relationships of power (e.g., between communities and higher-level government agencies) are central to governance and opportunities for collective action (Blythe et al. 2018; Bennett et al. 2019; Temper et al. 2018).

Transformative change, especially when it emerges from the bottom up, is hard to define, sensitive to external influences (political, biophysical) beyond a community, and difficult to sustain. Despite the challenges in applying the concept (see Blythe et al. 2018), a transformation lens provides a useful way to understand the important role of communities in reshaping how we interact with our oceans and coasts in Canada. Of particular note, a transformation perspective helps to shift the focus away from assessments of coastal change that focus only on threats and vulnerabilities, and that often emphasize the barriers to, or limited capacity of, coastal communities seeking to foster positive change. Instead, it emphasizes positive potential and proactive actions to promote change (Armitage, Charles, and Berkes 2017).

Our aim in this chapter is to identify some of the ingredients needed for communities to transform how they interact with and govern their ocean resources and coasts in the context of change, and in ways that sustain social and ecological systems. There are many ways to address this challenge. In this chapter, we identify and reflect on ingredients for transformative change through the lens of "community." We outline our approach to consider the role of communities in fostering transformative change in the governance of Canada's oceans and coasts, reflect on several examples of positive change and community-based innovations in seeking better outcomes for people and ecosystems, and synthesize lessons that can inform policy and lead to a future ocean for Canada that sustains healthy coastal communities.

INGREDIENTS FOR TRANSFORMATIVE CHANGE: A PERSPECTIVE FROM THE BOTTOM UP

Communities are not homogeneous – some scholars even argue that all communities are in fact "imagined" (Poteete and Ostrom 2004; Agrawal and Gibson 1999). It is true that as a social construct, the concept of "community" can mask all manner of social and political dynamics (Titz et al. 2018). However, the concept of coastal communities does reflect a particular vibrancy and also provides a lens for understanding how those closest to ocean and coastal resources must be central in shaping how we transform institutions of governance and resource management practices (Charles 2017). Indeed, the value of community involvement in coastal decision making is well established, and such involvement is an important antidote for hierarchical and/or colonial forms of decision making (see Silver et al. 2022).

There are promising examples of initiatives where coastal communities foster positive change, globally and in Canada (Charles et al. 2020; Armitage, Charles, and Berkes 2017). Transformations involve a significant reordering of and/or challenge to existing institutional and political structures of decision making (governance) in an effort to foster novel, environmentally sustainable, and more equitable outcomes (Wolfram 2019; Bennett et al. 2019; O'Brien 2012; Shah et al. 2018). Indeed, the notion of transformation is "normative" in that it reflects a desirable future and assumptions about the benefit of alternative futures (e.g., through technological innovations, institutional reforms, behavioural shifts). Our collective capacity for transformation is shaped by a range of factors and attributes (e.g., political, social, and technological) that coalesce across multiple scales. In this regard, Walker and colleagues (2006) outline several core aspects of transformative capacity based on a synthesis of cases, and provide a framework for our analysis here (see also Moore et al. 2014).

The first is the presence of "incentives to change" and sense of "agency" (i.e., belief in the capacity or power to effect change) (Westley et al. 2013). These incentives may emerge because of the failure (real or perceived) of previous policies to support positive social or ecological outcomes, a resource crisis (e.g., collapse of fisheries stocks), or changes in what people and communities value. The second dimension of transformative capacity involves a collective awareness and shared narrative of the need for solutions among a core group of actors (e.g., community leaders, community-based organizations) and in ways that help to form social networks of collective action across relevant scales. This takes transformation into the realm of governance and the need for governance processes and networks that are collaborative and adaptive (Folke et al. 2005; Kearney et al. 2007). The third aspect of transformative capacity is a "willingness to experiment" (an outcome of leadership capacity – see Davies et al. 2020), along with mechanisms to document the social learning that follows. Risks are inevitable and there may be costs for those seeking transformations where conditions are untenable. Leaders willing to experiment with alternative decision-making processes, power-sharing arrangements, and even livelihoods often emerge in the context of transformative change, and learning through these processes is recognized as crucial (Loeber 2007). Fourth, the capacity for transformation also depends on the presence of adequate stocks or reserves of human, social, natural, and built capital (Whitney et al. 2017). The implications for transformative change are often unclear, and for many communities there will be significant uncertainty as they seek alternative social and ecological outcomes.

These dimensions of transformative capacity are not fixed in time or space. In this regard, Olsson, Bodin, and Folke (2010) outline transformations as a social-ecological process involving a *preparation* phase, in which the initial conditions for transformation are established; the *navigation* phase, in which a transformation emerges given the right conditions (i.e., a window of opportunity) and takes shape; and a final phase, in which the *resilience* of the new system (alternative future or system trajectory) is fostered. Each of the dimensions of transformative capacity will manifest at different times in these three phases, and understanding when and how that happens can point researchers and policy-makers toward opportunities to leverage more durable social and ecological outcomes.

The material and process dimensions of transformative capacity are influenced by the actors involved. Here, we refer to actors as the community-based individuals (e.g., local leaders within communities) and/or place-based organizations catalyzing transformative change; inevitably, they reflect diverse interests, roles and responsibilities, and capacities (e.g., to participate, to take action). In this regard, our emphasis is very much on the agency exhibited by community-based actors, even as we recognize the importance of higher-level actors in supporting (e.g., through resources, sharing of power, etc.), or undermining, governance innovations from the bottom up. Ultimately, these material, process, and "actor" dimensions of capacity yield particular social and ecological outcomes associated with transformative change.

Finally, the capacity of local communities and actors to take transformative actions is influenced by external contextual factors beyond the community. These include, for example, the relative impacts of climate change on resources and the community, broader policies and governance processes that shape community rights and participation in environmental decision making, and changing markets and economic opportunities. Community actors are constantly contending with these factors as they navigate transformations.

In combination, these various dimensions provide a general framework for the analysis of our cases and an entry point for action in which coastal communities in Canada are leading efforts to transform how we govern our oceans and coasts (Figure 9.1). We examine several of these examples below.

LESSONS IN TRANSFORMATIVE CHANGE FROM CANADA'S THREE OCEANS AND COASTS

Canada's oceans and coasts, and the communities they support, are ecologically and culturally diverse. Yet, the ingredients for transformative change are likely to share a number of commonalities. We draw on the experiences in three Canadian cases to examine how these ingredients for transformative change can emerge from the bottom up, and the role that communities and community-based actors have in reshaping how we interact with and govern our coasts and ocean resources.

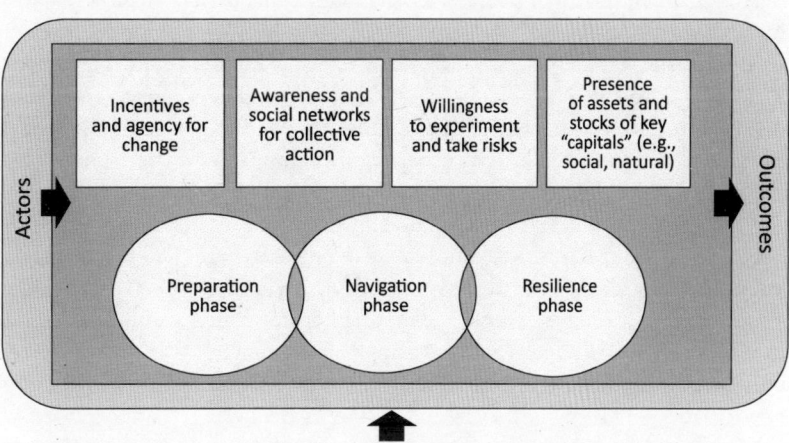

Figure 9.1 A general framework for assessing capacity for transformation.

We also explore the way external factors – including national policies and governance processes, as well as climate change – influence community efforts to transform ocean governance and how community actors contend with these situational factors during transformation processes.

Case 1 – Atlantic

Sea-cage finfish aquaculture in Canada commonly takes place in the sheltered bays, coves, and inlets of the coastal zone. A long list of potential impacts of sea-cage aquaculture on coastal environments have been identified. These range from impacts associated with released nutrients, chemicals, pathogens, and escaped fish to changes in habitat quality and migratory routes, as well as effects on human health, traditional fisheries, and world fish supplies (see reviews in DFO 2005; Sapkota et al. 2008; Burridge et al. 2010; Edwards 2015; Sundberg et al. 2016).

Sea-cage aquaculture was slow to develop in Nova Scotia. Production of Atlantic salmon (*Salmo salar*), the principal farmed species, was 630 tonnes in 1995, compared with 14,490 tonnes for the same year in the adjacent province of New Brunswick (DFO 2018). In 1995, a sea-cage fish farm was granted a licence (but not for salmon) to operate by the Nova Scotia Department of Agriculture and Fisheries (now the Nova Scotia Department of Fisheries and Aquaculture, or NSDFA) in traditional lobster fishing territory near Spectacle Island in Port Mouton Bay, a small bay (~56 km²) located on the southwest coast of Nova Scotia.

In early 2005, the Nova Scotia government released a discussion document seeking public input for the further development of the province's aquaculture industry (NSDFA 2005). The following year, the province announced that a second fish farm was under review for Port Mouton Bay. By then, residents and fishers were observing environmental changes to the bay – such as increased occurrence of nuisance green algae fouling beaches and lobster traps, losses of eelgrass (*Zostera marina*) habitat, and declining numbers of clams, scallops, mussels, and periwinkles – that many fishers and residents believed were associated with the fish farm (Gilbert 2007). For many in the community, these were clear indicators of ecological impact. Their observations, coupled with a shared sense of history, created incentives and a sense of

agency that oriented the community toward imagining policy change (Loucks et al. 2017), and therefore emerging *incentives and a sense of agency* that changes in policy could be made. What has followed has been an ongoing effort by the communities around Port Mouton Bay to foster transformative change and influence in direct ways the governance of its coastal and ocean resources.

Fishing, particularly lobster fishing, has been and continues to be a significant part of the social, cultural, and economic well-being of communities around Port Mouton Bay, along with herring bait fishing and Irish moss and rockweed harvesting, coastal tourism, and recreational harvesting of mussels, scallops, and clams (FPMB 2019a). However, unable to engage provincial and federal scientists to examine the potential environmental impacts observed from the existing fish farm, and concerned about the potential for additional impacts from a second fish farm, the citizens and fishers around Port Mouton Bay initiated a community network for collective action called the Friends of Port Mouton Bay (FPMB). The FPMB reached out to university and retired scientists and began a series of self-funded studies to examine the changes they were observing in their bay. One study included applying a decision support system (Marine Finfish Aquaculture Decision Support System, or MFADSS) developed by federal government scientists to assist Department of Fisheries and Oceans (DFO) habitat managers in making decisions on farm site location when evaluating finfish lease applications (Doucette and Hargrave 2002) – an evaluation tool that had not been applied to the proposed fish farm by either the federal or provincial government. The results indicated that the shallow water depths and low current velocities at both the existing farm and the proposed new lease location, combined with the presence of sills that create depositional basins that retain settled organic waste from sea cages, made both locations in Port Mouton Bay unsuitable for salmon aquaculture (Hargrave 2009).

At the same time, the FPMB held rallies, petitioned the provincial government, and made numerous submissions to various levels of government requesting that the second site not be approved. On March 12, 2009, the premier of Nova Scotia announced an "indefinite" moratorium on more finfish aquaculture operations in Port Mouton Bay (FPMB 2009). Having secured protection from a second fish farm, the FPMB turned their research focus on the existing farm site.

Over the next 10 years, research by the FPMB in collaboration with university and retired researchers resulted in the publication of eight peer-reviewed environmental, policy, and social science studies as well as more than two dozen reports (FPMB 2019b). An 11-year lobster study (2007–17) found that during periods when the Port Mouton Bay fish farm was actively raising fish (a feed period), market lobster catches in Port Mouton Bay dropped by 42% on average and female berried lobster counts dropped 56% on average, compared with when the fish farm was not in production (a fallow period) (Milewski et al. 2018). Moreover, both market and berried lobster tended to be lower nearest the fish farm, and higher furthest away from the farm. Eelgrass, a species designated in 2009 as an ecologically significant species but not monitored near fish farms, was found to be impacted by the Port Mouton Bay fish farm (Cullain et al. 2018). Results indicated lower mean eelgrass shoot density, cover, and biomass closer to the fish farm, compared with a reference site and other Nova Scotia sites (Cullain et al. 2018). Copper used in fish feed and as antifoulants in sea-cage nets and structures was being transported from the seabed below the sea cages to the sea surface, where concentrations exceeded the level considered safe for the protection of marine life (Loucks et al. 2012). Overall, the fish farm was the largest source of nitrogen (15%) to the bay compared with all land-based sources (2%), and the second-largest source compared with atmospheric deposition (83%) (McIver et al. 2018). Anthropogenic nitrogen loading is one of the principal causes of degradation of coastal ecosystems worldwide and potentially can lead to shifts in local food web structures and ecological simplification (McIver et al. 2018). Despite the federal government's long-standing interest in increasing knowledge for regulatory purposes about the impacts of aquaculture operations (DFO 2002, 2005), these studies represented the first of their kind in Canada.

To render a *shared narrative* around perceived community resources, the FPMB hosted a community asset–mapping workshop whose results were later implemented as an online community asset "story map" featuring plain language versions of local scientific literature (FPMB 2018, 2019a), thereby increasing collective awareness of the issues challenging the resilience of Port Mouton Bay. Displaying a willingess to experiment with diverse information sources, the FPMB and the community continue to engage in social and scientific studies with experts internal or external to the community on local issues such as the dune and beach conditions at Carter's Beach, Indigenous archeological history, eelgrass mapping, and multi-partner ecosystem monitoring (see also Charles et al. 2020).

The Port Mouton Bay fish farm was only one of more than a dozen marine finfish farms in Nova Scotia and more than 100 farms in Atlantic Canada. By 2012, broad public discontent over a lack of public disclosure regarding fish farm disease outbreaks, drug and pesticide use, escaped fish, and, in some jurisdictions, environmental monitoring data and the absence of provincial or federal planning and procedural systems that would allow for equitable, meaningful, and effective participation by citizens in coastal zone decision making had resulted in the formation of the Atlantic Coalition for Aquaculture Reform (ACAR). In June 2012, the FPMB joined over 100 groups at a press conference calling on the Nova Scotia government to ban marine open-net pen fish farms (*CBC News* 2012). By the fall of 2012, and in response to requests by individuals and community groups, the NSDFA began releasing historical and current environmental monitoring data on open-net pen finfish operations. The data were compiled by ACAR and publicly released in 2013. The unique capacity of the FPMB and its partners to mobilize critical assets (e.g., intellectual and human resource assets, limited financial assets) was starting to bring about greater transparency. Although the FPMB exhibited a willingness to take greater organizational and economic risk (e.g., community-driven science and monitoring, social and natural capital inventories), acceptance by government and their role in local natural resource decision making and economic development would prove to be a more challenging step in the transformation process.

Overall, the ACAR report indicated that environmental quality associated with Nova Scotia finfish sites – measured by only one parameter, sediment sulfides – had decreased between 2006 and 2011 compared with data in a 2006 provincial report covering the 2003–05 monitoring period (Milewski 2013). The report and subsequent sulfide monitoring data revealed that numerous individual sampling stations at the Port Mouton Bay fish farm during production periods (2007–09, 2013–14) had sediment sulfide levels above levels the federal and provincial governments consider harmful to fish habitat (>3,000 μm [micrometre]) (DFO 2006; NSDFA 2014). Depending on the production year, sediment sulfides ranged from 5,000 to 10,000 μm; in 2007 and 2014, mean sediment sulfides were >4,000 μm (NSDFA 2007, 2014).

Increasing province-wide protests over salmon farming prompted the Nova Scotia government to impose a moratorium on new fish and shellfish aquaculture sites in 2013 and to appoint an independent panel to review Nova Scotia's regulatory framework for aquaculture (*CBC News* 2015a). The 159-page panel report released in 2015 called for a complete overhaul in the way aquaculture was regulated in the province, and recommended that the overhaul follow seven guiding principles: effectiveness, openness, transparency, accountability, proportionality, integration, and precaution (Doelle and Lahey 2014). To improve the efficiency, effectiveness, and fairness of the regulatory process, the report recommended the development of a classification system that identifies sites suitable or unsuitable for finfish aquaculture, and sensitive areas based first on biophysical conditions, such as water depth, current speed, oceanographic and benthic circulation patterns, proximity to salmon rivers, and other biogeophysical conditions (Doelle and Lahey 2014).

The Doelle and Lahey (2014) report also referred to the need for industry to secure a social licence to operate (SLO) as a means of establishing more accessible and transparent decision-making processes between industry and the communities in which they operate: a view also expressed in the 2010 DFO National Aquaculture Strategic Action Plan (DFO 2010). The social conditions for meaningful community engagement in SLO negotiations require knowledge, credibility, power, and trust (Mather and Fanning 2019; Leith, Ogier, and Haward 2014; Parsons and Moffat 2014). SLO is not a legally binding agreement but an undefined process, and the NSDFA, DFO, and industry have yet to develop objective measures to assess when, and if, SLO has been achieved (Milewski and Smith 2019).

In 2015, the Nova Scotia Minister of Fisheries and Aquaculture held a press conference to announce that he was prepared to lift the moratorium on aquaculture in the province in 2016 and that an Aquaculture Regulatory Advisory Committee composed of community, First Nations, industry, and municipal government representatives would be appointed to provide advice on regulating aquaculture in Nova Scotia (*CBC News* 2015a). A new process for applying for aquaculture sites would be put in place that emphasized more opportunities for the public to comment on proposed developments and increased transparency and objectivity in decision making (NSDFA 2015). Since then, several major global salmon fish farm companies have signalled their interest in fish farm expansion in Nova Scotia (Withers 2019a, 2019b). No announcement has been made about the development of a coastal planning process that would establish a classification system for coastal development, including aquaculture, in Nova Scotia, and the new process for applying and reviewing aquaculture licence applications would be put to the test.

In the meantime, the existing farm site in Port Mouton Bay has been fallow since 2015. The lease for the site was set to expire in 2020 (NSDFA 2017). The FPMB had written repeatedly to the provincial Minister of Fisheries and Aquaculture to share the results of their studies and request that the lease not be renewed (FPMB 2018). The local municipal council consistently supported the community's request for a ban on fish farming in the bay (FPMB 2006, 2012, 2019b). DFO and NSDFA scientists initiated studies in the vicinity of the Port Mouton Bay fish farm to examine the late summer/fall movement of lobster, model the flow of nitrogen from the fish farm, and estimate benthic organic loading rates under different farm production scenarios (personal communication by C. McKindsey, S. Robinson, and G. Reid).

After a public consultation period in early 2020, the NSDFA issued a 10-year licence and 20-year lease renewal for the Port Mouton Bay fish farm on April 8. A publicly released "Findings and Decision" document included copies of public submissions solicited by the province and the factors considered in its decision-making process (NSDFA 2020). Missing from the document were submissions made by the author and co-author of some of the peer-reviewed published studies done in Port Mouton Bay (e.g., Loucks et al. 2012; Loucks, Smith, and Fisher 2014; Cullain et al. 2018; McIver et al. 2018; Milewski et al. 2018), despite their having gotten an acknowledgement from the province that their submissions had been received. Submissions by other community members were also missing from the decision document. Any references to, or discussion of, the findings from these published studies – with one exception – were also absent. The Findings and Decision document made general reference to the concerns raised by the public regarding the decrease in lobster catch rates associated with the operation of the fish farm (NSDFA 2020), findings published in two studies (Loucks, Smith, and Fisher 2014; Milewski et al. 2018). The province claimed that several other studies offered different conclusions, yet no references for these studies were presented, nor were any details on the results of these purported studies provided. Without exception, all of the comments made by the public and accepted by the province expressed unanimous opposition to the renewal of the licence and lease for the Port Mouton Bay fish farm (NSDFA 2020). Except for the studies by Milewski and colleagues (2018) and by Loucks, Smith, and Fisher (2014), no other peer-reviewed studies on lobster catch rates in relation to the fish farm in Port Mouton Bay have been published to date.

Case 2 – Arctic

Inuit have long anticipated the need for integrated ocean management, and indeed have initiated and launched a range of actions aimed at securing the health, protection, and security of the Arctic Ocean as a regional sea. These efforts have taken place in a variety of fora, ranging from the National Energy Board hearings of 1974–75 concerning marine impacts of offshore mineral and oil and gas drilling and exploration in Lancaster Sound, to the Berger Commission hearings during 1975–76 on northern pipeline development, to recent court cases (e.g., Isabella Bay). They have also taken place at United Nations and International Labour Organization fora. As well, marine and offshore issues have been subjects of federal negotiations of rights and resources, with varying degrees of success (see Chapter 14).

These efforts are cross-scale. For example, they range from the international circumpolar perspective of the Inuit Circumpolar Council (ICC; formed in 1978, formerly known as the Inuit Circumpolar Conference), representing Inuit in Canada, Greenland, and Alaska, to the local and regional struggles of Inuit communities across the Arctic to confront shipping as it threatens marine-based livelihoods, sustainability, marine safety, and environmental protection. In 1986, the ICC published the Inuit Arctic Policy, featuring objectives for Arctic Ocean protection, and a host of policy recommendations to governments to recognize human rights and ensure sustainability. In conjunction with the Inuit Arctic Policy, the Inuit Regional Conservation Strategy was also adopted. It took an international polar perspective centred on the Arctic Ocean and rooted in the World Conservation Strategy, and embraced all Inuit communities in Canada, Greenland, and Alaska. Action on this strategy was overshadowed, however, by the eruption of the issue that came to be identified as "Northern Contaminants" (i.e., the effort to ban or phase out persistent organic pollutants [POPs] and heavy metals).

With the struggles for recognition of Inuit interests and rights to participate, the Northern Contaminants issue put Canada at the forefront of global and regional ecosystem science and policy formation, which resulted in administrative, legal, and governmental change (e.g., the *Stockholm Convention on Persistent Organic Pollutants*). These struggles also precipitated the participation of Inuit and other circumpolar Indigenous Peoples' Organizations (IPOs) in the Rovaniemi Process (Finland), which culminated in the creation of the Arctic Council in 1996, with permanent IPO observers (Government of Canada 2017). The Arctic Council is composed of the eight circumpolar nations (Canada, Denmark, Finland, Iceland, Norway, Russia, Sweden, and the United States), and a growing contingent of nation-state observers. This is currently the state of the art of Arctic Ocean governance as it applies to Canada's role internationally. Historically, however, Canada played key roles at crucial moments in the evolution of Arctic and national and international oceanic legal systems, specifically the *United Nations Convention on the Law of the Sea* (UNCLOS) and the recognition of Canada's interests through the clauses aimed at "Protection of Ice-infested Waters."

Importantly, we also see the emergence of bilateral international agreements, such as the *Agreement on the Conservation of Polar Bears*, and new intra-Inuit regional cooperation agreements (e.g., the Lancaster Sound Region Study [Green Paper] successor, the Tallurutiup Imanga

National Marine Conservation Area [NMCA]). Although we do not see a place of priority or privilege assigned to the Arctic Ocean in the domestic concerns addressed by the new Arctic and Northern Policy Framework (ANPF) released by the Government of Canada in September 2019, it appears prominently in the framework's "International Chapter," which notes that an extensive international legal framework applies to the Arctic Ocean, including UNCLOS (Government of Canada 2019).

In keeping with the focus of this chapter on transformation from below, we present an "Inuit-first" perspective on the origins of policy change and management initiatives, not to the exclusion of the achievements of other Indigenous Peoples (particularly the Māori) but taking a Canada-centric view of geopolitics. The North has played a unique role in Canada's ocean strategy by virtue of the extent to which Inuit and other northern Indigenous Peoples have contributed proactively to policy development and lawmaking. Much of the real work of "made in the North" policy has been done by Indigenous organizations, and insofar as ocean health and protection of the Arctic marine environment are concerned, Inuit remain vocal agents for action at local, regional, and international levels. Yet, there is still much to do to transform ocean governance from the bottom up. Community needs for participation in regulation of marine activities that affect them are still largely unmet.

For instance, the hamlet of Chesterfield Inlet is an Inuit community in the Kivalliq Region of Nunavut, west of Hudson Bay, and functions within the Government of Nunavut framework as a municipality. The Kivalliq Region has been home to a number of mines in recent history, and active exploration efforts continue. In Nunavut, communities are consulted about new mining projects through the Nunavut Impact Review Board, and have the capacity to participate in negotiation of benefit agreements. Often, mining companies contribute to regional organizations such as the Kivalliq Inuit Association to ensure that neighbouring communities share in the benefits of this resource extraction. Communities that are considered to be directly impacted by mining operations in their area are compensated for this disturbance; however, determining who is directly impacted is contentious for Chesterfield Inlet.

Ships transport supplies to the mines near Baker Lake dock in Chesterfield Inlet's harbour for a week at a time; supplies are offloaded onto smaller boats that travel along the inlet to the mine. Elders and hunters report that the presence of large ships in the harbour and smaller ships along the inlet results in noticeable changes in the locations where seals and other marine wildlife can be found in the area. These observations are part of a collective awareness and shared narrative within the community, which leaders began sharing at community meetings with the mining company, government officials, and other organizations, such as the World Wildlife Fund (*CBC News* 2015b). As a result, community hunters need to travel further and are less able to harvest "country food" – an important source of both food security and cultural well-being. These negative impacts on community health created the incentives and agency for change.

However, the community has remained in the preparation phase as these impacts are not considered "direct," according to the mining company responsible for the shipping. Chesterfield Inlet community leaders have advocated for changes to shipping traffic management that will support the mining needs and limit the impact on wildlife and hunters. A willingness to experiment is evidenced in their suggested changes. Community leaders have suggested that ships travel at a reduced speed along the inlet to limit the sound, which has been found to impact marine mammals, and that they include Inuit hunters on board to navigate around sensitive areas for local wildlife, that marine traffic during important harvesting times be limited, and that a deep-sea port in Chesterfield Inlet be constructed to enable cargo to be offloaded immediately and transported on land to the mine, reducing the time large ships spend in the harbour and the traffic along the inlet (Newell et al. 2020; Lightfoot 2020). Advocating for the construction of the deep-sea port locally is one way by which community leaders in Chesterfield Inlet are aiming to increase their stocks of built capital in order to mitigate and adapt to the impacts that increased resource development in the area are having on community health. This is part of an overall push by community leaders to change the governance model to one in which local voices are included in decision making beyond the point of approving resource development through the Nunavut Impact Review Board.

The inclusion of local voices in governance is important to ensure that the community's needs are balanced with economic interests in the area; local stakeholders' voices, values, and knowledge must not only be present at the table but respected and considered equal to those of other participants in the decision-making process. The creation of the Government of Nunavut in 1999 marked a new beginning in federal-territorial relations but the real work of implementation clearly continues at regional and local scales, as evidenced by the example from recent experience in Chesterfield Inlet (Newell et al. 2020). As climate change expands opportunities for economic development and resource extraction in the Canadian Arctic, it is important that governance models include community involvement, not just at the beginning stages but throughout. Community involvement throughout the process will ensure that the dynamic nature of development impacts on the community are considered and that adjustments are made when unforeseen impacts occur. As the mayor of Chesterfield Inlet expressed in a presentation to government officials and mining and shipping company executives: when the community originally approved of this mine, they had no idea of the magnitude of the impact that increased shipping would have on the community, and while the Chesterfield Inlet is not against continuing resource development, it should not come at the cost of the community's food security (Lightfoot 2020).

Case 3 – Pacific

The landscape currently known as the Central Coast of British Columbia represents a complex and dynamic site of governance collaboration, negotiation, and conflict. Indigenous leaders have asserted their *agency* through claims for inherent rights, responsibilities, and authority to manage ancestral territories based in their own knowledges and expertise. The tools and strategies of *collective action* to transform ocean governance are emerging, but that collective action is manifested through a complex mix of confrontation, negotiation, litigation, collaboration, ceremony, and celebration (Frid, McGreer, and Stevenson 2016; Jones, Rigg, and Pinkerton 2017; Klain, Beveridge, and Bennett 2014; von der Porten, Corntassel, and Mucina 2019). In this context, eulachon (*Thaleichthys pacificus*), a small ecologically and culturally important fish, represents a case of contested jurisdiction and potential governance transformation.

Because of the nature of their life cycle, eulachon occupy two distinct bodies of water, living their adult lives in the Pacific Ocean and returning to spawn in glacier-fed rivers along the coast (Moody 2008; Schweigert et al. 2012). This results in some governance complexity: while the marine environment is managed by the Canadian state, de facto management by *Nuxalkmc* (Nuxalk people) and other Indigenous Peoples has been uninterrupted on much of the coast (Hilland 2013; Moody 2008). Considering the cultural role of eulachon and the lack of state investment in protection of the species, eulachon remains, in the eyes of many, an "Indigenous" fish. Prior to colonial contact, *Nuxalkmc* had a thriving relationship with eulachon, or *sputc*, an anadromous smelt that spawns in glacial-fed rivers of coastal British Columbia, including those of Nuxalk territory. A cultural keystone species, eulachon remain vital to Nuxalk well-being, culture, and identity, supporting ancestral systems of knowledge and governance (Beveridge 2019; Hilland 2013).

Once abundant throughout British Columbia, eulachon failed to return to rivers in the Central Coast region in 1999, and have not returned in harvestable numbers since (Moody 2008). While the reasons for this loss may be characterized as complex (Schweigert et al. 2012), *Nuxalkmc* experts observe that the timing of the decline directly followed eulachon bycatch associated with an expansion of a shrimp trawl fishery into the Queen Charlotte Sound management area (Hilland 2013; Moody 2008). Subsequently, DFO did close the area to shrimp trawling and imposed mandatory bycatch reduction devices in 2000 (Schweigert et al. 2012), but further conservation action has been slow. Thus, *Nuxalkmc* see the federal fisheries management system as having failed to protect eulachon in the marine environment, impacting management rights, resource access, and related benefits (Hilland 2013; Turner et al. 2013).

The rapid disappearance of eulachon and the government's failure to address it has provided strong impetus for community-based action. In this context, a community-engaged project, the Nuxalk Sputc Project, sought to strengthen community authority related to eulachon from the ground up. As a part of the "preparatory activity" (see Figure 9.1) for transformative change, a key goal of the project was to generate a shared narrative related to management priorities and related governance mechanisms. Initiated by the Nuxalk Stewardship Office, over the course of four years the project employed an iterative, community-engaged methodology informed by Nuxalk ways of knowing (Beveridge et al. 2020). This process offered Nuxalk leadership the knowledge, background, and tools to speak strongly on behalf of eulachon, with the full support and authority of *Nuxalkmc* and Nuxalk laws. In so doing, the project addressed key elements of governance transformation – in this case, a movement toward increased Indigenous authority and related resistance to current power structures.

Through the Nuxalk Sputc Project, we learned that the ingredients required to move toward bottom-up governance transformation (in this context) included: (1) enhancing local and community capacity for knowledge documentation, articulation, and representation; (2) upholding community governance processes and collective decision-making practices; and (3) community engagement (Beveridge et al. 2020). Strengthening Nuxalk management authority through the project required engaging local governance institutions and related decision-making processes. Articulating Nuxalk knowledge in an accessible and relevant manner was integral to regenerating community-level consensus and responsibility around management priorities, including cultural and spiritual practices often not included in standard governance frameworks (Beveridge et al. 2020). Given the distributed nature of Nuxalk governance structures and decision-making processes (Hilland 2013; McIlwraith 1992), a community engagement process leading to collective ownership of knowledge was key to upholding local management legitimacy.

Our experience confirmed that articulating and sharing Nuxalk knowledge for internal purposes alone required extensive internal *capacities,* often restricted by limited human, financial, and educational *assets* (Bowie 2013; Pinkerton and John 2008; von der Porten, de Loë, and Plummer 2015). Addressing Nuxalk eulachon management priorities necessarily requires collaboration with actors at other levels of jurisdiction. While these may include nonstate actors (industry, NGOs, a supportive public, and other First Nations), interactions with the state remain central to any potential transformation of this system. The importance of power relationships with nonstate actors is recognized as key to strengthening

management authority (Bowie 2013; Jones, Rigg, and Pinkerton 2017; von der Porten, Corntassel, and Mucina 2019; von der Porten et al. 2016). Thus far, however, extractive commercial interests have predominated over Nuxalk priorities related to ocean management and research (e.g., area closures, ocean range research, consideration of genetic diversity), and First Nations are only peripherally included in decision making.

For *Nuxalkmc,* engaging in an external (state) management system will require even more local resources than those required by the internally oriented Nuxalk Sputc Project. While funding constraints and human resources capacity may play a role in First Nations' collaborative potential, external power dynamics and institutional biases are also essential factors (Bowie 2013; von der Porten et al. 2016). In a political and legal context that aspires to reconciliation and holds an increasing expectation of First Nations' collaboration in environmental management, it is the responsibility of the state to "level the playing field" to mediate respectful and responsible relationships (von der Porten, de Loë, and Plummer 2015; von der Porten et al. 2016). Such a transformation would be supported by court decisions (e.g., *Delgamuukw, Tsilhq'otin*[1]) (Hoehn 2016; Kotaska 2013), international agreements (e.g., *United Nations Declaration on the Rights of Indigenous Peoples* [UNDRIP]), and reconciliation agreements and frameworks (Low 2018; McGee, Cullen, and Gunton 2010). This would promote management led by community initiatives and priorities, rather than by consultation with First Nations as "stakeholders" or actors with equal legal standing to other groups (Castleden et al. 2017; Singleton 2009; von der Porten and de Loë 2015). The case of eulachon therefore presents the Canadian state with a transformative opportunity to align with inherent Indigenous rights and responsibilities, and to embrace collaborative, nation-to-nation management approaches.

DISCUSSION AND CONCLUSION

Transforming how we govern Canada's oceans and coasts will require sustained effort. In this chapter, we have examined several cases in which the capacity for change has coalesced to foster significant shifts in governance and the potential for transformation. These experiences are ongoing and they will not always yield clearly positive social or ecological outcomes – trade-offs and challenges are inevitable, and favourable contexts necessary.

The dimensions of transformative capacity we outline in each case (e.g., those related to agency, networks, assets, and capacity; see Figure 9.1) are not fixed in time or space, nor is it easy to conclude that a transformation has been achieved. Still, the capacity or latent potential for transformation across Canada's coasts reflects a phase of *preparation* (where initial conditions for transformation are established), a phase of *navigation* (where evidence of a governance transformation emerges), and a phase of *resilience* (in which a new system or alternative future trajectory is fostered). Each of the dimensions of transformative capacity will manifest at different times in these three phases, and understanding when and how that happens can point researchers and policy-makers toward opportunities to leverage more durable social and ecological outcomes.

The cases we have explored appear to be largely situated in the preparation and navigation phases, yet with the potential to become more entrenched, and with opportunities for those involved to build the resilience of new system configurations. For example, in the case of eulachon, despite legal precedents and rhetorical attention to collaborative decision-making institutions and reconciliation, there remain significant limitations in the state's apparent willingness and capacity for nation-to-nation engagement. This is instructive for coastal communities and governments of all levels across Canada. For instance, in the face of settler-colonial impacts, Indigenous legal scholars suggest that First Nations rearticulate how their particular intellectual processes inform both formal and informal management systems, including decision-making processes, cultural practices and ethics, roles and responsibilities, relationships and kinship networks (Friedland and Napoleon 2015; Napoleon and Friedland 2016; Napoleon and Overstall 2007). This is, in part, what the Nuxalk Sputc Project set out to accomplish. The importance of relationships with nonstate actors, including industry, NGOs and supportive public, and other First Nations, is also recognized as key to strengthening management authority (Bowie 2013; Jones, Rigg, and Pinkerton 2017; von der Porten, Corntassel, and Mucina 2019; von der Porten et al. 2016). However, state willingness to

consider power sharing, joint decision-making, and systemic change processes remains essential to transformation (see also Silver et al. 2022).

In Port Mouton Bay, a transformation in the governance of coastal and ocean resources has shown less progress. For instance, the development of an effective coastal planning process remains incomplete, and the process for applying and reviewing aquaculture licence applications is opaque, lacking transparency and fairness. A transformation has not been achieved, yet the experiences documented by the Friends of Port Mouton Bay, and the novel approach to research and action they have developed, provide crucial models moving forward for other coastal communities seeking to prepare, navigate, and ultimately transform relationships of power with governments and industry. Creating governance changes that have been driven by community-based actors remains a significant challenge.

The Arctic constitutes the most unfamiliar of the ocean contexts for consideration here: colonized within living memory, and subjected to regulatory and administrative forms of governance that were based on racial profiling and pervasive discrimination. Inuit resilience is embedded in culture, and is being articulated in ongoing forms of negotiation in each generation. This offers a new window through which to view the model in Figure 9.1: one where resilience remains fully alive in living memory, actively pursuing restoration. The recent past constitutes a history of various forms of expropriation of rights, resources, and territories in right of Canada, and the restoration of Indigenous rights begins with recognition (see Chapters 2 and 14). The transcendent importance of culture as a source of agency and capacity inspires broader conceptions of community, to include transnational community and international nongovernmental agency. The Arctic Ocean community has done its work. It remains for the non-Arctic Canadian partners to develop the awareness, willingness, and equity needed to make it work. The external drivers of climate and markets will not cease. Different approaches to assessing capacity are needed, making the present approach timely.

In this chapter, we have shared insights from three cases about: (1) what processes, relationships, and capacities are required to support governance transformation from the ground up; and (2) the interjurisdictional

Table 9.1

Synthesis of lessons and policy recommendations

Framework element	*Lesson/policy implication*
External drivers of change	Ocean resources are known to be at risk on a global scale, with multiple drivers of change impacting a wide range of fisheries.
	Management agencies have insufficient local ecological and social knowledge required for decision making on local-level marine resource allocation/use.
	Industry and higher-level (municipal, regional, provincial, federal) marine resource management goals often differ from the goals of community members.
Incentives and agency for change	Infringements on resource rights and access, and/or evidence of resource decline, are common incentives for engagement in transformative action.
	International, federal, provincial, and local legislation and decision-making authorities with transformative potential are already in place (e.g., UNDRIP).
	Federal oceans management is widely recognized to be operating at low capacity and/or with conflicting goals.
	The political and legal context in Canada aspires to reconciliation, holds an expectation of local/Indigenous participation in marine management, and supports Indigenous right to manage lands and waters.
Development of social networks for collective action	Examples of collective action are increasingly available to support ocean governance transformation.
	Community-engaged processes lend legitimacy to local management planning/action.
	Local management legitimacy requires clear knowledge articulation and representation, upholding of local governance institutions and decision-making processes, and consideration of nuanced roles and responsibilities.
	Negotiation of ocean resource rights and responsibilities may be more successful if they are networked and cross-scale.
	Collective knowledge documentation is an important first step to action.
Willingness to experiment and take risks	Transformative opportunities for sharing of power already exist, but require willingness to yield control, to recognize Indigenous rights, and to experiment (e.g., Tina Ngata, https://tinangata.com/).
	Community-driven science takes time and human and financial resources.
	Funding for the development and implementation of local-level ecological monitoring and ecosystem assessment must be part of marine resource management.
Presence of assets and stocks of key "capitals" (e.g., social, natural)	Federal agencies require increased capacity and resources to respectfully engage Indigenous decision makers and decision-making processes on a nation-to-nation basis.
	Local managers require additional resources to address community management needs while interfacing with interjurisdictional (e.g., nation-to-nation) management contexts at other scales.
	Engagement of local knowledge systems is key to unlocking transformative change.
	Equitable engagement and decision-making process that integrates the results of community science into management process must be developed.

engagement, leadership, and knowledge (including notably Indigenous leadership) needed to move through phases of transformative change (see Table 9.1). As these examples show, collective understanding and consensus are necessary to maintain legitimacy and authority, and this is especially so in Indigenous legal systems characterized by decentralized institutions and interactive processes (Napoleon and Overstall 2007). Enacting governance institutions, decision-making protocols, and knowledge-sharing practices requires engagement with both political and cultural bases of authority. Ultimately, the cases reveal a shift in emphasis from technical approaches to consistent, trusting relationships (Bowie 2013), and involve working with Indigenous Peoples (leadership, decision makers, and knowledge holders) rather than extracting Indigenous knowledges (McGregor 2009), along with non-Indigenous coastal communities. In the context of coastal management, this kind of change is fundamentally systemic and relational.

NOTE

1 *Delgamuukw v British Columbia,* 1997 SCC 3; *Tsilhqot'in Nation v British Columbia,* 2014 SCC 44.

REFERENCES

Agrawal, A., and C.C. Gibson. 1999. "Enchantment and Disenchantment: The Role of Community in Natural Resource Conservation." *World Development* 27 (4): 629–49.

Armitage, D., A. Charles, and F. Berkes, eds. 2017. *Governing the Coastal Commons: Communities, Resilience and Transformation.* London and New York: Taylor and Francis.

Ban, N.C., H.M. Alidina, and J.A. Ardron. 2010. "Cumulative Impact Mapping: Advances, Relevance and Limitations to Marine Management and Conservation, Using Canada's Pacific Waters as a Case Study." *Marine Policy* 34: 876–86.

Bennett, N., J. Blythe, A.M. Cisneros-Montemayor, G.G. Singh, and U.R. Sumaila. 2019. "Just Transformations to Sustainability." *Sustainability* 11 (14): 3881.

Berkes, F. 2015. *Coasts for People: Interdisciplinary Approaches to Coastal and Marine Resource Management.* New York and London: Routledge.

Beveridge, R. 2019. "Standing Up for Sputc: The Nuxalk Sputc Project, Eulachon Management, and Well-Being." PhD dissertation, University of Victoria. http://hdl.handle.net/1828/10830.

Beveridge, R., M. Moody, G. Murray, and C.T. Darimont. 2020. "The Nuxalk Sputc (Eulachon) Project: Strengthening Indigenous Management Authority through Community-Driven Research." *Marine Policy* 119: 103971; https://doi.org/10.1016/j.marpol.2020.103971.

Blythe, J., J. Silver, L. Evans, D. Armitage, N.J. Bennett, M.L. Moore, and K. Brown. 2018. "The Dark Side of Transformation: Latent Risks in Contemporary Sustainability Discourse." *Antipode* 50 (5): 1206–23.

Bowie, R. 2013. "Indigenous Self-Governance and the Deployment of Knowledge in Collaborative Environmental Management in Canada." *Journal of Canadian Studies/Revue d'études canadiennes* 47 (1): 91–121.

Burridge, L., J.S. Weis, F. Cabello, J. Pizarro, and K. Bostick. 2010. "Chemical Use in Salmon Aquaculture: A Review of Current Practices and Possible Environmental Effects." *Aquaculture* 306: 7–23.

Castleden, H., D. Martin, A. Cunsolo, S. Harper, C. Hart, P. Sylvestre, R. Stefanelli, L. Day, and K. Lauridsen. 2017. "Implementing Indigenous and Western Knowledge Systems (Part 2): 'You Have to Take a Backseat' and Abandon the Arrogance of Expertise." *International Indigenous Policy Journal* 8 (4). https://doi.org/10.18584/iipj.2017.8.4.8.

CBC News. 2012. "Groups Rally against Ocean-Based Salmon Farming." *CBC News,* June 4. https://www.cbc.ca/news/canada/nova-scotia/groups-rally-against-ocean-based-salmon-farming-1.1253832.

–. 2015a. "New Nova Scotia Aquaculture Regulations 'More Accountable.'" *CBC News,* October 26. https://www.cbc.ca/news/canada/nova-scotia/nova-scotia-aquaculture-fish-farms-moratorium-1.3288661.

–. 2015b. "Shipping Route Affecting Marine Mammals: Chesterfield Inlet Mayor." *CBC News,* January 30. https://www.cbc.ca/news/canada/north/shipping-route-affecting-marine-mammals-chesterfield-inlet-mayor-1.2937831.

Charles, A. 2001. *Sustainable Fishery Systems.* Oxford: Wiley-Blackwell.

–. 2012. "People, Oceans and Scale: Governance, Livelihoods and Climate Change Adaptation in Marine Social-Ecological Systems." *Current Opinion in Environmental Sustainability* 4: 351–57.

–. 2017. "The Big Role of Coastal Communities and Small-Scale Fishers in Ocean Conservation." In *Conservation* for the Anthropocene Ocean: Interdisciplinary Science in Support of Nature and People, edited by P.S. Levin and M.R. Poe, 447–61. London: Academic Press/Elsevier.

Charles, A., L. Loucks, F. Berkes, and D. Armitage. 2020. "Community Science: A Typology and Its Implications for Governance of Social-Ecological Systems." *Environmental Science and Policy* 106: 77–86.

Cheung, W.W.L., R. Watson, and D. Pauly. 2013. "Signature of Ocean Warming in Global Fisheries Catch." *Nature* 497: 365–68.

Cullain, N., R. McIver, A.L. Schmidt, I. Milewski, and H.K. Lotze. 2018. "Potential Impacts of Finfish Aquaculture on Eelgrass (*Zostera marina*) Beds and Possible Monitoring Metrics for Management: A Case Study in Atlantic Canada." *PeerJ* 6: e5630.

Davies, K., K.T. Fisher, G. Couzens, A. Allison, E.I. van Putten, J.M. Dambacher, M. Foley, and C.J. Lundquist. 2020. "Trans-Tasman Cumulative Effects Management: A Comparative Study." *Frontiers in Marine Science* 7. https://doi.org/10.3389/fmars.2020.00025.

DFO (Fisheries and Oceans Canada). 2002. *DFO's Aquaculture Policy Framework.* Ottawa: Fisheries and Oceans. http://publications.gc.ca/collections/collection_2012/mpo-dfo/Fs23-411-2002-eng.pdf.

–. 2005. *Assessment of Finfish Cage Aquaculture in the Marine Environment.* DFO Canadian Science Advisory Secretariat Research Document 2005/034. Ottawa: Fisheries and Oceans Canada. https://publications.gc.ca/collections/collection_2011/mpo-dfo/Fs70-6-2005-034-eng.pdf.

–. 2006. "Science Expert Opinion on the Effects of Free Sulfides in Marine Sediments on Macrobenthic Infauna Biodiversity." February 23. Maritime Region Expert Opinion 2006/001.

–. 2010. *National Aquaculture Strategic Action Plan Initiative (NASAPI) (2011–2015).* Ottawa: Fisheries and Oceans Canada. http://waves-vagues.dfo-mpo.gc.ca/Library/343981.pdf.

–. 2018. "Fisheries Statistics." https://www.dfo-mpo.gc.ca/stats/aqua/aqua95-eng.htm.

Doelle, M., and W. Lahey. 2014. *A New Regulatory Framework for Low-Impact/High-Value Aquaculture in Nova Scotia. The Final Report of the Independent Aquaculture Regulatory Review for Nova Scotia [The Doelle-Lahey Panel].* Halifax: Province of Nova.Scotia. https://novascotia.ca/fish/aquaculture/laws-regs/docs/regulatory-review-final-report.pdf.

Dolan, A.H., M. Taylor, B. Neis, R. Ommer, J. Eyles, D. Schneider, and B. Montevecchi. 2005. "Restructuring and Health in Canadian Coastal Communities." *EcoHealth* 2: 195–208.

Doucette, L.I., and B.T. Hargrave. 2002. *A Guide to the Decision Support System for Environmental Assessment of Marine Finfish Aquaculture.* Canadian Technical Report of Fisheries and Aquatic Sciences 2426. Moncton: Fisheries and Oceans Canada, Gulf Region.

Ecotrust Canada. 2004. *CATCH-22: Conservation, Communities and the Privatization of B.C. Fisheries.* Vancouver: Ecotrust Canada.

–. 2009. *Briefing: A Cautionary Tale about ITQs in BC Fisheries.* Vancouver: Ecotrust Canada.

–. 2018. *Fisheries for Communities Gathering: Proceedings Report.* Vancouver: Ecotrust Canada.

Edwards, P. 2015. "Aquaculture Environment Interactions: Past, Present and Future Trends." *Aquaculture* 447: 2–14.

Folke, C., T. Hahn, P. Olsson, and J. Norberg. 2005. "Adaptive Governance of Social-Ecological Systems." *Annual Review of Environment and Resources* 30: 441–73.

FPMB (Friends of Port Mouton Bay). 2006. "Mayor Leefe Writes to Minister of Fisheries: Region of Queens against Fish Farm Proposal." August 23. http://friendsofportmoutonbay.ca/letters.html.

–. 2009. "Moratorium on Finfish Aquaculture Announced!" https://www.friendsofportmoutonbay.ca/news.html.

–. 2012. "Region of Queens Asks the Province to: 'Designate Port Mouton Bay as a Closed Area Not Suitable for Aquaculture.'" https://www.friendsofportmoutonbay.ca/docs/Regions-of-Queens-Motion-Port-Mouton-Bay-Closed-for-Aquaculture-Mar-19-2012.pdf.

–. 2018. "Second Call to Close Spectacle Island Fish Farm (Lease #0835) in Port Mouton Bay." November 19. http://www.friendsofportmoutonbay.ca/news.html.

–. 2019a. "Port Mouton Bay Launches On-line Asset Map: The 'Story' of Our Community, Our Assets and Climate Change Challenges: by Lydia Ross." February 11. https://www.friendsofportmoutonbay.ca/news.html.

–. 2019b. "Our Documents." http://www.friendsofportmoutonbay.ca/documents.html.

Friedland, H., and V. Napoleon. 2015. "Gathering the Threads: Developing a Methodology for Researching and Rebuilding Indigenous Legal Traditions." *Lakehead Law Journal* 1: 29.

Frid, A., M. McGreer, and A. Stevenson. 2016. "Rapid Recovery of Dungeness Crab within Spatial Fishery

Closures Declared under Indigenous Law in British Columbia." *Global Ecology and Conservation* 6: 48–57.

Gilbert, G. 2007. "Survey of Fishers and Mossers in Port Mouton Bay." http://www.friendsofportmoutonbay.ca/fishersurvey.html.

Government of Canada. 2017. "Declaration on the Establishment of the Arctic Council (Ottawa, Canada, 1996)." https://www.international.gc.ca/world-monde/international_relations-relations_internationales/arctic-arctique/declaration_ac-declaration_ca.aspx?lang=eng&_ga=2.108256002.1093978593.1570050859%E2%80%931%20442566674.1570050859.

–. 2019. "International Chapter." In *Canada's Arctic and Northern Policy Framework.* Ottawa: Crown-Indigenous Relations and Northern Affairs Canada. https://www.rcaanc-cirnac.gc.ca/eng/1562867415721/1562867459588.

Hargrave, B. 2009. "Application of a Traffic Light Decision System for Marine Finfish Aquaculture Siting Assessment in Port Mouton, Nova Scotia. Report Submitted to Friends of Port Mouton Bay." http://www.friendsofportmoutonbay.ca/docs/Summary-MFADSS-application-Port-Mouton-20090423-HargraveBT.pdf.

Hilland, A. 2013. "Extinguishment by Extirpation: The Nuxalk Eulachon Crisis." LLM thesis, University of British Columbia.

Hoehn, F. 2016. "Back to the Future – Reconciliation and Indigenous Sovereignty After Tsilhqot'In." *UNBJL* 67: 109–45.

Jones, R., C. Rigg, and E. Pinkerton. 2017. "Strategies for Assertion of Conservation and Local Management Rights: A Haida Gwaii Herring Story." *Marine Policy* 80: 154–67.

Kearney, J., F. Berkes, A. Charles, E. Pinkerton, and M. Wiber. 2007. "The Role of Participatory Governance and Community-Based Management in Integrated Coastal and Ocean Management in Canada." *Coastal Management* 35: 79–104.

Klain, S., R. Beveridge, and N. Bennett. 2014. "Ecologically Sustainable but Unjust? Negotiating Equity and Authority in Common-Pool Marine Resource Management." *Ecology and Society* 19 (4): 52.

Kotaska, J.G. 2013. "Reconciliation 'at the End of the Day': Decolonizing Territorial Governance in British Columbia after *Delgamuukw.*" PhD dissertation, University of British Columbia.

Leith, P., E. Ogier, and M. Haward. 2014. "Science and Social License: Defining Environmental Sustainability of Atlantic Salmon Aquaculture in South-Eastern Tasmania, Australia." *Social Epistemology* 28: 277–96.

Lightfoot, P. 2020. "Chesterfield Inlet Mayor Pitches Deep-Sea Port to Ease Effects of Increased Shipping." *Nunatsiaq News,* February 11.

Loeber, A. 2007. "Designing for Phronèsis: Experiences with Transformative Learning on Sustainable Development. *Critical Policy Studies* 1: 389–414.

Loucks, L., F. Berkes, D. Armitage, and T. Charles. 2017. "Emergence of Community Science as a Transformative Process in Port Mouton Bay, Canada." In *Governing the Coastal Commons: Communities, Resilience and Transformation,* edited by D. Armitage, A. Charles, and F. Berkes, 43–58. London: Routledge.

Loucks, R.H., R.E. Smith, C.V. Fisher, and B. Fisher. 2012. "Copper in the Sediment and Sea Surface Microlayer Near a Fallowed Open-Net Fish Farm." *Marine Pollution Bulletin* 64: 1970–73.

Loucks, R.H., R.E. Smith, and E.B. Fisher. 2014. "Interactions between Finfish Aquaculture and Lobster Catches in a Sheltered Bay." *Marine Pollution Bulletin* 88: 255–59.

Low, M. 2018. "Practices of Sovereignty: Negotiated Agreements, Jurisdiction, and Well-Being for Heiltsuk Nation." PhD dissertation, University of British Columbia.

Mather, C., and L. Fanning. 2019. "Social License and Aquaculture: Towards a Research Agenda." *Marine Policy* 99: 275–82.

McGee, G., A. Cullen, and T. Gunton. 2010. "A New Model for Sustainable Development: A Case Study of the Great Bear Rainforest Regional Plan." *Environment, Development and Sustainability* 12: 745–62.

McGregor, D. 2009. "Linking Traditional Knowledge and Environmental Practice in Ontario." *Journal of Canadian Studies* 43 (3): 69–100.

McIlwraith, T.F. 1992. *The Bella Coola Indians.* 2 vols. Toronto: University of Toronto Press.

McIver, R., I. Milewski, R. Loucks, and R. Smith. 2018. "Estimating Nitrogen Loading and Far-Field Dispersal Potential from Background Sources and Coastal Finfish Aquaculture: A Simple Framework and Case Study in Atlantic Canada." *Estuarine, Coastal and Shelf Science* 205: 46–57.

Milewski, I. 2013. "Nova Scotia Environmental Monitoring Program for Finfish Aquaculture: An Update (2006–2011)." Prepared for the Atlantic Coalition for Aquaculture Reform, Fredericton, NB.

Milewski, I., R.H. Loucks, B. Fisher, R.E. Smith, J.S.P. McCain, and H.K. Lotze. 2018. "Sea-Cage Aquaculture

Impacts Market and Berried Lobster (*Homarus americanus*) Catches." *Marine Ecology Progress Series* 598: 85–97.

Milewski, I., and R.E. Smith. 2019. "Sustainable Aquaculture in Canada: Lost in Translation." *Marine Policy* 107: 103571. https://doi.org/10.1016/j.marpol.2019.103571.

Moody, M. 2008. "Eulachon Past and Present." MSc thesis, University of British Columbia.

Moore, M.L., O. Tjornbo, E. Enfors, C. Knapp, J. Hodbod, J.A. Baggio, and D. Biggs. 2014. "Studying the Complexity of Change: Toward an Analytical Framework for Understanding Deliberate Social-Ecological Transformations." *Ecology and Society* 19 (4): 54.

Napoleon, V., and H. Friedland. 2016. "An Inside Job: Engaging with Indigenous Legal Traditions through Stories." *McGill Law Journal* 61: 725–54.

Napoleon, V., and R. Overstall. 2007. *Indigenous Laws: Some Issues, Considerations and Experiences.* Aboriginal Policy Research Consortium International (APRCI).

Newell, S., N.C. Doubleday, and Community of Chesterfield Inlet, Nunavut. 2020. "Sharing Country Food: Connecting Health, Food Security and Cultural Continuity in Chesterfield Inlet, Nunavut." *Polar Research* 39: 3755.

NSDFA (Nova Scotia Department of Fisheries and Aquaculture). 2005. *Growing Our Future: Long-Term Planning for Aquatic Farming in Nova Scotia.* Halifax: NSDFA. https://novascotia.ca/fish/documents/Growing-our-future.pdf.

–. 2007. "Environmental Monitoring Program. Level II Monitoring Results for Port Mouton Lease #0835 July 01 2007." NSDFA, Halifax.

–. 2014. "Environmental Monitoring Program. Level II Monitoring Results for Port Mouton Lease #0835 July 16 2014." NSDFA, Halifax.

–. 2015. "Transparent and Accountable: A New Approach to Regulating Aquaculture." Op-ed by K. Colwell, Minister of Fisheries and Aquaculture. https://novascotia.ca/news/release/?id=20151028003.

–. 2017. "Information for the Public." https: novascotia.ca/fish/aquaculture/ public-information/.

–. 2020. "Findings and Decision – Renewal of Ocean Trout Farm Inc. for AQ#0835." https://novascotia.ca/fish/aquaculture/public-information/public-notice/Decision_Document-AQ0835-2020.04.07.pdf.

O'Brien, K. 2012. "Global Environmental Change II: From Adaptation to Deliberate Transformation." *Progress in Human Geography* 36: 667–76.

Olsson, P., Ö. Bodin, and C. Folke. 2010. "Building Transformative Capacity for Ecosystem Stewardship in Social-Ecological Systems." In *Adaptive Capacity and Environmental Governance,* edited by D. Armitage and R. Plummer, 263–85. Berlin: Springer.

Ommer, R. 2007. *Coasts under Stress: Restructuring and Social-Ecological Health.* Montreal and Kingston: McGill-Queen's University Press.

Parsons, R., and K. Moffat. 2014. "Constructing the Meaning of Social Licence." *Social Epistemology* 28 (3–4): 340–63.

Pinkerton, E., and L. John. 2008. "Creating Local Management Legitimacy." *Marine Policy* 32 (4): 680–91.

Poteete, A.R., and E. Ostrom. 2004. "Heterogeneity, Group Size and Collective Action: The Role of Institutions in Forest Management." *Development and Change* 35: 435–61.

Sapkota, A., A.R. Sapkota, M. Kucharski, J. Burke, S. McKenzie, P. Walker, and R. Lawrence. 2008. "Aquaculture Practices and Potential Human Health Risks: Current Knowledge and Future Priorities." *Environment International* 34: 1215–26.

Schweigert, J., C. Wood, D. Hay, M. McAllister, J. Boldt, B. McCarter, et al. 2012. *Recovery Potential Assessment of Eulachon (*Thaleichthys pacificus*) in Canada.* DFO Canadian Science Advisory Secretariat Research Document 2012/098. Ottawa: Fisheries and Oceans Canada.

Shah, S.S., L. Rodina, J.M. Burt, E.J. Gregr, M. Chapman, S. Williams, N.J. Wilson, G. McDowell. 2018. "Unpacking Social Ecological Transformations: Conceptual, Ethical and Methodological Insights." *Anthropocene Review* 5 (3): 250–65.

Silver, J.J., D.K. Okamoto, D. Armitage, S.M. Alexander, C. Atleo (Kam'ayaam/Chachim'multhnii), J. Burt, R. Jones (Nang Jingwas), et al. 2022. "Fish, People, and Systems of Power: Understanding and Disrupting Feedback between Colonialism and Fisheries Science." *American Naturalist* 200 (1): 168–80. https://doi.org/10.1086/720152.

Silver, J.J., and J. Stoll. 2019. "How Do Commercial Fishing Licences Relate to Access?" *Fish and Fisheries* 20 (5): 993–1004.

Singleton, S. 2009. "Native People and Planning for Marine Protected Areas: How 'Stakeholder' Processes Fail to Address Conflicts in Complex, Real-World Environments." *Coastal Management* 37 (5): 421–40.

Stephenson, R.L., S. Paul, M. Wiber, E. Angel, A. Benson, A. Charles, O. Chouinard, et al. 2018. "Evaluating and Implementing Social-Ecological Systems: A Comprehensive Approach to Fisheries Management." *Fish and Fisheries* 19 (5): 853–73.

Stocks, A. 2016. *State of Coastal Communities in British Columbia 2016.* Vancouver: T. Buck Suzuki Environmental Foundation.

Sundberg, L., T. Ketola, E. Laanto, H. Kinnula, J.K.H. Bamford, R. Penttinen, and J. Mappes. 2016. "Intensive Aquaculture Selects for Increased Virulence and Interference Competition in Bacteria." *Proceedings of the Royal Society B* 283: 20153069.

Temper, L., M. Walter, I. Rodriguez, A. Kothari, and E. Turhan. 2018. "A Perspective on Radical Transformations to Sustainability: Resistances, Movements and Alternatives." *Sustainability Science* 13: 747–64.

Titz, A., T. Cannon, and F. Krüger. 2018. "Uncovering 'Community': Challenging an Elusive Concept in Development and Disaster Related Work." *MDPI Societies* 8 (3): 71.

Turner, N.J., F. Berkes, J. Stephenson, and J. Dick. 2013. "Blundering Intruders: Extraneous Impacts on Two Indigenous Food Systems." *Human Ecology* 41 (4): 563–74.

von der Porten, S., J. Corntassel, and D. Mucina. 2019. "Indigenous Nationhood and Herring Governance: Strategies for the Reassertion of Indigenous Authority and Inter-Indigenous Solidarity Regarding Marine Resources." *AlterNative: An International Journal of Indigenous Peoples* 15 (1): 62–74.

von der Porten, S., R. de Loë, and R. Plummer. 2015. "Research Article: Collaborative Environmental Governance and Indigenous Peoples: Recommendations for Practice." *Environmental Practice* 17 (2): 134–44. https://doi.org/10.1017/S146604661500006X.

von der Porten, S., D. Lepofsky, D. McGregor, and J.J. Silver. 2016. "Recommendations for Marine Herring Policy Change in Canada: Aligning with Indigenous Legal and Inherent Rights." *Marine Policy* 74: 68–76. https://doi.org/10.1016/j.marpol.2016.09.007.

Walker, B., C.S. Holling, S.R. Carpenter, and A. Kinzig. 2004. "Resilience, Adaptability and Transformability in Social-Ecological Systems." *Ecology and Society* 9 (2): 5. http://www.ecologyandsociety.org/vol9/iss2/art5/.

Walker, B.H., L.H. Gunderson, A.P. Kinzig, C. Folke, S.R. Carpenter, and L. Schultz. 2006. "A Handful of Heuristics and Some Propositions for Understanding Resilience in Social-Ecological Systems." *Ecology and Society* 11 (1): 13. http://www.ecologyandsociety.org/vol11/iss1/art13/.

Westley, F.R., O. Tjornbo, L. Schultz, P. Olsson, C. Folke, B. Crona, and Ö. Bodin. 2013. "A Theory of Transformative Agency in Linked Social-Ecological Systems." *Ecology and Society* 18 (3): 27. https://doi.org/10.5751/es-05072-180327.

Whitney, C.K., N.J. Bennett, N.C. Ban, E.H. Allison, D. Armitage, J.L. Blythe, J.M. Burt, et al. 2017. "Adaptive Capacity: From Assessment to Action in Coastal Social-Ecological Systems." *Ecology and Society* 22 (2): 22. https://doi.org/10.5751/ES-09325-220222.

Withers, Paul. 2019a. "Cooke Unveils Ambitious Fish Farm Expansion Plans in Nova Scotia." *CBC News,* March 29. https://www.cbc.ca/news/canada/nova-scotia/cooke-aquaculture-fish-farm-plans-expansion-1.5076206.

–. 2019b. "Global Salmon Farming Company Eyes $500M Expansion along Nova Scotia Coast." *CBC News,* April 3. https://www.cbc.ca/news/canada/nova-scotia/nova-scotia-fish-farms-mitsubishi-cermaq-1.5083342.

Wolfram, M. 2019. "Assessing Transformative Capacity for Sustainable Urban Regeneration: A Comparative Study of Three South Korean Cities." *Ambio* 48 (5): 478–93.

10

Knowledge Co-Production and Enhanced Governance Fit in Rebuilding Canada's Coastal Fisheries

Evan J. Andrews, Graham Epstein, Jennifer L. Silver, Ella-Kari Muhl, and Derek Armitage

Recovering the abundance, structure, and function of marine life is a "global grand challenge" (Duarte et al. 2020). Rebuilding fisheries reflects efforts to halt overfishing, reverse stock decline, and restore marine habitats to a state that supports population stability (Garcia et al. 2018). In Canada, interest in coastal fisheries rebuilding is gaining momentum. For example, Baum and Fuller's report (2016, 54) on the status of marine resources indicated that approximately 80% of fish stocks on the east coast and 90% of fish stocks on the West Coast are overfished or depleted. Meanwhile, a report of the Office of the Auditor General of Canada (2016), *Sustaining Canada's Major Fish Stocks,* identifies 15 fish stocks that fall within the critical zone of the Fisheries and Oceans Canada (DFO) Sustainable Fisheries Framework, which requires the development of rebuilding plans (DFO 2019b). Now, DFO is beginning to write and implement rebuilding plans that are meant to address gaps identified in the report of the Office of the Auditor General of Canada (2016) (see DFO 2020b).

Recent legislative change also has implications for fisheries rebuilding (see the text box "The policy context for fisheries rebuilding in Canada"). In 2019, the federal *Fisheries Act* was updated when Bill C-68 was passed and received royal assent.[1] This was not the first time that the *Fisheries Act* had been modified, but the latest amendments broaden the Fisheries Minister's discretion to integrate consideration of Indigenous rights, social and cultural dimensions, and stock rebuilding into decision making. It will take time for DFO to develop policy

frameworks and implement the *Fisheries Act* amendments (DFO 2020b). Expanded governance objectives have led to novel and modified policy frameworks that, in turn, require new knowledge and analytical capacities (Howlett 2009).

This chapter examines the links among coastal fisheries rebuilding, knowledge, and governance fit – defined here as the extent to which rules for the use and management of fisheries are adjusted to local social and ecological conditions (see Epstein et al. 2015). Specifically, we assess how diverse knowledge types and knowledge co-production processes can catalyze governance arrangements that better fit the challenges of fisheries rebuilding, with a particular emphasis on Atlantic and Pacific contexts. We define knowledge co-production as a process in which researchers, other knowledge holders, and knowledge users collaborate to co-create information that is "actionable in decision-making" (Mach et al. 2020, 30).

We draw upon case studies from selected Canadian coastal fisheries to identify three critical roles for knowledge in improving governance fit and addressing challenges associated with fisheries rebuilding. First, different forms of evidence (e.g., from Indigenous and local knowledge, natural and social sciences) are needed to assess and anticipate changes to fish stocks and ecosystems, and to understand their implications for sustainable livelihoods and coastal communities. Second, diverse knowledge systems and associated viewpoints can generate a greater range of "alternative futures" or scenarios that may facilitate adaptive governance in a context of change and uncertainty. Third, diverse knowledge systems provide a foundation for alternative governance processes for rebuilding because they reflect the values, interests, rights, and interests in which knowledge is embedded.

GOVERNANCE FIT AND KNOWLEDGE

Governance fit is a way to assess the extent to which actors, decisions, processes, and practices are tailored to match what is known about the complexity and dynamism of a fishery, particularly in a local context (Pittman et al. 2015). In this regard, governance reflects the systems of knowledge, principles, institutions, decision processes, and social processes through which rebuilding processes must occur (see Garcia et al. 2018; Khan and Chuenpagdee 2014). For Canadian coastal fisheries, this system includes the networks of actors (e.g., politicians, policy-makers,

community and organizational leaders, and scientists) that interact to develop and assess a suite of decisions (e.g., goals, policies, and management plans), and implement those decisions through a pattern of practices (e.g., stock assessments, management strategy evaluation [MSE], policy processes, public referrals) to control use of fish stocks and allocate benefits to different groups in society.

Knowledge is a critically important condition for governance fit (Epstein et al. 2015) and the development of strategies to rebuild coastal fisheries (Khan and Neis 2010). However, knowledge is produced by diverse sources, such as the natural and social sciences, and local and Indigenous knowledge types. These different types and sources of knowledge are often contested and shaped by the political, structural, and financial realities embedded in governance systems (Young et al. 2018). Further,

THE POLICY CONTEXT FOR FISHERIES REBUILDING IN CANADA

Policies that interpret rebuilding objectives in Canada's amended *Fisheries Act* are described in DFO's Sustainable Fisheries Framework (DFO 2019b) and operationalized through the precautionary approach and rebuilding plans, all incorporated into integrated fisheries management planning (DFO 2009, 2016). Rebuilding is specified in relation to objectives and the precautionary approach. Decisions about stocks in the critical zone (depleted) are guided by conservation objectives explicitly, whereas stocks in the cautious or healthy zones implicate diverse objectives, as implied by Section 6.2 of the *Fisheries Act*. Fisheries monitoring is guided by the principles found in the Fishery Monitoring Policy (DFO 2019a) and specified for the Pacific (DFO 2012).

Rebuilding planning is shaped by operational knowledge procedures for coastal fisheries on both coasts. Biophysical assessments, models, and key methods are assessed through the Canadian Science Advisory Secretariat (DFO 2020a). DFO regional scientists and managers communicate model assumptions, anticipated harvest rates, and potential harvest decisions (e.g., quotas, opening and closing dates) related to harvest control rules to the Minister. The Minister revises or approves the decision, a decision level that depends on the status of the stock and the potential societal implications of the stock decision (Andrews 2020).

Insights into social and decision processes can guide development and implementation of rebuilding plans with broader governance implications and opportunities. For example, stock rebuilding is described as the need to maintain healthy fish stocks and rebuild weakened fish stocks in ways that mitigate adverse "socio-economic and cultural impacts" (*Fisheries Act,* Sections 6.1[2] and 6.2[2]). Rebuilding can therefore provide recommendations to guide how ecosystem-based fisheries management (EBFM) and management strategy evaluation are implemented, given diverse governance objectives (Wetzel and Punt 2016). Further attention to governance fit holds considerable promise for guiding the social and decision-making processes that are used to rebuild fisheries (see Garcia et al. 2018). However, the promise of improved governance requires that more attention be paid to the types, forms, and application of diverse knowledge systems.

certain knowledge is prioritized over others in different decision-making settings (e.g., quantitative models favoured by Western science can be prioritized over Indigenous knowledge). This affects how operational decisions about fish stocks, and strategic decisions about how governance is arranged and how trade-offs among objectives for rebuilding, are navigated (Garcia et al. 2018).

In Canadian fisheries governance, biophysical science and some economic assessments have key roles in rebuilding. However, the use of different types of social science and local and Indigenous knowledges are most often restricted to planning committees and advisory boards (Andrews 2020). At the strategic level, it is uncertain how different knowledge systems (e.g., information, viewpoints, values, interests, and demands) are incorporated into management evaluations, rule-making systems and policy frameworks, and decision-making processes (Soomai 2017), although local and Indigenous knowledges play central roles in strategic thinking for co-managed fisheries (e.g., Snook, Cunsolo, and Morris 2018). Furthermore, recent empirical research on the incorporation of social science in fisher behaviour in Newfoundland suggests that the knowledge base can be broadened at the strategic level in DFO's Atlantic regions if current organizational barriers are removed (Andrews 2020).

Knowledge co-production can be a catalyst or mechanism for enhancing governance fit (Cash et al. 2006; Armitage et al. 2011; Mach et al. 2020). At present, Western forms of knowledge and quantitative models typically drive fisheries and marine governance processes both in Canada and globally. These governance processes and the knowledge that informs them frequently emphasize ecological and/or economic efficiency objectives, and often at aggregate scales (Armitage et al. 2019). However, place-based and Indigenous knowledge, experience, and community values are needed to make context-sensitive

Table 10.1

Governance factors that shape knowledge co-production

Factor	Description
Partnerships	This refers to the presence of opportunities for actors to work together on an issue in a way in which different forms of knowledge are integrated into aspects of a decision process to produce new and varied outcomes and ways of producing science (Wyborn et al. 2019).
Institutional relationships	This refers to the extent to which knowledge co-production can be facilitated by institutions, and/or knowledge co-production as a means to shift institutional relationships, governing power, and knowledge, and the subsequent relationships between society and science, and state and citizen (Wyborn et al. 2019).
Interactions and feedback loops	Knowledge and societies co-evolve through feedbacks. Governance capacity is needed to recognize and address feedbacks that influence the way knowledge is shaped and moulded through repeated interactions among scientists, their organizational culture, and broader society (Jasanoff 2004).
Power and equity	Whose knowledge is chosen and who decides what knowledge is used and for what purpose involve questions of power. Therefore, when knowledge producers are involved, all users need to be engaged as equally and fairly as possible, while considering cultural differences and power dynamics (Ascher, Steelman, and Healy 2010).
Authority of expertise	Experts have privileged (but often invisible) influence over the decision process. Knowledge co-production provides influence to a range of "experts" who hold different types of knowledge (Ascher, Steelman, and Healy 2010).

decisions, especially where efforts to address trade-offs among diverse objectives (cultural, economic, and ecological) require more understanding of spatial social-ecological sensitivities (Armitage et al. 2011; Norström et al. 2020).

Knowledge co-production and exchange processes are situated in broader political, social, economic, and cultural realities (Wyborn et al. 2019), and thus can be enabled or constrained by several factors and vary across different contexts (see Table 10.1). Emerging challenges, particularly in the field of marine and fisheries governance, relate to complex vertical and horizontal feedback mechanisms and cross-scale drivers (Epstein et al. 2015). To address these challenges, multiple groups of people with different forms of knowledge relative to governance issues need to collaborate to successfully address social, political, ecological, and behavioural uncertainty relative to fisheries rebuilding, especially because rebuilding efforts often involve feedback across different scales (see also Cash et al. 2006). Table 10.1 highlights five governance factors that intersect with knowledge processes and opportunities to foster governance fit.

KNOWLEDGE, GOVERNANCE, AND FIT: LESSONS FROM CANADA'S FISHERIES

We identify three main pathways through which diverse knowledges and knowledge co-production processes can catalyze governance arrangements that better fit the challenges of fisheries rebuilding: (1) developing a diverse evidence base to assess and anticipate changes associated with rebuilding; (2) fostering space for the viewpoints needed to appraise management interventions and rebuilding strategies; and (3) ensuring that diverse values, interests, rights, and perspectives are reflected in rebuilding policy.

Evidence to Assess and Anticipate Changes Associated with Rebuilding

Anticipating change and fostering rebuilding require a well-developed and interdisciplinary evidence base. A diversity of evidence can reveal opportunities to fit scientific advice and other forms of evidence to conditions of complexity and uncertainty. In Canada, there is strong indication that evidence is needed on the changing ocean and ecosystem conditions that are contributing to rapid changes in fish communities (Mullowney et al. 2014).

However, these changes are complex because they also shape patterns of use and distribution of benefits in coastal and Indigenous communities. For most marine resources, changes in the abundance and distribution of species are tracked through peer-reviewed stock assessments with the principle of maximum sustainable yield (MSY), along with key reference points to guide decisions and management interventions (Soomai 2017). Within this current approach, fish stock depletion, past and potential, points to some key opportunities to strengthen the evidence base for rebuilding (see the text box "Examples of past and potential species depletion reveal knowledge needs for rebuilding").

The "Examples" text box points to the limits of Western scientific knowledge and economic models alone for rebuilding, and the potential benefits of incorporating more diverse evidence to enhance rules for rebuilding. The Pacific herring example highlights the need to develop a better understanding of the structure and function of the marine ecosystem to explicitly manage complex interactions among ecosystem components and fisher behaviour. The Arctic example, meanwhile, highlights the potential benefits of linking local and Indigenous ecological knowledge to anticipate long-term changes, such as rapid sea-ice retreat, and understand their implications for local communities. The restructuring after the Atlantic cod collapse highlights the need for a systems approach to proactively develop potential social-ecological responses to collapsing marine resource systems, with shifting patterns of access, livelihoods, and distribution of benefits to coastal communities (Perry et al. 2011). To take these opportunities, these recommendations need to be discussed in the context of policy, particularly relating to decision rules and expertise that can evaluate and reconcile diverse knowledge types. Concerns exist in DFO about how to evaluate and reconcile local and Indigenous knowledge, as indicated in the Pacific Region fisheries management plan for northern Pacific salmon: "The challenge for resource managers is how to engage knowledge holders and how to ensure that the information can be accessed and considered in a mutually acceptable manner, by both knowledge holders, and the broader community of First Nations, stakeholders, managers, and policy makers involved in the fisheries" (DFO 2020d, 3).

EXAMPLES OF PAST AND POTENTIAL SPECIES DEPLETION REVEAL KNOWLEDGE NEEDS FOR REBUILDING

On the Pacific coast, the commercial Pacific herring (*Clupea pallasii*) fishery – which was established in the late 19th century – suffered multiple species depletions and recoveries before the onset of early modern fisheries governance in the 1970s and the subsequent emergence of more sophisticated scientific processes. The fishery is managed with five major and two minor stock areas, with a harvest control rule that predicts level of precaution to be implemented in Total Allowable Catch (TAC). This approach has had limited success, however, as three of the five major stocks have experienced persistent declines. These declines have been attributed to combinations of difficulty in modelling of the ecosystem-specific drivers, including pressures from predation (top-down) and changing food supplies (bottom-up) (Schweigert et al. 2010), and the influence of fishing behaviour, including the range of responses from fishers to new harvest rules, and the outcomes of those responses on ecosystem drivers of change (Cleary et al. 2015; Schweigert et al. 2010). Further, species depletions have been attributed to institutional arrangements for allocation that intensified impacts on fishing and tensions over access (Sporer 2001).

In the Arctic, drivers of potential species depletion are largely external to the system, e.g., climate change, which has resulted in the rapid retreat of sea ice, as harvest pressures are relatively muted compared with the Pacific and Atlantic coasts (Moore and Reeves 2018). Yet, rapid sea-ice retreat has and can continue to manifest in very uncertain ways with heterogeneous impacts on populations of benthic, groundfish, and cetacean communities and ultimately the Arctic communities whose livelihoods and lifeways depend on these species (Moore and Reeves 2018).

In the Atlantic, massive restructuring occurred following the Atlantic northern cod collapse, with a general shift toward the exploitation of invertebrates. In Newfoundland and Labrador, this pressure was placed mostly on northern shrimp (*Pandalus borealis*) and snow crab (*Chionoecetes opilio*) (Mullowney et al. 2014). Now both stocks are declining, with corresponding cuts to quotas and catches affecting the livelihoods of harvesters and the communities where they reside. The long-term effects of these changes are largely unknown because of difficulties in modelling interactions among resources through food webs and uncertainty over the influence of change across value chains, including the influence of fluctuating prices on various activities in the system. For example, high prices can mitigate the impacts of lower catches and encourage fishers to remain in the fishery, but those prices do little for volume, which is a driver of employment in processing.

Viewpoints for Appraising Management Interventions and Strategies

Practices that engage multiple viewpoints on alternative rebuilding scenarios and their impacts are needed. Policy-making for rebuilding invariably involves considerable uncertainty regarding the nature and likelihood of alternative future states, and is often articulated through several different scenarios of change (Wetzel and Punt 2016; Teh and Sumaila 2020). A key aspect of policy, then, is the assessment of ecosystem change with multiple scenarios of social-ecological feedback and corresponding indicators to facilitate monitoring, evaluation, and intervention (Wetzel and Punt 2016; Teh and Sumaila 2020). Fit can be improved when scenarios and their policy implications are developed in collaboration with stakeholders with different views of strategies and potential social-ecological outcomes (Österblom et al. 2013).

Stephenson and colleagues (2019) suggest that the scope of policies is currently broad enough to realize multiple objectives through the *Fisheries Act,* but that there is a problem of implementation for rebuilding. In particular, there is a lack of *ex ante* prioritization concerning how management systems may anticipate and

FACILITATING VIEWPOINTS ON THE NORTHERN SHRIMP FISHERY OF NEWFOUNDLAND AND LABRADOR

The collapse and subsequent closure of Atlantic cod fisheries in Newfoundland and Labrador was followed soon after by a relatively rapid rise in shellfish populations, including northern shrimp (*Pandalus borealis*) (Lilly, Parsons, and Kulka 2000). Now northern shrimp stocks are declining rapidly (Le Corre et al. 2019), with corresponding impacts on livelihoods and lifeways in coastal communities. Environmental change has had considerable socio-cultural, political, and economic consequences (e.g., outmigration, school closures, and community closures) in Newfoundland and Labrador, although impacts have varied widely across different communities (Perry et al. 2011). Yet once again new opportunities for fish harvesting are emerging, including an anticipated boom in Acadian redfish (*Sebastes fasciatus*) (DFO 2020e). It is critical that we learn from past and ongoing experiences with species depletions across communities in Newfoundland and Labrador about rebuilding marine systems, as well as the knowledge, policies, and decision making that are needed to understand and address rapid change (Kahn and Chuenpagdee 2014).

In Newfoundland and Labrador, growing shrimp stocks presented an unexpected opportunity for government stakeholders to address high levels of unemployment and support economic development. Fishers who had been affected by the closure of the cod fishery could obtain a licence if they chose to gear up for shrimp (Foley, Mather, and Neis 2015). For a while, times were good as shrimp stocks continued to expand, supporting both the large offshore and smaller inshore fleets as well as special allocations to community organizations to support social and economic development (Foley and Mather 2016). Nonetheless, shrimp stocks have been on the decline as part of a more general shift in the species composition of the North Atlantic ecosystem (DFO 2018), contributing to political conflicts over the allocation of fishing rights in the shrimp fishery (Foley and Mather 2018) and once again threatening the well-being of coastal communities.

While new emerging stocks such as Acadian redfish present an opportunity for rebuilding these fish-harvesting systems, the case of the northern shrimp fishery in Newfoundland and Labrador offers several cautionary points. The first relates to the response to rapid changes in the abundance of species. The rapid growth of shrimp stocks in the mid to late 1990s gave government stakeholders an unexpected opportunity to address social and economic challenges in the province. Licences were provided freely to affected cod fishers, who were nonetheless required to invest heavily in the boats and equipment that are required for shrimp trawling. Shrimp licence holders incurred and continue to hold considerable debt (FFAW 2019), which creates economic challenges for households and the communities in which they reside as stocks decline, and exacerbates broader social and political conflicts.

Second, there is a large and growing body of evidence suggesting that high levels of dependence on individual fisheries pose a significant threat to the well-being of fishers and coastal communities (e.g., Epstein et al. 2018). As a result, it is critically important that stakeholders approach emerging opportunities by drawing upon diverse knowledge systems to develop an understanding of long-term resource dynamics and potential future states, and viewpoints that propose strategies for addressing conflicts. Perhaps most importantly, incorporating this understanding and these viewpoints to explore future resource dynamics can reveal diversified opportunities. Further, these knowledge practices can help avoid linking the future of fishers and communities to the conditions of any individual stock and the vulnerability that comes with this high level of dependence.

address societal responses when systems show signs of weakening. We turn to an example on the Atlantic coast in the text box "Facilitating viewpoints on the northern shrimp fishery of Newfoundland and Labrador."

The text box demonstrates that one important consideration in this process is how harvesters and coastal communities interpret different scenarios and how their potential behavioural responses affect management strategy evaluation. Another consideration is how impacts from other marine activities are suspected to influence the effectiveness of management interventions, as reflected in an ecosystem-based fisheries management approach. While the precautionary approach and EBFM provide regular opportunities to manage rebuilding, MSE provides an opportunity for long-term trial-and-error learning. Chief among the challenges for MSE is to have the appropriate evidence base from which to appraise management interventions (e.g., current strategies), and new proposals that reflect alternative strategies (Wetzel and Punt 2016). Diverse evidence types can support the former, but viewpoints are necessary for the latter.

Values, Interests, Rights, and Perspectives Addressed in Policy Development

Knowledge co-production can also serve an important role in fisheries policy development by providing a mechanism for incorporating the values, interests, rights, and perspectives of different stakeholders and rights holders (Khan and Neis 2010). Knowledge in the context of fisheries rebuilding is often construed as an "objective" representation of a system and its components, linkages, and trends. However, co-production processes may allow policy-makers to address key factors that motivate the underlying behaviours that shape conservation, economic, social, and institutional outcomes (Andrews 2020). Further, particular values, interests, and perspectives reflect different commitments to involve coastal and Indigenous communities, along with other partners, in the scope and scale of decision making (Ommer 2007). As such, knowledge co-production fosters fit by ensuring that policies and their implementation address human-centred problems, activities, and goals in coastal fisheries that manifest impacts on fish stocks and ecosystems. There are lessons about how colonial histories and Indigenous values and ways of knowing can enhance

strategic decision making for rebuilding through knowledge co-production (see the text box "The Pacific clam fishery and lessons for decision making").

The text box indicates that knowledge systems and efforts to co-produce knowledge enhance our appreciation for colonial histories and understanding of Indigenous values and ways of knowing; in turn, more pluralistic objectives for any rebuilding effort may be devised. It also demonstrates how knowledge co-production can facilitate interactions among knowledge systems embedded in cultural histories and Western forms of scientific management, provide opportunities for different expertise on system change to emerge, and make it possible to engage viewpoints that influence decision making at regional scales.

Mechanisms for knowledge co-production should reflect how formal authority and informal influence are distributed among actors. Co-management arrangements provide a general model for how to structure decision making with community and industry partners. Co-management partnerships can enhance the legitimacy and accountability of partners because they "literally have to live with [the] consequences" of their decisions that reflect their values, interests, and demands (Brunner 2010, 322). New partnership models (i.e., those that go beyond conventional planning committees and advisory boards) can cultivate shared authority for co-producing and implementing policies, coordinate with multiple users, and develop monitoring and appraisals of policies and sharing of resources necessary for rebuilding.

DISCUSSION AND CONCLUSIONS

Rebuilding is a key imperative for Canada's coastal fisheries and is also a regulatory requirement. Fisheries managers are expected to return depleted stocks to a healthy status and maintain healthy stocks so that they are not depleted (see the text box "Examples of past and potential species depletion reveal knowledge needs for rebuilding"). Yet these stocks are fished by multiple fleets in different institutional contexts, and through different governance models (governmental authority, advisory partnerships, and shared authority through co-management). An important opportunity is to consider how governance can fit this complicated reality in Canadian coastal fisheries.

THE PACIFIC CLAM FISHERY AND LESSONS FOR DECISION MAKING

The phrase "when the tide is out, our table is set" is common among First Nations up and down the coast of British Columbia. Indigenous oral histories and the archaeology of this place tell us that intertidal species and spaces have been integral to trade and food security for millennia (Deur et al. 2015). Indigenous management practices, including modifying habitat to support species such as butter clams, cockles, and sea cucumbers, helped maintain abundant stocks in front of villages and at other important harvest sites (Groesbeck et al. 2014). Indigenous governance systems – refined and reinforced over generations – articulate roles and responsibilities for individuals, families, and hereditary leaders (Menzies 2010).

It is instructive to put the BC wild-harvest commercial clam fishery into conversation with Indigenous practices and governance systems from the nearshore and intertidal areas. DFO asserted management authority over clam harvesting and began keeping province-wide records of clam landings in 1951. Participation in the clam fishery was initially open and did not require a licence. Two species – butter clam (*Saxidomus giganteus*) and razor clam (*Siliqua patula*) – predominated in the early decades. In the 1970s, foreign market demand grew for smaller "steamer-sized" species such as the Manila clam (*Venerupis phlippinarum*). Thereafter, BC seafood processors prioritized the buying of Manila clams from harvesters because they sought to build and expand their reach into US and European markets.

The number of participants in the clam fishery increased rapidly through the mid-1980s (Mitchell 1995/96). The landed volume of clams peaked in 1988, then between 1988 and 1990, landings declined by an estimated 29% (Silver 2010). DFO began to reduce and stagger openings in an attempt to create continuous supply and stabilize the price that BC processors could get. It also invested more resources and scientist time in Manila clam stock assessment, and introduced individual licences that forced harvesters to select and harvest in one of several areas (Pinkerton and Silver 2011). Entry remained open in the sense that anyone could apply for a clam licence and the annual registration fee remained under Cdn$100. In 1997, however, DFO implemented a plan to limit the total number of clam licences. To be granted a licence moving forward, an individual had to have been licensed and active five out of six years between 1989 and 1994 (Mitchell 1995/96).

The clam fishery illustrates a pattern common to commercial fisheries around the world: state assertion of governance authority; emergence and exploitation of an export opportunity; rapid growth in participation and landings; and forms of "fisheries rationalization," such as licence limitation (Carothers and Chambers 2012). There are many things to be gleaned from the clam fishery example and to reflect critically upon in the context of the broader pattern (and its tendency toward stock depletion). Three are especially relevant to this chapter.

First, the example reminds us that management strategies and governance systems are often already in place and that they are usually collectively devised and practised. When state claims to governance authority emerge, they are commonly made on a jurisdiction-wide basis and in concert with the assertion that scientific monitoring, stock assessment, and decision making have to operate at that scale.

The second point is that harvest practices and management approaches are often tailored to maximizing the aggregate economic value of one or more export-oriented fisheries. This is different from thinking of commercial fisheries as locally oriented and/or regionally oriented with multiple objectives, including seasonal income for harvesters; local food security; opportunities to pass down fishing knowledge and practices to younger generations; and sharing access to the resource rather than limiting it to individuals deemed to be the most "productive."

Finally, and relevant to the question of rebuilding, the example shows that long-standing ways of doing things can and should inform new ways of thinking. Indigenous harvesting and fishing techniques that are adapted to place – including clam gardens and fish traps – are gaining a great deal of attention in British Columbia and beyond (for more, visit the Clam Garden Network at http://www.clamgarden.com). Moreover, since the late 1990s to early 2000s, DFO has been experimenting with clam licence arrangements that grant First Nations collective use and management rights to specific beaches within their territory. It has also been supporting at least one "Community Clam Management Board" on the West Coast of Vancouver Island.

We recognize that the challenge of rebuilding is too broad and multi-faceted to be addressed fully here (see Garcia et al. 2018). Rather, we have focused on the roles of knowledge in enhancing governance fit and promoting rebuilding. An emphasis on knowledge and governance supports the implementation and tracking of governance objectives for rebuilding in Canada's coastal fisheries (Stephenson et al. 2019). By applying theory to case studies and examples in coastal fisheries across Canada's coasts, we have identified and examined three roles for knowledge and the importance of knowledge co-production in support of rebuilding efforts:

- to develop diverse types of evidence to assess and anticipate changes associated with rebuilding
- to foster space for the viewpoints needed to appraise management interventions and rebuilding strategies
- to ensure that diverse values, interests, rights, and perspectives are included in rebuilding policy.

Empirical research shows how coastal fisheries rebuilding is situated within a milieu of social, ecological, economic, and political conditions (including reconciliation) that shape fish stock depletion. Based on these conditions, the research argues, rebuilding solutions need to tailored, or "fit," to the local context (Khan and Neis 2010; Khan and Chuenpagdee 2014). Indeed, Ommer (2007) identifies a "misalignment" between the scope of goals for governance and the scope of challenges and burdens experienced by coastal and Indigenous communities in Canada. She argues that "blockages, rigidities, and wrong turns in decisions" have resulted in the "misalignment or asymmetry in the scale of situations, interests, knowledge, and/or power between those making key decisions concerning the exploitation of key resources and those exposed to the consequences of these decisions" (28). This example highlights how knowledge is a key component of advancing alignment of governance in rapidly changing fisheries (434).

Fish stocks and ecosystems are changing rapidly, indicating a need for enhanced knowledge and knowledge co-production processes to understand and respond to these changes. The case studies presented in the previous section demonstrate how knowledge can be co-produced to enhance governance fit at operational and strategic levels. Key insights include the promotion of multiple knowledge types in advancing ecosystem-based science across regions about fish stock change and corresponding changes in fishing pressure with biophysical sciences, social sciences, and local and Indigenous knowledges (see the text box "Examples of past and potential species depletion reveal knowledge needs for rebuilding"). Our examples highlight how diverse evidence can foster an understanding of complex change processes and reveal the types of rules likely to be effective in navigating change (see Epstein et al. 2015). Examples from fisheries across all three coasts reveal that knowledge about change includes the implications to livelihoods and coastal communities from changing fish stocks and management strategies for rebuilding.

In addition to monitoring changes to the fish stocks, our example of the northern shrimp fishery of Newfoundland and Labrador (see also Foley and Mather 2018) indicates the importance of facilitating diverse viewpoints on different management strategies. In this context, allocation and access decisions benefit from such diverse viewpoints on existing management strategies and on the proposal of alternative management strategies (see the text box "Facilitating viewpoints on the northern shrimp fishery of Newfoundland and Labrador"). Diverse viewpoints can integrate insights from long-term trial-and-error processes for rebuilding and contribute to participants' feelings of autonomy regarding decision making that affects their fishery (see Weeks et al. 2014). The case of the Newfoundland northern shrimp fishery reveals that viewpoints are also important in addressing and mitigating conflict as fish stocks decline, extending the potential for understanding implications of different management strategies in, for example, an MSE approach.

At the strategic level, opportunities exist to support knowledge co-production for rebuilding through further development and implementation of Canada's amended *Fisheries Act* (Stephenson et al. 2019). Lessons for decision making from a Pacific clam fishery indicate that there are opportunities to foster knowledge co-production by learning from traditional harvesting and managing practices (see the text box "The Pacific clam fishery and lessons for decision making"). Furthermore, the story of this fishery shows how those practices are deeply embedded in a social, institutional, and historical context that differs markedly from the interests and values that guide most fisheries management processes in Canada.

Knowledge co-production can be promoted through existing co-management or delegation agreements in which authority is shared for developing, assessing, and monitoring rebuilding strategies that complement a precautionary approach. There are co-management arrangements with Indigenous fisheries and some non-Indigenous fisheries in the Atlantic – for example, with the Torngat Fish Producers Co-operative in Labrador (Snook, Cunsolo, and Morris 2018) – that may be better situated to advance knowledge co-production than other partnerships, and especially where access, standing, and influence of management decisions are unclear or inconsistent across regions and management areas (see Andrews 2020). For most of the commercial stocks, knowledge is shared through industry participation in stock assessment peer review processes, annual management decision consultations, and integrated fisheries management planning committees (DFO 2016, 2020a). Empirical analyses are needed to understand the extent to which these partnerships can manifest the three roles for knowledge to anticipate and address change in Canadian coastal fisheries.

Table 10.2

Contextual factors for knowledge co-production

Factor	Pacific	Atlantic	Knowledge challenges
Partnerships	Various, including knowledge in co-management with shared authority; industry partnerships	Various, including knowledge in co-management with shared authority for some fisheries, and consultative and advisory roles for representatives for others	There are differing opportunities for knowledge and knowledge systems. Consultative and advisory roles limit the scope of influence or autonomy for values, practices, and interests related to alternatives in the governing systems.
Institutional relationships	Communal licensing and individual transferable licences	Communal licensing, owner-operator model (inshore and midshore), and enterprise allocations (offshore)	In Pacific, corporate ownership limits actors that can provide knowledge and viewpoints. In Atlantic, representation shapes who can provide knowledge, with knowledge in Newfoundland and Labrador shaped through a strong union presence.
Interactions and feedback loops	Interactions restricted to authority provided in partnerships (e.g., advisory and consultative versus co-management)	Interactions restricted to authority provided in partnerships (e.g., advisory and consultative versus co-management)	Broader understanding of change to livelihoods and communities and viewpoints on management strategies and new proposals are limited in current approach.
Power and equity	Exclusion of actors, e.g., non-corporate licence holders	No representation for actors in Newfoundland and Labrador	Diverse viewpoints are constrained by partnership structures that limit power for some industry and community actors. Knowledge is limited about how to mitigate conflicts.
Authority of expertise	DFO science remains the key authority for expertise on rebuilding	DFO science along with Northwest Atlantic Fisheries Organization (NAFO) in some regions are the key authority for expertise on rebuilding	Knowledge for rebuilding is needed beyond the precautionary approach and licensing implications. Expertise is needed for broader understanding of social and decision processes needed to advance rebuilding.

The development of mechanisms for knowledge co-production also needs to reflect differences in the institutional contexts of Canada's fisheries (Ommer 2007). Access and allocation decisions for Atlantic and Pacific coastal fisheries are shaped through different licensing models, patterns of ownership, and extent of influence in management decisions provided to industry. The examples and cases presented in this chapter shed light on certain important contextual factors along the Pacific and Atlantic coasts that influence the nature of challenges for knowledge co-production (Table 10.2). The focus here primarily on the Pacific and Atlantic coast draws attention to knowledge co-production in a commercial fishing context, with high levels of fishing pressure and diverse fleets shaped by different values, interests, and rights seeking long-term well-being through fishing. This is an ongoing situation that is distinct from fisheries, fishing, and harvesting of marine mammals off the Arctic coast, which includes large regional differences in fishing in eastern and western Arctic coastal areas, lower levels of fishing pressure overall, and distinct concerns in Indigenous communities about climate change impacts (see the text box "Examples of past and potential species depletion reveal knowledge needs for rebuilding"). Future research can continue to explore the three roles for knowledge in different regional settings along the Arctic coast by revealing and drawing on contextual factors shaping knowledge co-production.

Important questions remain as to how knowledge co-production processes can catalyze governance arrangements that better fit the challenges of fisheries rebuilding. Several potentially valuable insights emerge from our analysis. First, diverse types of knowledge need to be interpreted, evaluated, and ultimately incorporated in Canadian coastal fisheries governance. In fact, there is an emerging opportunity to incorporate diverse types of knowledge through the development of interdisciplinary indicators of change across marine, community, and governance aspects of ecosystem-based management of coastal fisheries. Second, recent research has proposed different methods that can be used to capture alternative ways of knowing and thinking about change, based on narratives and stories (Lowery et al. 2020), typologies of fishers, their behaviour and corresponding pressures on fish stocks (Andrews 2020), and assessment of inter-

actions that shape change processes (Khan and Chuenpagdee 2014). Third, there is a need for consistent strategic-level policy direction at the national level that envisions diverse viewpoints as a core component of rebuilding, and allows for different actors' perspectives, including community and industry partners and those outside of the consultative and advisory processes. Both evidence and viewpoints can be envisioned in the context of knowledge co-production. Further, an emphasis is needed on the legal and policy opportunities to advance knowledge co-production, and the governing systems to facilitate the incorporation of values, cultural practices, and interests about alternative social and decision processes to advance rebuilding. If we hope to rebuild fisheries and recover the abundance of marine life in line with the "global grand challenge," we require diverse types of knowledge to better fit governance to contexts of change and uncertainty.

NOTE

1 *Fisheries Act,* RSC 1985, c F-14.

REFERENCES

Andrews, E. 2020. "Fisher Behaviour and Its Implications for the Governability of Canada's Atlantic Inshore Fisheries." PhD dissertation, University of Waterloo.

Armitage, D., F. Berkes, A. Dale, E. Kocho-Schellenberg, and E. Patton. 2011. "Co-Management and the Co-Production of Knowledge: Learning to Adapt in Canada's Arctic." *Global Environmental Change* 21 (3): 995–1004.

Armitage, D.R., D.K. Okamoto, J.J. Silver, T.B. Francis, P.S. Levin, A.E. Punt, I.P. Davies, et al. 2019. "Integrating Governance and Quantitative Evaluation of Resource Management Strategies to Improve Social and Ecological Outcomes." *BioScience* 69 (7): 523–32.

Ascher, W., T. Steelman, and R.G. Healy. 2010. *Knowledge and Environmental Policy: Re-Imagining the Boundaries of Science and Politics.* Cambridge, MA: MIT Press.

Baum, J.K., and S.D. Fuller. 2016. *Canada's Marine Fisheries: Status, Recovery Potential and Pathways to Success.* Toronto: Oceana Canada.

Brunner, R.D. 2010. "Adaptive Governance as a Reform Strategy." *Policy Sciences* 43 (4): 301–41.

Carothers, C., and C. Chambers. 2012. "Fisheries Privatization and the Remaking of Fishery Systems." *Environment and Society* 3 (1): 39–59.

Cash, D.W., W.N. Adger, F. Berkes, P. Garden, L. Lebel, P. Olsson, L. Pritchard, and O. Young. 2006. "Scale and Cross-Scale Dynamics: Governance and Information in a Multilevel World." *Ecology and Society* 11 (2): 8.

Cleary, J., A. Benson, N. Taylor, and S. Cox. 2015. *Stock Assessment and Management Advice for BC Pacific Herring: 2015 Status and 2016 Forecast.* DFO Canadian Science Advisory Secretariat Science Response 2015/038. Ottawa: Fisheries and Oceans Canada. https://www.researchgate.net/publication/288670300_Stock_Assessment_and_Management_Advice_for_BC_Pacific_Herring_2015_Status_and_2016_Forecast.

DFO (Fisheries and Oceans Canada). 2009. "A Fishery Decision-Making Framework for Incorporating the Precautionary Approach." https://www.dfo-mpo.gc.ca/reports-rapports/regs/sff-cpd/precaution-eng.htm.

– 2012. *Strategic Framework for Fishery Monitoring and Catch Reporting in the Pacific Fisheries.* Ottawa: Fisheries and Oceans Canada. https://www.pac.dfo-mpo.gc.ca/fm-gp/docs/framework-monitoring-cadre-surveillance-eng.html.

– 2016. "Policy and Operational Framework for Integrated Management of Estuarine, Coastal, and Marine Environments in Canada." https://www.dfo-mpo.gc.ca/oceans/publications/cosframework-cadresoc/page01-eng.html.

– 2018. "Northern Shrimp and Striped Shrimp – Shrimp Fishing Areas 0, 1, 4–7, the Eastern and Western Assessment Zones, and North Atlantic Fisheries Organization (NAFO) Division 3M." https://www.dfo-mpo.gc.ca/fisheries-peches/ifmp-gmp/shrimp-crevette/shrimp-crevette-2018-002-eng.html.

– 2019a. "Fishery Monitoring Policy." https://www.dfo-mpo.gc.ca/reports-rapports/regs/sff-cpd/fishery-monitoring-surveillance-des-peches-eng.htm.

–. 2019b. "Sustainable Fisheries Framework." https://www.dfo-mpo.gc.ca/reports-rapports/regs/sff-cpd/overview-cadre-eng.htm.

–. 2020a. "Canadian Science Advisory Secretariat (CSAS)." https://www.dfo-mpo.gc.ca/csas-sccs/index-eng.htm.

–. 2020b. "Departmental Plan 2020–2021." https://www.dfo-mpo.gc.ca/rpp/2020-21/fin-eng.html.

–. 2020c. "Integrated Fisheries Management Plans." https://www.dfo-mpo.gc.ca/fisheries-peches/ifmp-gmp/index-eng.html.

–. 2020d. "Integrated Fisheries Management Plan June 1, 2020 – May 31, 2021 Salmon Northern BC." https://waves-vagues.dfo-mpo.gc.ca/library-bibliotheque/40881398.pdf.

–. 2020e. *Redfish (Sebastes fasciatus and S. mentella) Stocks Assessment in Units 1 and 2 in 2019.* DFO Canadian Science Advisory Secretariat Science Advisory Report 2020/019. Ottawa: Fisheries and Oceans Canada. https://www.dfo-mpo.gc.ca/csas-sccs/Publications/SAR-AS/2020/2020_019-eng.html.

Deur, D., A. Dick, K. Recalma-Clutesi, and N.J. Turner. 2015. "Kwakwaka'wakw 'Clam Gardens': Motive and Agency in Traditional Northwest Coast Mariculture." *Human Ecology* 43 (2): 201–12.

Duarte, C.M., S. Agusti, E. Barbier, G.L. Britten, J.C. Castilla, J. Gattuso, R.L. Fulweiler, et al. 2020. "Rebuilding Marine Life." *Nature* 580 (7801): 39–51.

Epstein, G., E. Andrews, D. Armitage, P. Foley, J. Pittman, and R. Brushett. 2018. "Human Dimensions of Ecosystem-Based Management: Lessons in Managing Trade-Offs from the Northern Shrimp Fishery in Northern Peninsula, Newfoundland." *Marine Policy* 97: 10–17.

Epstein, G., J. Pittman, S.M. Alexander, S. Berdej, T. Dyck, U. Kreitmair, K.J. Rathwell, S. Villamayor-Tomas, J. Vogt, and D. Armitage. 2015. "Institutional Fit and the Sustainability of Social-Ecological Systems." *Current Opinion in Environmental Sustainability* 14: 34–40.

FFAW (Food, Fish and Allied Workers Union). 2019. "LPR Review for Northern Shrimp Step in Right Direction for Inshore Fishery." https://ffaw.ca/the-latest/news/lrp-review-northern-shrimp-step-right-direction-inshore-fishery/.

Foley, P., and C. Mather. 2016. "Making Space for Community Use Rights: Insights from 'Community Economies' in Newfoundland and Labrador." *Society and Natural Resources* 29 (8): 965–80.

–. 2018. "Ocean Grabbing, Terraqueous Territoriality and Social Development." *Territory, Politics, Governance* 7 (3): 1–19.

Foley, P., C. Mather, and B. Neis. 2015. "Governing Enclosure for Coastal Communities: Social Embeddedness in a Canadian Shrimp Fishery." *Marine Policy* 61: 390–400.

Garcia, S.M., Y. Ye, J. Rice, and A. Charles. 2018. *Rebuilding of Marine Fisheries. Part 1: Global Review.* FAO Fisheries and Aquaculture Technical Paper No. 630/1. Rome: Food and Agriculture Organization.

Groesbeck, A.S., K. Rowell, D. Lepofsky, and A.K. Salomon. 2014. "Ancient Clam Gardens Increased Shellfish Production: Adaptive Strategies from the Past Can Inform Food Security Today." *PLOS One* 9 (3): e91235.

Howlett, M. 2009. "Policy Analytical Capacity and Evidence-Based Policy-Making: Lessons from Canada." *Canadian Public Administration* 52 (2): 153–72.

Jasanoff, S., ed. 2004. *States of Knowledge: The Co-Production of Science and the Social Order.* London: Routledge.

Khan, A., and R. Chuenpagdee. 2014. "An Interactive Governance and Fish Chain Approach to Fisheries Rebuilding: A Case Study of the Northern Gulf Cod in Eastern Canada." *Ambio* 43 (5): 600–13.

Khan, A., and B. Neis. 2010. "The Rebuilding Imperative in Fisheries: Clumsy Solutions for a Wicked Problem?" *Progress in Oceanography* 87 (1): 347–56.

Le Corre, N., P. Pepin, G. Han, Z. Ma, and P.V. Snelgrove. 2019. "Assessing Connectivity Patterns among Management Units of the Newfoundland and Labrador Shrimp Population." *Fisheries Oceanography* 28: 183–202.

Lilly, G., D. Parsons, and D. Kulka. 2000. "Was the Increase in Shrimp Biomass on the Northeast Newfoundland Shelf a Consequence of a Release in Predation Pressure from Cod?" *Journal of Northwest Atlantic Fishery Science* 27: 45–62.

Lowery, B., J. Dagevos, R. Chuenpagdee, and K. Vodden. 2020. "Storytelling for Sustainable Development in Rural Communities: An Alternative Approach." *Sustainable Development* 28: 1813–26.

Mach, K.J., M.C. Lemos, A.M. Meadow, C. Wyborn, N. Klenk, J.C. Arnott, N.M. Ardoin, et al. 2020. "Actionable Knowledge and the Art of Engagement." *Current Opinion in Environmental Sustainability* 42: 30–37.

Menzies, C.R. 2010. "Dm sibilhaa'nm da laxyuubm Gitxaała: Picking Abalone in Gitxaała Territory." *Human Organization* 69 (3): 213–20.

Mitchell, D.A. 1995/96. "Management of the Intertidal Clam Resource: A British Columbia Experiment in Limited Entry and Local Participation." *Western Geography* 5/6: 45–73.

Moore, S.E., and R.R. Reeves. 2018. "Tracking Arctic Marine Mammal Resilience in an Era of Rapid Ecosystem Alteration." *PLOS One Biology* 16 (10): e2006708.

Mullowney, D.R., E.G. Dawe, E.B. Colbourne, and G.A. Rose. 2014. "A Review of Factors Contributing to the Decline of Newfoundland and Labrador Snow Crab (*Chionoecetes opilio*)." *Reviews in Fish Biology and Fisheries* 24 (2): 639–57.

Norström, A.V., C. Cvitanovic, M.F. Löf, S. West, C. Wyborn, P. Balvanera, A.T. Bednarek, et al. 2020. "Principles for Knowledge Co-Production in Sustainability Research." *Nature Sustainability* 3: 1–9 .

Office of the Auditor General of Canada. 2016. *Sustaining Canada's Major Fish Stocks.* Ottawa: Office of the Auditor General of Canada.

Ommer, R.E. 2007. *Coasts under Stress: Restructuring and Social-Ecological Health.* Montreal and Kingston: McGill-Queen's University Press.

Österblom, H., A. Merrie, M. Metian, W.J. Boonstra, T. Blenckner, J.R. Watson, R.R. Rykaczewski, et al. 2013. "Modeling Social-Ecological Scenarios in Marine Systems." *BioScience* 63 (9): 735–44.

Perry, I.R., R.E. Ommer, M. Barange, S. Jentoft, B. Neis, and U.R. Sumaila. 2011. "Marine Social-Ecological Responses to Environmental Change and the Impacts of Globalization." *Fish and Fisheries* 12: 427–50.

Pinkerton, E., and J. Silver. 2011. "Cadastralizing or Coordinating the Clam Commons: Can Competing Community and Government Visions of Wild and Farmed Fisheries Be Reconciled?" *Marine Policy* 35 (1): 63–72.

Pittman, J., D. Armitage, S. Alexander, D. Campbell, and M. Alleyne. 2015. "Governance Fit for Climate Change in a Caribbean Coastal-Marine Context." *Marine Policy* 51: 486–98.

Schweigert, J.F., J.L. Boldt, L. Flostrand, and J.S. Cleary. 2010. "A Review of Factors Limiting Recovery of Pacific Herring Stocks in Canada." *ICES Journal of Marine Science* 67 (9): 1903–13.

Silver, J.J. 2010. "Seeking Certainty: A Political Ecology of Shellfish Aquaculture Expansion on the West Coast of Vancouver Island, British Columbia." PhD dissertation, University of British Columbia.

Snook, J., A. Cunsolo, and R. Morris, R. 2018. "A Half Century in the Making: Governing Commercial Fisheries through Indigenous Marine Co-Management and the Torngat Joint Fisheries Board." In *Arctic Marine Resource Governance and Development,* edited by N. Vestergaard, B.A. Kaiser, L. Fernandez, and J. Nymand Larsen, 53–73. Cham, Switzerland: Springer International Publishing.

Soomai, S. 2017. "The Science-Policy Interface in Fisheries Management: Insights about the Influence of Organizational Structure and Culture on Information Pathways." *Marine Policy* 81: 53–63.

Sporer, C. 2001. "Initial Allocation of Transferable Fishing Quotas in Canada's Pacific Marine Fisheries." In *Case Studies in the Allocation of Transferable Quota Rights in Fisheries.* FAO Technical Paper 411, 222–303. Rome: Food and Agriculture Organization.

Stephenson, R.L., M. Wiber, P. Stacey, E. Angel, A. Benson, A. Charles, O. Chouinard, et al. 2019. "Integrating Diverse Objectives for Sustainable Fisheries in Canada." *Canadian Journal of Fisheries and Aquatic Sciences* 76: 480–96.

Teh, L.S.L., and U.R. Sumaila. 2020. "Assessing Potential Economic Benefits from Rebuilding Depleted Fish Stocks." *Ocean and Coastal Management* 195: 105289.

Weeks, R., P.M. Aliño, S. Atkinson, P. Beldia, A. Binson, W.L. Campos, R. Djohani, A.L. Green, R. Hamilton, and V. Horigue. 2014. "Developing Marine Protected Area Networks in the Coral Triangle: Good Practices for Expanding the Coral Triangle Marine Protected Area System." *Coastal Management* 42: 183–205.

Wetzel, C.R., and A.E. Punt. 2016. "The Impact of Alternative Rebuilding Strategies to Rebuild Overfished Stocks." *ICES Journal of Marine Science* 73 (9): 2190–2207.

Wyborn, C., A. Datta, J. Montana, M. Ryan, P. Leith, B. Chaffin, C. Miller, and L. Van Kerkhoff. 2019. "Co-Producing Sustainability: Reordering the Governance of Science, Policy, and Practice." *Annual Review of Environment and Resources* 44: 319–46.

Young, O.R., D.G. Webster, M.E. Cox, J. Raakjær, L.O. Blaxekjær, N. Einarsson, R.A. Virginia, et al. 2018. "Moving beyond Panaceas in Fisheries Governance." *Proceedings of the National Academy of Sciences* 115 (35): 9065–73.

11 Bright Spots in Integrated Management of Canada's Oceans and Coasts

Sondra Eger, Natalie C. Ban, Chelsea Boaler, Irene Brueckner-Irwin, Simon Courtenay, Graham Epstein, Carie Hoover, Jim McIsaac, and Robert L. Stephenson

There is a large and growing consensus that environmental management must move toward an integrated approach to govern human activities in coastal and marine areas (e.g., Armitage, Charles, and Berkes 2017; Bennett et al. 2018; Stephenson et al. 2019, 2021). Current centralized and sector-based approaches have proven inadequate in addressing the externalities of individual sectors, resulting in a range of unintended consequences from social and biophysical feedbacks (Pikitch et al. 2004; Cury et al. 2011). Furthermore, sector-based management often neglects the consideration and informed decision making of trade-offs across different actors, activities, and objectives, as well as an understanding of the cumulative effects of human activities. Despite the explicit need for integrated management (IM), it remains relatively rare in Canada. This chapter highlights progress in changing governance to respond to changing oceans and access to resources as seen in previous chapters. Additionally, it contributes to the theory and practice of IM by examining the conditions in which IM approaches have emerged in Canada's coastal and marine areas.

The literature on IM has evolved over the past 30 years to encompass diverse approaches toward understanding and classifying different types of management integration (Underdal 1980; Cicin-Sain 1993; Thia-Eng 1993; Cicin-Sain and Knecht 1998; Lockwood et al. 2010; Eger et al. 2021). For instance, Dickinson, Rutherford, and Gunton (2010) suggest that integration involves linkages and coordination across different sectors (horizontal), levels of government (vertical), spaces, and sustainable development objectives. In general, IM refers to a comprehensive, collaborative, and coordinated approach toward planning and managing human activities in ecosystems. The intention of IM is to move beyond traditional sector-based approaches to coordinate the governance of human activities in socio-ecological systems and to provide mechanisms to reconcile diverse interests and values (e.g., Sainsbury and Sumaila 2003; Foley et al. 2010; Foster, Haward, and Coffen-Smout 2005; Stephenson et al. 2019).

Institutions, which reflect the rules, norms, and conventions that provide a guide for human interactions and decision making, are especially salient (Kooiman and Bavinck 2005; Armitage et al. 2009). Environmental governance in Canada is largely organized around a traditional sector-based approach in which different departments and branches of government are responsible for different activities. Such an approach tends to neglect externalities and cumulative effects of human activities, and lacks mechanisms for adequate, transparent, equitable consideration of trade-offs and evaluation of cumulative performance (Stephenson et al. 2019). In contrast, IM requires a "whole-of-government approach" to effectively integrate management across different departments and levels (e.g., municipal, provincial, federal, Indigenous). Examples of IM include the New South Wales Marine Estate (Brooks et al. 2020), Great Barrier Reef Marine Park (Commonwealth of Australia 2018), the Barents Sea (Olsen et al. 2016), and the North Sea (Interdepartmental Directors' Consultative Committee North Sea 2015).

Canada was once recognized globally as a leader in promoting IM, largely through innovative legislation such as the *Oceans Act*[1] and accompanying strategies (see Government of Canada 2002; Bailey et al. 2016). The *Oceans Act* called for the development of IM plans and recognized the need for a framework to guide future activities and decisions. However, while the *Oceans Act* may have enabled managers to advance IM plans, it lacked accompanying regulations or prescriptions to guide and compel action, potentially contributing to the limited implementation of IM plans. Although five pilot projects were initiated to implement IM in Large Ocean Management Areas (LOMAs) in the Pacific, Atlantic, and Arctic coasts, these efforts have largely stalled (e.g., McCuaig and Herbert 2013; Ricketts and Hildebrand 2011; Bailey et al. 2016) and only the Pacific North Coast Integrated Management Area (PNCIMA) and the Beaufort Sea Integrated Management Area (BSIMA) initiatives have endorsed plans that are active. Additionally, several smaller-scale IM planning activities (including the Southwest New Brunswick Marine Resources Planning Initiative; see Parlee and Wiber 2018) have been only partially developed or short-lived. IM has been difficult to implement as a result of pre-existing limits on the scope of management authority and insufficient consideration of governance systems and processes (Stephenson et al. 2019, 2023; Vince et al. 2015).

Although Canada has had very limited success in developing and implementing a comprehensive integrated management system, this chapter presents a few "bright spots" in IM. These are cases where one or more of the central features of IM have been established. These include efforts to advance our understanding of the structure and function of coastal systems and/or address the broader effects of one or more human activities on these systems. The cases were selected by drawing upon the research of members of the OceanCanada Partnership (OCP) working along Canada's three coasts. We identify initiatives that have led to, or are anticipated to lead to, positive social, economic, and/or ecological outcomes, with a particular emphasis on the maintenance of ecosystem integrity and community well-being. We focus on cases that range considerably in terms of scope from single-sector management actions that address broader ecosystem effects to attempts that fully integrate management

within a defined area. This chapter contributes to the theory and practice of IM by evaluating these cases using a recently developed theoretical framework (Stephenson et al. 2019) to understand the conditions in which IM might emerge and persist. In the process, we also generate insights to support further integration of human activities in Canada's coastal areas.

A LENS FOR EVALUATION OF INTEGRATED MANAGEMENT

There is a vast literature outlining a wide range of conditions that might facilitate or undermine integration within coastal and marine systems (Dickinson et al. 2010; O'Boyle and Jamieson 2006; Carvalho and Fidélis 2013; Brooks and Fairfull 2017; Stojanovic, Ballinger, and Lalwani 2004). Stephenson and colleagues (2019) synthesized this literature to develop an analytical framework for evaluating key features that support the implementation of IM through five critical phases of the management process. In what follows, we first introduce the case studies and continue by applying this framework to analyze each phase of the implementation process and the extent to which nine hypothesized key features of IM are present in each case.

CASE STUDIES

To take stock of a variety of experiences with IM across different contexts, we chose a diverse set of case studies from Atlantic Canada to the Arctic and Pacific coasts (Figure 11.1). We focused on cases where we have had personal experience and knowledge, and that represent several aspects of the evolution of IM in Canada at present. The case studies are not comprehensive, but rather illustrate various approaches taken and how they are contributing to IM in Canada at the time of this exercise in 2021.

Hawke Box, Atlantic

The Hawke Box in Newfoundland and Labrador (Shrimp Fishing Area [SFA] 6/Northwest Atlantic Fisheries Organization [NAFO] Division 2J) was first established in 2002 as a pilot project at the request of crab fishers, to address concerns about the impacts of shrimp trawling on crab populations. After consultation with actors (considered here to be stakeholders and rights holders) and a review of available information, Fisheries and Oceans

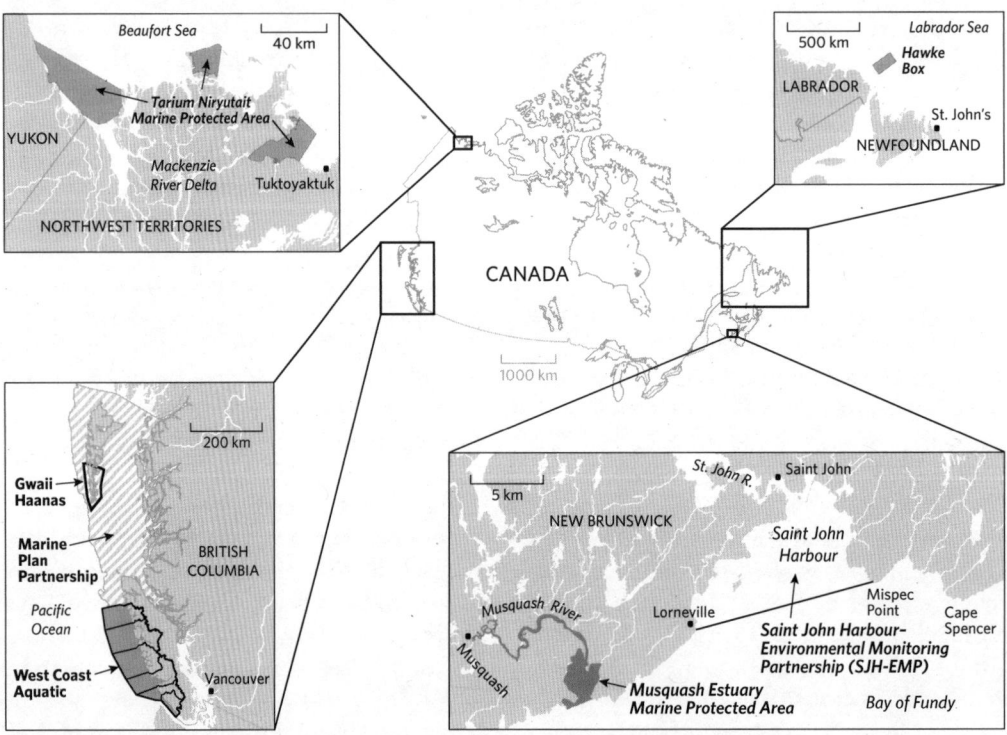

Figure 11.1 Map showing the various case study areas across Canada's three coasts.

Canada (DFO) implemented a 400-square-mile "no-trawl/no-gillnetting" study area in NAFO Division 2J. The study area was later expanded to cover 6,400 km² in July 2003. The Hawke Box was designed with the goal of protecting crab from potentially harmful interactions with shrimp fishing gear and to provide a refuge for aggregations of cod and other species of conservation concern. Crab fishers played a key role in pressuring DFO for the change, with support from their union, the Fish, Food and Allied Workers Union (FFAW-Unifor). Relatively large and growing shrimp quotas at the time may have reduced opposition from shrimp fishers. Although debates remain regarding the impacts of the closure on crab (Dawe et al. 2007; Mullowney et al. 2012, 2014; Nguyen et al. 2014), crab landings have declined more slowly within the closure than outside, and several species, including cod, have increased in abundance and biomass within the closure (Kincaid and Rose 2017). Crab fishers in the area continue to strongly support the closure

(Kincaid and Rose 2014), while shrimp fishers who were most affected by the closure would prefer to see the area opened to trawling (Epstein et al. 2018). Monitoring and evaluation have been in place in some capacity since 2002.

Saint John Harbour–Environmental Monitoring Partnership, Atlantic

The Saint John Harbour–Environmental Monitoring Partnership (SJH-EMP) was the pilot project for the Canadian Watershed Research Consortium (CWRC), led by the Canadian Water Network (CWN) from 2010 to 2016, which acknowledged the need for integrated monitoring nationally and locally. The objective of the CWRC was to develop monitoring frameworks in support of cumulative effects assessment at the geographic scale of regions or watersheds. The cumulative effects assessment was deemed necessary because a separate review of environmental assessments for recent major projects revealed that information collected for other purposes (e.g., graduate student theses) was not being used. The intent was to coordinate and integrate monitoring to (1) identify information needs of decision makers in and around the harbour, (2) encourage data collection and information sharing to reduce total costs, and (3) identify and address gaps in monitoring. The SJH-EMP partnership for conducting, regulating, or benefiting from monitoring in the Saint John Harbour region dissolved when CWN funding expired, but some of the partners have continued to fund and carry out elements of the recommended monitoring program (Canadian Water Network 2016; Guerin et al. 2023).

Musquash Estuary Marine Protected Area, Atlantic

The Musquash Estuary Marine Protected Area (MEMPA) was formally proposed to DFO by a local environmental nongovernmental organization (ENGO) and fishing association in 1998 and was officially designated as an *Oceans Act* Marine Protected Area (MPA) in 2006. The MEMPA is 7 km² and covers a salt marsh–estuary system in the Bay of Fundy (DFO 2017). Although the local community championed designation of the MEMPA in order to steward and protect the marsh from potential future industrial development, some community members felt the designation process was too slow on the part of DFO due to shifting government priorities and staff turnover (Dehens and Fanning 2018). The MEMPA is managed by DFO with guidance from the multi-actor Musquash Advisory Committee. The management of the intertidal area was transferred to DFO through an agreement between the provincial and federal governments (DFO 2017). Government, First Nations, NGOs, and community groups are all represented on the advisory committee. Actors have generally expressed support for the MEMPA, and significant monitoring and research efforts continue in the area, where evaluation and adaptive management are ongoing (Cooper, Jones, and Blanchard 2023; DFO 2017, 2023).

Tarium Niryutait Marine Protected Area, Arctic

The Beaufort Sea Beluga Management Plan, established by DFO in 1991, led to the establishment of the Beaufort Sea Integrated Management Planning Initiative (BSIMPI) working group, who assessed the potential for an MPA taking into consideration the renewal of oil and gas exploration in the early 2000s and the Inuvialuit concerns for protection of beluga whales and their habitats (Fisheries Joint Management Committee 2013). This integrated initiative ultimately led to the designation of Canada's first Arctic MPA, the Tarium Niryutait Marine Protected Area (TNMPA) in 2010. It covers approximately 1,800 km², across three individual sub-areas of the Mackenzie River Delta and Estuary in the Inuvialuit Settlement Region (ISR). The TNMPA was co-created by Inuvialuit and DFO, with involvement from other rights holders (communities, hunters, and trappers), co-management boards, and private industry. It is currently co-managed by the Fisheries Joint Management Committee (FJMC), an Inuvialuit co-management board, and DFO, with an established governance structure, management plan, and monitoring plan (DFO 2013a, 2013b). The TNMPA has been fully implemented, with evaluation of the MPA monitoring activities against the conservation objective being assessed every five years. The first five-year review was completed in 2018 and evaluated the progress of co-management toward the goals and conservation objective of the MPA (Brewster et al. 2021).

Marine Plan Partnership, North Pacific

The Marine Plan Partnership (MaPP) (http://mappocean.org; Diggon et al. 2019) is a collaboration between the Province of British Columbia and 16 First Nations governments. The aim of MaPP is to conduct and implement coastal and marine planning on the North Pacific coast in order to inform decisions regarding the sustainable economic development and stewardship of British Columbia's coastal marine environment. Provincial and First Nations governments had previously been working with the federal government in the Pacific North Coast Integrated Management Area initiative (PNCIMA), a fully integrated marine planning initiative under Canada's *Oceans Act,* when the federal government unilaterally reduced the scope of the initiative to a strategic planning process. The province and First Nations, coming off a successful collaboration on the Great Bear Rainforest on land, wanted to expand this collaboration to the adjacent marine area and continue with the original idea from PNCIMA of integrated marine planning. The partners initiated MaPP through a letter of intent (LOI). MaPP began in November 2011, following the LOI, with the creation of five multi-sectoral actor advisories – one regional and four subregional tables – to inform plan development and decision making. Four subregional plans were drafted by governments and actors from 2011 to 2015, and were endorsed by the respective governments in 2015. Since then, annual work plans and implementation reports have been drafted for each subregion, with an ecosystem-based management (EBM) monitoring framework still in development. Multi-sectoral advisories meet regularly and are more engaged in some subregions than others. The four subregional plans will undergo review in 2024, after ten years of implementation.

West Coast Aquatic, Pacific

The West Coast Aquatic (WCA) Governance Board is a forum for coastal communities and other persons and bodies affected by aquatic resource management. Its aim is to explore how affected parties can participate more fully with governments in all aspects of the IM of aquatic resources in the management area, including decision making. The vision is for the aquatic resources of the West Coast of Vancouver Island to be managed by people working together for the benefit of current and future generations of aquatic resources, people, and communities. Canada, British Columbia, the Nuu-chah-nulth Tribal Council, local governments, and eight marine sectors initially ratified the WCA Terms of Reference in 2001, and amended them in 2013. In 2012, the WCA Coastal Strategy – which outlined shared values, vision, goals and objectives, and priority action items – was finalized by the board and endorsed by all governments and actor sectors. This integrated plan, however, did not include an implementation agreement or funding for implementation. External foundation funding was secured for marine spatial planning, which morphed into the Marine Ecosystem Reference Guide in 2018. As of 2023, with no core funding for integrated marine management, the West Coast Aquatic Management Association, which served as the secretariat for the board, has pivoted to focus on designing and facilitating collaborative initiatives in natural resource management.

Gwaii Haanas, Pacific

A combination of legal structures was used to create the Gwaii Haanas National Park Reserve, National Marine Conservation Area Reserve, and Haida Heritage Site, collectively referred to as Gwaii Haanas. The Council of the Haida Nation (CHN) used a Haida Heritage Site designation to integrate social and cultural values into managing the area in 1985. In 1993, the CHN and Canada agreed to jointly manage the terrestrial area through a new Archipelago Management Board (AMB) under a Parks Canada designation. For Canada, the *Canada National Parks Act* in 2000 and the *Canada National Marine Conservation Areas (NMCA) Act* in 2002 were used to bring the terrestrial and marine areas into the co-management arrangement.[2] Regardless of the Canadian designations, the CHN does not consider the area a park. Under the *NMCA Act,* Marine Protected Areas are designated as either "fully protected" or "sustainable use." The sustainable-use designation allows tourism and commercial fishing uses. The Land-Sea-People Plan, an IM plan, was finalized and endorsed in 2018 by the CHN and Canada. The AMB, made up of six representatives, provides leadership and is supported by a diverse actor advisory. The two governments provide resources and funding to implement the plan.

PHASES OF INTEGRATED MANAGEMENT

We apply the IM framework of Stephenson and colleagues (2019) to trace the development of IM in each of the cases by examining (1) the preconditions or drivers that motivated actors to pursue new management relationships among actors, (2) the process by which IM links have been explored and developed, and (3) factors that might have enabled or undermined the formation of new institutional arrangements. While the specific drivers of change differed across the seven cases, there were also several similarities in terms of the high-level features that compelled actors to consider institutional change. These include general recognition of the need to integrate management of multiple activities, to protect valued components of the ecosystem, and to increase collaboration among governmental (municipal, provincial, and federal governments, and Indigenous authorities/communities) and nongovernmental actors in management.

The institutional design and (re)arrangement for each case study is quite different. Here, we consider institutions broadly to include governance models, structures (formal and informal), and relationships (e.g., statutory requirements, approvals, and adaptive elements) (Stephenson et al. 2019). IM typically involves a complex network of formal and informal institutions that influence how actors interact and make decisions within a system (e.g., multi-actor groups, committees, advisory boards). Some scholars have highlighted the need for a transformation in marine governance to integrate policies across different sectors, resources, and interests (Pikitch et al. 2004; Bailey et al. 2016), whereas others have argued that more effective management of individual species or sectors may be sufficient and has a greater likelihood of success (Hilborn 2011). In the Atlantic cases, these initiatives were largely undertaken within existing institutional arrangements, taking a stepwise approach to address gaps and externalities on a case-by-case basis. For example, in the case of Saint John Harbour, indicators were developed to monitor trends in ecosystem conditions (simple measure of complex systems); and the Hawke Box closure, which places restrictions on two technologies (gillnets and trawling), provides benefits to both target and nontarget species. In contrast, cases on the West Coast often involved a reorganization of institutional structures and arrangements to integrate the perspectives of diverse actors and interests in the governance of human activities in coastal areas. West Coast Aquatic is an IM collaboration between Canada, British Columbia, Nuu-chah-nulth, and local governments to participate in shared marine decision making; the Marine Plan Partnership is a collaboration between the Province of British Columbia and 16 First Nations governments, covering all areas of provincial marine jurisdiction; and Gwaii Haanas involves co-management between Canada (through Parks Canada and DFO) and the CHN, covering Canadian and Haida jurisdiction. Similar to Gwaii Haanas, the Tarium Niryutait Marine Protected Area is an equal partnership established by Inuvialuit (Inuit) and the federal government, and operates under a specific co-management structure.

Faced with a desire for change, the transition toward a more integrated approach in each of the cases was facilitated by the presence of enabling factors, including leadership, pressure applied to both state agencies and elected officials, and adequate funding. In fact, a lack of sustained funding was highlighted as a core barrier to building upon early progress in the case of both Saint John Harbour and West Coast Aquatic. In general, it appears that programs with existing resources are more likely to persist, whereas those that rely upon short-term grants are likely to struggle to maintain momentum through intermittent funding cycles. Further, the lack of clear and consistent political direction that balances short-term government priorities with a longer-term vision has created uncertainty concerning the value of contributing to, and investing in, processes to develop IM systems. In the case of West Coast Aquatic, federal authorities have limited traction and implementation of the integrated initiative.

Although our synthesis identifies relatively few barriers, likely due to the sampling of "bright spots," the literature has identified several factors that appear to limit success with regard to IM (Rothwell and VanderZwaag 2006; Leslie and McLeod 2007; Marshak et al. 2017; Kong, Yang, and Sun 2018; Kelly, Ellis, and Flannery 2019). For example, through a study in the Bay of Fundy in Atlantic Canada, Eger and Courtenay (2021) identify five main impediments to operationalizing IM relating to (1) capacity to sustain interventions; (2) commitment from legal authorities; (3) engagement of diverse actor groups; (4) vertical integration of policies; and (5) informal structures that facilitate horizontal integration. Although the Bay of Fundy is unique in its jurisdictional complexity, these impediments are likely relevant to other coastal areas in Canada.

KEY FEATURES OF INTEGRATED MANAGEMENT

Table 11.1 provides a general evaluation of the cases by considering the extent to which nine key features of IM (Stephenson et al. 2019) are present in each case. Features were coded by first developing a common understanding of the definitions for each feature, and then turning to one or more authors with knowledge of each case to determine whether a feature is absent (0), present (1), or partially present (0.5).

The case studies reveal several enabling factors and a range of different processes through which actors have been able to achieve some measure of IM in these case studies. Actors in each case study *recognized the need* for some type of integration to advance understanding and management of the coastal area. West Coast Aquatic, for instance, was specifically organized around the development of an IM plan, while the Marine Plan Partnership initiative was developed to create marine plans, although

Table 11.1

Evaluations of seven case studies across the nine features of integrated management (summary)

Case study title	Hawke Box	Saint John Harbour	Musquash Estuary MPA	Tarium Niryutait MPA
1. Recognition of need for IM	Partially present – recognized by crab fishers who felt that northern shrimp fishing had a negative impact on snow crab health and abundance	Present/partially present – development of a common integrated monitoring framework; no recognition of IM benefits	Present –recognition of need to work across sectors and actors for an integrated approach to marine protection	Present – recognized desire for the MPA from Inuvialuit and federal government, an other groups
2. Shared vision for IM	Partially present – clear intention to achieve identified objectives	Present – consensus achieved on information needs for cumulative effects assessment decision making and vision	Partially present – shared objective to create MPA but various motivations among groups of participants	Present – long-term vision c protecting Inuvialuit harves rights and limiting oil and g in MPA management plans
3. Legal frameworks to support IM	Partially present – some existing mechanisms (integrated fisheries management plan) and ad hoc operational framework, but nothing formalized	Partially present – monitoring priorities tied to regulatory requirements for some participants (i.e., cumulative effects assessment) but no other governance pieces	Present – rooted in the *Oceans Act* and has a governance structure that includes an advisory committee	Present – multiple legal fram works to support IM, includi the *Inuvialuit Final Agreem* and *Oceans Act* as well as o informal working groups
4. Effective participation	Partially present – some groups were not adequately engaged (e.g., shrimp fishers were engaged initially and then not engaged throughout)	Present – excellent engagement, governance, and meeting structures during funding period	Present – bottom-up participatory process with ongoing involvement of many groups	Present – multiple structure and pathways for actors to b involved
5. Integrate common objectives across sectors	Partially present – ecological and conservation objectives are a priority, but social considerations are lacking	Partially present – partners shared objectives for a healthy harbour but no direct socio-economic aspects were considered	Partially present – ecological integration is a priority and occurred across sectors, but less emphasis on economic and socio-cultural objectives	Present – 82 comprehensive indicators in monitoring pla conservation objectives, but to monitor them are still bei developed and/or revised
6. Consider trade-offs and cumulative effects of activities	Partially present – clear winners and losers, trade-offs not addressed, some cumulative effects considered on crab habitat	Present –monitoring framework specifically designed to measure ecological cumulative effects, but no explicit consideration of trade-offs	Partially present – cumulative effects and trade-offs between ecological and socio-economic factors discussed, but no formal framework or decision-making tool in place to systematically address cumulative effects or trade-offs	Trade-offs present; cumulati effects partially present – tra offs with regard to exclusion oil and gas for harvesting are were considered, though cu tive effects evolved to becor a priority in the region and t MPA later on
7. Process flexible to changing conditions	Partially present – an experiment that evolved and became formalized	Present – semi-annual meetings to adjust work plan through actor feedback	Partially present – perception of inflexibility of MPAs despite principles of adaptive management within MPA plan and ongoing opportunities for diverse input	Present/partially present – management plans, governa structure, and monitoring pl were flexible; spatial bounda and MPA regulations less so
8. Process for ongoing review and refinement	Present – monitoring to ensure progress	Present – opportunities for review, monitoring at semi-annual meetings (during funding period)	Present – management and monitoring reviews every five years	Present – process for review was outlined from the onset the MPA to be reviewed ever five years
9. Effective resources, capacity, and tools	Present – uses existing tools and resources	Present – well resourced by Canadian Water Network during research phase but not after as it lacked a champion/coordinator	Partly present – existing resources but lengthy designation process and DFO staff turnover hindered initial designation progress	Present/partially present – sufficient funding, but needs more high-level Inuvialuit an local capacity and better too for implementing managem
Total score (x/9)	5–6	6.5–8	6.5	7–9

Note: Scores are the sum of values for each feature and case study, rated as 0 (absent), 0.5 (partially present), or 1 (fully present).

Marine Plan Partnership	West Coast Aquatic	Gwaii Haanas	Total score (x/7)
Present – coastal and marine planning through collaborative decision making was the purpose of the initiative	Present – IM of aquatic resources is the stated purpose	Present – sustainable use and mixed objectives acknowledged across the differing acts and laws	6–6.5
Present – shared vision of Province of British Columbia and First Nations for land and marine spatial planning	Present – stated as the purpose in terms or reference and Coastal Strategy	Present – shared by federal government, CHN, and actors	6
Partially present – created through a letter of intent, working within various existing legislation but unable to implement actions within federal authority	Partially present – terms of reference and Coastal Strategy were supported by all levels of government, albeit weakly by DFO	Present –multiple agreements and plans to support initiative; integrated objectives framed by the Land-Sea-People Plan (2018)	5
Partially present – designed to have effective participatory structures, but participation varied (e.g., lack of federal government participation and inclusion of First Nations vision)	Partially present – strong participation from some groups but not all (e.g., Province and Parks Canada have been strong, DFO has been poor)	Present – strong participation throughout (e.g., Parks Canada and CHN participation has been strong; DFO support is growing)	5.5
Present – promote ecosystem-based fisheries management (EBFM) and integrate across activities that mirror the diverse EBFM objectives of PNCIMA	Present – clearly articulated integrated objectives within terms of reference and Coastal Strategy that are supported across governments and actors	Present – goals of integrated governance and sustainable use are common across sectors	5.5–6
Present – plans explicitly acknowledge both cumulative effects and trade-offs with a framework developed to link with an existing EBFM framework	Partially present – considered indirectly through conflicts and compatibilities that led to the development of priorities (e.g., InVEST analysis tool)	Present/ partially present – trade-offs were considered and cumulative effects considered somewhat; explicit goal to reduce negative cumulative effects	4.5–5.5
Present – required flexibility during change from PNCIMA to MaPP, where participation has evolved	Present – Governance Board has added sectors from 2013 terms of reference review/renewal; staffing has changed	Partially present – 10-year review window	5–5.5
Present – adaptive approach to implementation, performance, and monitoring	Partially present- terms of reference were amended in 2013	Partially present – Land-Sea-People Plan reviewed and renewed every 10 years	6
Partially present – relatively well supported with resources from external foundations and ongoing resources from Province and First Nations	Partially present/absent – no secure funding, initial government support for governance, funding has decreased over past decade	Present – clear and constant funding, mainly from Parks Canada, local resources, CHN, and actors; growing engagement from DFO staff	4–5
7.5	5.5–6.5	7.5–8	

without federal involvement. Gwaii Haanas was supported by diverse instruments to enable co-management on land and in the sea, including agreements and plans. The Hawke Box closure and Saint John Harbour, meanwhile, were less ambitious in terms of the scope of activities, focusing on the integration of fisheries plans and monitoring activities, respectively. In the MPA cases, the recognized need was for greater integration of objectives in order to effectively protect the environment. In the case of the Tarium Niryutait MPA, the need to preserve traditional harvesting grounds in the face of potential increases in oil and gas exploration led to an integrated approach that included industry in talks regarding the establishment of an MPA.

The case studies further demonstrated the importance of developing a *common vision* regarding the high-level objectives of IM. Indeed, participants in the vast majority of cases either held or developed a common, though not necessarily unanimous, vision of the objectives they meant to achieve. For the Marine Plan Partnership, West Coast Aquatic, Tarium Niryutait MPA, and Gwaii Haanas, the vision was for full IM, whereas for the other cases, the vision incorporated only a portion of IM (i.e., greater marine protection through MPAs, more comprehensive and collaborative monitoring, and resolution of perceived conflict between fisheries activities).

Many of the cases have managed to make progress toward IM in the absence of supportive *legal frameworks*. While the MPAs (Musquash Estuary MPA, Tarium Niryutait MPA, and Gwaii Haanas) were supported by the *Oceans Act* or the *NMCA Act,* other cases tended to not have any direct or specific legal framework of support. Instead, actors in these cases pursued innovative strategies, including ad hoc adaptation of existing legislation, formal letters of intent, and informal agreements. For instance, the Saint John Harbour partnership emerged as a response to uncertainty among decision makers across Canada regarding their ability to fulfill legal obligations to monitor cumulative effects. Also, the Tarium Niryutait MPA applied the *Oceans Act* in a novel manner to protect an area for the traditional harvesting of a culturally significant species.

Participation and collaboration were strong features of nearly all the cases. While all cases included diverse actors, not all initiatives implemented participatory

processes effectively. The nature of participation in the Hawke Box, for example, has been criticized by some for neglecting the perspectives of shrimp fishers and Indigenous communities. With regard to the Tarium Niryutait MPA, the establishment of the *Inuvialuit Final Agreement* (IFA) outlined four key pillars, one of which is protecting wildlife for future generations. The co-management structure under the IFA alleviated the need for a new management protocol to establish and promote many of the same values DFO is used to supporting in the establishment of MPAs.

Although each case resulted in the development of a more *diverse set of objectives,* the Hawke Box, Saint John Harbour, and Musquash Estuary MPA tended to focus more on ecological outcomes, while the Marine Plan Partnership, West Coast Aquatic, Gwaii Haanas, and Tarium Niryutait MPA included social (including economic and cultural) aspects. Furthermore, consideration of trade-offs and cumulative effects was not consistently addressed across cases. For example, the Hawke Box generated clear winners (crab fishers) and losers (shrimp fishers) and generally failed to explicitly acknowledge this result and develop strategies for mediating potential impacts. Meanwhile, the MPA cases – Musquash Estuary MPA and Tarium Niryutait MPA – specifically considered *cumulative ecological impacts* in at least some respects. The Marine Plan Partnership and Saint John Harbour clearly established monitoring that would directly inform cumulative impact evaluation. With regard to acknowledging trade-offs, all cases at least partially considered them. For example, in the Tarium Niryutait MPA, cumulative impacts were not directly considered initially, whereas *trade-offs* were considered extensively. Additionally, a combination of both social and ecological indicators suggests that many dimensions are important to the long-term success of the MPA, although measuring these regularly remains a challenge.

Flexibility was a key feature of many of the reviewed cases, providing opportunities to adjust strategies based upon input from participatory and evaluation processes. However, most cases are in the relatively early stages of implementation and important questions remain as to whether this flexibility will be retained to allow for further refinement based upon the experiences of actors. The MPAs and Marine Plan Partnership explicitly require performance reviews. Finally, the evaluation of *effective resourcing* of knowledge, capacity, and tools is somewhat mixed. While all cases had some additional resourcing to support integration activities, some cases (e.g., the MPAs and Marine Plan Partnership) were provided with relatively substantial resources, while others struggled to maintain momentum and administrative duties as seed funding expired.

LESSONS FROM CANADA'S INTEGRATED MANAGEMENT "BRIGHT SPOTS"

Using a common framework (Stephenson et al. 2019), we sought to generate insights into IM in Canada through the evaluation of diverse case studies. In summary, two cases (Tarium Niryutait MPA and Musquash Estuary MPA) aligned with federal commitments to establish MPAs, while two others emerged from the evolution of fisheries management (Hawke Box) and monitoring (Saint John Harbour) processes. The cases in British Columbia, such as the Marine Plan Partnership, which is an offshoot of the previous IM Large Ocean Management Area initiative (PNCIMA), and Gwaii Haanas, are the most fully fleshed out IM initiatives of the group of cases. In the Pacific and Arctic initiatives, Indigenous Peoples were core and essential partners, highlighting an important avenue for supporting the transition toward IM in Canada, as well as broader reconciliation efforts.

Several cases demonstrated the beneficial function of innovative institutional structures, such as multi-actor partnerships for management and decision making. With regard to IM elements, there was a general recognition of the need and a vision for IM, as well as the need for a process of ongoing review and refinement across the seven cases. The weakest elements were consideration of trade-offs and cumulative effects, legal frameworks to support IM, and resourcing. Furthermore, although the cases pursued multiple objectives, these tended to focus on ecological aspects and neglected the social (including economic and cultural) aspects.

Each of the cases discussed in this chapter has pursued a different pathway toward better outcomes and provided insights that could potentially be scaled up or replicated. We acknowledge that these bright spots do not encompass all features of IM, but they all demonstrate some level of successful integration that provides valuable insights for the design and implementation of systems to better support IM. Here, we summarize the key lessons that emerged from the case studies, and highlight ways to promote IM in the future.

Nongovernmental actors appear to play a critical role in IM, specifically, in applying pressure on existing management actors and elected officials for changes in policies and programs, and in investing time and resources to ensure implementation. The introduction of the Hawke Box in Newfoundland and Labrador, for instance, resulted from the coupling of sustained actor pressure from crab fishers and muted opposition from shrimp fishers during a period of relative shrimp abundance (Kincaid and Rose 2014; Epstein et al. 2018). Second, participants in planning processes generally recognized the need for some level of IM, and held a shared vision with respect to the high-level objectives they meant to achieve through integration, although more specific objectives and motivations differed in some cases. For example, general support for the concept of the Musquash Estuary MPA did not reflect varying motivations and assumptions behind creation of the protected area, which could lead to disconnected perceptions about the effectiveness of the MPA and IM moving forward. Third, all case studies embody a change in conventional sector-based management. In each case, nongovernmental actors have been involved in supporting, developing, and modifying governance arrangements required for integration.

Making progress toward IM is not necessarily a linear process. The set of key features we used as a lens is seen as a checklist of the major requirements that are likely to result in successful IM. Many factors, however, must align to allow IM to be fully implemented. It requires time and effort to build trust and collaborations among diverse groups and to shape processes to include the key features. Further, attempts to achieve integration require long-term investments of resources and capacity from multiple actor groups and authorities to address the information requirements of integrated management and negotiate trade-offs among diverse interests. In some of the case studies, there had been considerable preparatory work by participants who had been working together for many years. For example, the ecosystem-based management framework in the Marine Plan Partnership involved over six years of discussion and negotiation to achieve consensus, and yet debates still remain concerning the indicators that will be used to track progress. Therefore, IM requires modification of sector participation and management institutions, and some consistency of process and leadership that extend beyond typical terms of elected political office (i.e., three to five years) and shifting governmental priorities. We suggest that, in addition to the technical challenge of how to implement IM, there is the reality of a social challenge of dealing with the question of who will be better or worse off under given IM scenarios.

There are many ways to approach IM. The cross-case comparison of the seven OCP-related cases shows that despite a lack of regulatory direction for IM specifically, some success has been seen in Canada. Canada attempted to introduce IM in Large Ocean Management Areas following introduction of the *Oceans Act,* but after several years of planning with actors, progress on integrated management is acknowledged to be very slow, with most of the pilot integrated management plans (e.g., LOMAs) never adopted or implemented (Office of the Auditor General of Canada 2005). Perhaps Canada lacked the traction needed because of the nature of the *Oceans Act* as enabling rather than regulatory legislation. Although we have seen some success from our bright spots toward IM through mechanisms catalyzed from both inside and outside of the *Oceans Act,* in order to make more substantial progress toward the integrated management of activities in Canada, a change from enabling regulatory legislation may be a way forward. Doing so may reduce flexibility of the process, however.

The transition toward IM of Canada's coastal regions has inherently been a legislative process, subject to election cycles and political influence. It involves a multiplicity of actors and competing interests, and a diversity of potential venues in which those actors might adopt a range of strategies to either support or resist changes to the status quo (Imperial 1999; Epstein et al. 2018). Some previous efforts to achieve integration in coastal areas have floundered as debates have emerged concerning the adequacy of indicators and scientific models to inform IM. There remains a need for administrative reform, including the following considerations: potential impacts on livelihoods; availability and adequacy of indicators and scientific models; suitability of existing management structures for implementing integrated management; lack of

appropriate processes, methods, and guidance for negotiating trade-offs among different actors and objectives; and complexities with weaving knowledge systems together for a comprehensive understanding of the management region (Murawski 2007; Patrick and Link 2015; Bailey et al. 2016; Sander 2018; Stephenson et al. 2023).

Bright spots should be supported and expanded, even as transformative approaches are also prioritized. Conceptually, there is a broad spectrum of potential approaches to the implementation of IM, ranging from large-scale replacement of existing management with a different paradigm to an incremental modification of single-sector management toward a more integrated approach. This debate exists within a broad range of disciplines. For example, Duinker and Greig (2006, 159) argue that "we need revolution ... not evolution" with regard to how we undertake cumulative effects assessment (see also Sinclair, Doelle, and Duinker 2017). Similar arguments are made in EBM and Marine Spatial Planning (Kelly, Ellis, and Flannery 2019). It is important to note, however, that large transformational changes in public policy are relatively rare, requiring the confluence of several enabling factors (Baumgartner and Jones 2009; Baumgartner et al. 2009), and that even small incremental changes can yield benefits. We acknowledge here that existing sector-based institutional arrangements are likely to be inadequate, and that there is value in transformative approaches to governance for aiding in the successful implementation of IM. We do, however, accept that this transformation may be incremental in some cases. An example suggested by Stephenson and colleagues (2019) is to revise sector-based plans to incorporate a broader spectrum of objectives and elicit a whole-of-government approach. We point to a broad range of beneficial actions, demonstrated within these studies, that are evidence of positive incremental change within existing systems toward IM in Canada.

CONCLUSION

Integrated management is used to understand progress in Canada toward aligning and adapting governance to changing oceans and changing access needs. It implies the integration of activities (sectors) in a comprehensive approach to govern social-ecological systems. While there is growing recognition of the need for IM, there have been few examples to date of successful implementation in Canada. Using a common lens of the essential features of integrated management, we looked across seven cases connected to the OceanCanada Partnership to better understand what has been successful with regard to achieving some degree of integration within coastal and marine systems that has led to positive social and ecological outputs. We found progress toward integration in development of fisheries management planning and monitoring, in MPA development, and in more explicit IM planning initiatives.

In evaluating these innovative case studies (or bright spots), we discussed the diverse features that are relevant to IM implementation. The seven key features – recognition of need; shared vision; legal framework; effective participation; evaluation of trade-offs and cumulative effects; flexibility to adapt, review, and refine; and effective resourcing – are important to the achievement, evaluation, and comparison of IM. Therefore, IM should link (integrate) planning, decision making, and management arrangements across sectors in a unified framework to enable a more comprehensive view of social-ecological sustainability and the consideration of cumulative effects and trade-offs (Stephenson et al. 2019). In particular, the use of the framework facilitated the following insights for amplifying efforts toward IM and changing governance in Canada:

- Nongovernmental actors appear to play a critical role in IM.
- Making progress toward IM is not necessarily a linear process.
- There are many ways to approach IM.
- The transition toward IM of Canada's coastal regions has inherently been a legislative process, subject to election cycles and political influence.
- Bright spots should be supported and expanded, even as transformative approaches are also prioritized.

As we transition from and/or transform the governance systems for coastal and marine systems (see Chapter 9), bright spots such as those highlighted in this chapter provide hope and encouragement that the changes needed can happen in real contexts. Although different in scope and history, these bright spots demonstrate a general movement toward IM and help draw our attention to the innovative practices and governance processes that can be replicated and scaled up to other contexts. Looking forward, we suggest research agendas to address the following areas of investigation to further facilitate change in governance: (1) how to raise the priority within government to address the lack of accompanying regulations to guide and compel action on IM in Canada; (2) circumstances in which change needs to be transformative versus incremental; and (3) examples of governance models or mechanisms that allow for a whole-of-government approach (in the context of participatory governance and shared responsibility, perhaps better termed a "whole-of-society approach") in support of IM, given the ever-changing nature of coastal and marine systems and dynamics.

ACKNOWLEDGMENTS

We would like to acknowledge all the contributors of case study content, especially Paul Macnab and Connie Blakeston for their extensive knowledge of the case studies presented here. We thank Rashid Sumaila and Derek Armitage for encouraging us and giving us the opportunity to reflect on and share experiences of integration taking place on all three coasts. Finally, we appreciate the feedback we received from two anonymous peer reviewers.

NOTES

1 *Oceans Act,* SC 1996, c 31.
2 *Canada National Parks Act,* SC 2000, c 32; *Canada National Marine Conservation Areas Act,* SC 2002, c 18.

REFERENCES

Armitage, D., T. Charles, and F. Berkes, eds. 2017. *Governing the Coastal Commons: Communities, Resilience and Transformations.* London: Routledge/Earthscan.

Armitage, D., R. Plummer, F. Berkes, R.I. Arthur, A. Charles, I. Davidson-Hunt, A. Diduck, et al. 2009. "Adaptive Co-Management for Social-Ecological Complexity." *Frontiers in Ecology and the Environment* 7 (2): 95–102.

Bailey, M., B. Favaro, S.P. Otto, A. Charles, R. Devillers, A. Metaxas, P. Tyedmers, et al. 2016. "Canada at a Crossroad: The Imperative for Realigning Ocean Policy with Ocean Science." *Marine Policy* 63: 53–60.

Baumgartner, F.R., C. Breunig, C. Green-Pedersen, B.D. Jones, P.B. Mortensen, M. Nuytemans, and S. Walgrave. 2009. "Punctuated Equilibrium in Comparative

Perspective." *American Journal of Political Science* 53 (3): 603–20.

Baumgartner, F.R., and B.D. Jones. 2009. *Agendas and Instability in American Politics.* 2nd ed. Chicago: University of Chicago Press.

Bennett, N.J., M. Kaplan-Hallam, G. Augustine, N. Ban, D. Belhabib, I. Brueckner-Irwin, A. Charles, et al. 2018. "Coastal and Indigenous Community Access to Marine Resources and the Ocean: A Policy Imperative for Canada." *Marine Policy* 87: 186–93.

Brewster, J.D., K. Hansen-Craik, L.A. Harwood, and C. Blakeston. 2021. *State of the Tarium Niryutait Marine Protected Areas (TNMPA) Report: Inventory of Monitoring from 2010–2016.* Canadian Manuscript Report of Fisheries and Aquatic Sciences 3301. https://publications.gc.ca/collections/collection_2021/mpo-dfo/Fs97-6-3301-eng.pdf.

Brooks, K., K. Barclay, R. Grafton, and N. Gollan. 2020. "Transforming Coastal and Marine Management: Deliberative Democracy and Integrated Management in New South Wales, Australia." *Marine Policy* 139: 104053. https://doi.org/10.1016/j.marpol.2020.104053.

Brooks, K., and S. Fairfull. 2017. "Managing the NSW Coastal Zone: Restructuring Governance for Inclusive Development." *Ocean and Coastal Management* 150: 62–72.

Canadian Water Network. 2016. *Synthesis of Learnings of the Canadian Watershed Research Consortium.* Waterloo, ON: Canadian Water Network. https://cwn-rce.ca/report/synthesis-of-learnings-of-the-canadian-watershed-research-consortium/.

Carvalho, T., and T. Fidélis. 2013. "The Relevance of Governance Models for Estuary Management Plans." *Land Use Policy* 34: 134–45.

Cicin-Sain, B. 1993. "Sustainable Development and Integrated Coastal Management." *Ocean and Coastal Management* 21 (1): 11–43.

Cicin-Sain, B., and R.W. Knecht. 1998. *Integrated Coastal and Ocean Management: Concepts and Practices.* Washington, DC: Island Press.

Commonwealth of Australia. 2018. "Reef 2050 Long-Term Sustainability Plan – July 2018." https://www.dcceew.gov.au/parks-heritage/great-barrier-reef/publications/reef-2050-long-term-sustainability-plan-2018.

Cooper, J.A., O. Jones, and M. Blanchard. 2023. "Review of Baseline Monitoring within the Musquash Estuary Marine Protected Area." Canadian Science Advisory Secretariat Research Document 2023/028. https://waves-vagues.dfo-mpo.gc.ca/library-bibliotheque/41117980.pdf.

Cury, P.M., I.L. Boyd, S. Bonhommeau, T. Anker-Nilssen, R.J. Crawford, R.W. Furness, J.A. Mills, E.J. Murphy, H. Österblom, and M. Paleczny. 2011. "Global Seabird Response to Forage Fish Depletion – One-Third for the Birds." *Science* 334: 1703–6.

Dawe, E.G., K.D. Gilkinson, S.J. Walsh, W. Hickey, D.R. Mullowney, D.C. Orr, and R.N. Forward. 2007. *A Study of the Effect of Trawling in the Newfoundland and Labrador Northern Shrimp (*Pandalus borealis*) Fishery on Mortality and Damage to Snow Crab (*Chionoecetes opilio*).* Canadian Technical Report of Fisheries and Aquatic Sciences 2752. St. John's: Fisheries and Oceans Canada.

Dehens, L.A., and L.M. Fanning. 2018. "What Counts in Making Marine Protected Areas (MPAs) Count? The Role of Legitimacy in MPA Success in Canada." *Ecological Indicators* 86: 45–57.

DFO (Fisheries and Oceans Canada). 2013a. *Tarium Niryutait: Marine Protected Area Management Plan.* Winnipeg: Fisheries and Oceans Canada/Fisheries Joint Management Committee.

–. 2013b. *Tarium Niryutait Marine Protected Area Monitoring Plan.* Winnipeg: Fisheries Joint Management Committee.

–. 2017. *Musquash Estuary: A Management Plan for the Marine Protected Area and Administered Intertidal Area.* Dartmouth, NS: Fisheries and Oceans Canada. http://publications.gc.ca/collections/collection_2018/mpo-dfo/Fs104-15-2017-eng.pdf.

–. 2023. "Proceedings of the Regional Advisory Meeting on the Review of Musquash Monitoring Plan and Assessment Framework; May 11–12, 2021." Canadian Science Advisory Secretariat Proceedings Series 2023/012. https://waves-vagues.dfo-mpo.gc.ca/library-bibliotheque/41117992.pdf.

Dickinson, M., M. Rutherford, and T. Gunton. 2010. "Principles for Integrated Marine Planning: A Review of International Experience." *Environments* 37 (3): 21–46.

Diggon, S., C. Butler, A. Heidt, J. Bones, R. Jones, and C. Outhet. 2019. "The Marine Plan Partnership: Indigenous Community-Based Marine Spatial Planning." *Marine Policy* 132: 103510. https://doi.org/10.1016/j.marpol.2019.04.014.

Duinker, P., and L. Greig. 2006. "The Impotence of Cumulative Effects Assessment in Canada: Ailments and Ideas for Redeployment." *Environmental Management* 37 (2): 153–61.

Eger, S., and S. Courtenay. 2021. "Integrated Coastal and Marine Management: Insights from Lived Experiences in the Bay of Fundy, Atlantic Canada." *Ocean and Coastal Management* 204: 105457. https://doi.org/10.1016/j.ocecoaman.2020.105457.

Eger, S., R. de Loë, J. Pittman, G. Epstein, and S.C. Courtenay. 2021. "Systematic Review of Governance in Recent Integrated Coastal and Marine Management Literature." *Marine Policy* 132: 104688. https://doi.org/10.1016/j.marpol.2021.104688.

Epstein, G., E. Andrews, D. Armitage, P. Foley, J. Pittman, and R. Brushett. 2018. "Human Dimensions of Ecosystem-Based Management: Lessons in Managing Trade-Offs from the Northern Shrimp Fishery in Northern Peninsula, Newfoundland." *Marine Policy* 97: 10–17.

Fisheries Joint Management Committee. 2013. *Beaufort Sea Beluga Management Plan.* 4th amended printing. Inuvik, NT: Fisheries Joint Management Committee. https://www.beaufortseapartnership.ca/wp-content/uploads/2015/04/Beaufort-Sea-Beluga-Management-Plan-2013.pdf.

Foley, M., B. Halpern, F. Micheli, M. Armsby, M. Caldwell, C. Crain, E. Prahler, et al. 2010. "Guiding Ecological Principles for Marine Spatial Planning." *Marine Policy* 34 (5): 955–66.

Foster, E., M. Haward, and S. Coffen-Smout. 2005. "Implementing Integrated Oceans Management." *Marine Policy* 29 (5): 391–405.

Government of Canada. 2002. *Canada's Oceans Strategy: Our Oceans, Our Future. Policy and Operational Framework for Integrated Management of Estuarine, Coastal and Marine Environments in Canada.* Ottawa: Fisheries and Oceans Canada. https://waves-vagues.dfo-mpo.gc.ca/library-bibliotheque/264678.pdf.

Guerin, A.J., K.A. Kidd, M.-J. Maltais, A. Mercer, and H.L. Hunt. 2023. "Temporal and Spatial Trends in Benthic Infauna and Potential Drivers in a Highly Tidal Estuary in Atlantic Canada." *Estuaries and Coasts* 46: 1612–31. https://doi.org/10.1007/s12237-023-01222-w.

Hilborn, R. 2011. "Future Directions in Ecosystem Based Fisheries Management: A Personal Perspective." *Fisheries Research* 108: 235–39.

Imperial, M.T. 1999. "Institutional Analysis and Ecosystem-Based Management: The Institutional Analysis and Development Framework." *Environmental Management* 24: 449–65.

Interdepartmental Directors' Consultative Committee North Sea. 2015. *Integrated Management Plan for the North Sea.* Netherlands: Rijkswaterstaat Noordzee.

Kelly, C., G. Ellis, and W. Flannery. 2019. "Unravelling Persistent Problems to Transformative Marine Governance." *Frontiers in Marine Science* 6 (213). https://doi.org/10.3389/fmars.2019.00213.

Kincaid, K.B., and G.A. Rose. 2014. "Why Fishers Want a Closed Area in Their Fishing Grounds: Exploring Perceptions and Attitudes to Sustainable Fisheries and Conservation 10 Years Post Closure in Labrador, Canada." *Marine Policy* 46: 84–90.

–. 2017. "Effects of Closing Bottom Trawling on Fisheries, Biodiversity, and Fishing Communities in a Boreal Marine Ecosystem: The Hawke Box off Labrador, Canada." *Canadian Journal of Fisheries and Aquatic Sciences* 74: 1490–1502.

Koen-Alonso, M., P. Pepin, M.J. Fogarty, A. Kenny, and E. Kenchington. 2019. "The Northwest Atlantic Fisheries Organization Roadmap for the Development and Implementation of an Ecosystem Approach to Fisheries: Structure, State of Development, and Challenges." *Marine Policy* 100: 342–52.

Kong, H., W. Yang, and Q. Sun. 2018. "Overcoming the Challenges of Integrated Coastal Management in Xiamen: Capacity, Sustainable Financing and Political Will." *Ocean and Coastal Management* 207: 104519.

Kooiman, J., and M. Bavink. 2005. *The Governance Perspective.* Amsterdam: Amsterdam University Press.

Leslie, H., and K. McLeod. 2007. "Confronting the Challenges of Implementing Marine Ecosystem-Based Management." *Frontiers in Ecology and the Environment* 5: 540–48.

Lockwood, M., J. Davidson, A. Curtis, E. Stratford, and R. Griffith. 2010. "Governance Principles for Natural Resource Management." *Society and Natural Resources* 23 (10): 966–1001.

Marshak, A.R., J.S. Link, R. Shuford, M.E. Monaco, E. Johannesen, G. Bianchi, M.R. Anderson, et al. 2017. "International Perceptions of an Integrated, Multi-Sectoral, Ecosystem Approach to Management." *ICES Journal of Marine Science* 74: 414–20.

McCuaig, J., and G. Herbert, eds. 2013. *Review and Evaluation of the Eastern Scotian Shelf Integrated Management (ESSIM) Initiative.* Canadian Technical Report of Fisheries and Aquatic Sciences 3025. Dartmouth, NS: Fisheries and Oceans Canada.

Mullowney, D.R., E.G. Dawe, E.B. Colbourne, and G.A. Rose. 2014. "A Review of Factors Contributing to the Decline of Newfoundland and Labrador Snow Crab (*Chionoecetes opilio*). *Reviews in Fish Biology and Fisheries* 24: 639–57.

Mullowney, D.R., C.J. Morris, E.G. Dawe, and K.R. Skanes. 2012. "Impacts of a Bottom Trawling Exclusion Zone on Snow Crab Abundance and Fish Harvester Behavior in the Labrador Sea, Canada." *Marine Policy* 36: 567–75.

Murawski, S.A. 2007. "Ten Myths Concerning Ecosystem Approaches to Marine Resource Management. *Marine Policy* 31: 681–90.

Nguyen, T.X., P.D. Winger, G. Legge, E.G. Dawe, and D.R. Mullowney. 2014. "Underwater Observations of the Behaviour of Snow Crab (*Chionoecetes opilio*) Encountering a Shrimp Trawl off Northeast Newfoundland. *Fisheries Research* 156: 9–13.

O'Boyle, R., and G. Jamieson. 2006. "Observations on the Implementation of Ecosystem-Based Management." *Fisheries Research* 79 (1): 1–12.

Office of the Auditor General of Canada. 2005. *Report of the Commissioner of the Environment and Sustainable Development to the House of Commons.* Ottawa: Office of the Auditor General.

Olsen, E., S. Holen, A.H. Hoel, L. Buhl-Mortensen, and I. Røttingen. 2016. "How Integrated Ocean Governance in the Barents Sea Was Created by a Drive for Increased Oil Production." *Marine Policy* 71: 293–300. https://doi.org/10.1016/j.marpol.2015.12.005.

Parlee, C.E., and M.G. Wiber. 2018. "Using Conflict over Risk Management in the Marine Environment to Strengthen Measures of Governance." *Ecology and Society* 23 (4): 5.

Patrick, W.S., and J.S. Link. 2015. "Myths That Continue to Impede Progress in Ecosystem-Based Fisheries Management." *Fisheries* 40: 155–60.

Pikitch, E., C. Santora, E. Babcock, A. Bakun, R. Bonfil, D. Conover, P. Dayton, et al. 2004. "Ecosystem-Based Fishery Management." *Science* 305: 346–47.

Ricketts, P., and L. Hildebrand. 2011. "Coastal and Ocean Management in Canada." *Coastal Management* 39 (1): 4–19.

Rothwell, D.R., and D. VanderZwaag. 2006. *Towards Principled Oceans Governance: Australian and Canadian Approaches and Challenges.* New York: Routledge.

Sainsbury, K., and U.R. Sumaila. 2003. "Incorporating Ecosystem Objectives into Management of Sustainable Marine Fisheries, Including 'Best Practice' Reference Points and Use of Marine Protected Areas." In *Responsible Fisheries in the Marine Ecosystem,* edited by M. Sinclair and G. Valdimarsson, 346–61. Rome: Food and Agriculture Organization of the United Nations.

Sander, G. 2018. "Ecosystem-Based Management in Canada and Norway: The Importance of Political Leadership and Effective Decision-Making for Implementation." *Ocean and Coastal Management* 163: 485–97.

Sinclair, J., M. Doelle, and P. Duinker. 2017. "Looking Up, Down, and Sideways." *Environmental Impact Assessment Review* 62: 183–94.

Stephenson, R.L., A.J. Hobday, C. Cvitanovic, K.A. Alexander, G.A. Begg, R. Bustamante, P.K. Dunstan, et al. 2019. "A Practical Framework for Implementing and Evaluating Integrated Management of Marine Activities." *Ocean and Coastal Management* 177: 127–38.

Stephenson, R.L., A.J. Hobday, E.H. Allison, D. Armitage, K. Brooks, A. Bundy, C. Cvitanovic, et al. 2021. "The Quilt of Sustainable Ocean Governance: Patterns for Practitioners." *Frontiers in Marine Science* 8: 630547. https://doi.org/10.3389/fmars.2021.630547.

Stephenson, R.L., A.J. Hobday, I. Butler, T. Cannard, M. Cowlishaw, I. Cresswell, C. Cvitanovic, et al. 2023. "Integrating Management of Marine Activities in Australia." *Ocean and Coastal Management* 234: 106465. https://doi.org/10.1016/j.ocecoaman.2022.106465.

Stojanovic, T., R. Ballinger, and C. Lalwani. 2004. "Successful Integrated Coastal Management: Measuring It with Research and Contributing to Wise Practice." *Ocean and Coastal Management* 47 (5): 273–98.

Thia-Eng, C. 1993. "Essential Elements of Integrated Coastal Zone Management." *Ocean and Coastal Management* 21 (1–3): 81–108.

Underdal, A. 1980. "Integrated Marine Policy: What? Why? How?" *Marine Policy* 4 (3): 159–69.

Vince, J., A.D.M. Smith, K.J. Sainsbury, I.D. Cresswell, D.C. Smith, and M. Haward. 2015. "Australia's Oceans Policy: Past, Present, and Future." *Marine Policy* 57: 1–8.

PART 5

Into the Future

12

Progress and Challenges in Making Canada's Ocean Laws and Policies Climate-Ready

Cecilia Engler, Phillip Saunders, and David L. VanderZwaag

Grasping Canadian ocean laws and policies is challenging for various reasons. Canadian marine environments along the Pacific, Arctic, and Atlantic coasts are naturally diverse, supporting distinct ecological niches, economic activities, and cultural values. Management of activities within Canadian waters has developed on a sectoral and regional basis in the context of shared federal, provincial, territorial, Aboriginal, and local responsibilities (Mageau et al. 2010). As a result, and despite the integrative aspiration of the *Oceans Act*,[1] the legal and regulatory framework is fragmented and tangled. Multiple global and local drivers of change contribute to this complexity. The greatest challenges are, arguably, climate change and ocean acidification, which will profoundly transform these interdependent coastal and marine social-ecological systems.

Climate change law and policy are evolving both internationally and domestically. Earlier scholarship focused on mitigation law, including tools and incentives to reduce greenhouse gas (GHG) emissions (Craig 2010; Ruhl 2010). Oceans were mostly absent from early climate change mitigation debates (Hoegh-Guldberg et al. 2019), but there is growing recognition of oceans' potential contribution to mitigation efforts. They include ocean-based renewable energies, enhanced blue carbon ecosystems, fertilization, alkalinization, or carbon sequestration (Gattuso et al. 2018).

Increasingly, attention on the part of practitioners and academia has shifted to adaptation law, which is a far more complex legal problem (Craig 2010). Climate change will impact all components of social-ecological systems, and adaptation efforts are not "owned" by any particular discipline, sector, or actor. Thus, climate change adaptation needs to be *integrated* in our legal systems as a whole, particularly in environmental and natural resources law.

Legal challenges of climate change adaptation are even more profound, however. Traditional environmental law has centred on stationary visions of ecosystems, i.e., the "idea that natural systems fluctuate within an unchanging envelope of variability" (Craig 2010). Anthropogenic impacts are considered unnatural and degrading but non-transformative and therefore generally reversible (Craig 2010). This understanding has informed the general goal of environmental law (ecosystem preservation and restoration) and the design of regulatory interventions (Craig 2010; Garmestani et al. 2019). However, climate change (alone and cumulatively and synergistically with other anthropogenic stressors) is pushing ecosystems to non-analogue realities where this regulatory assumption and the associated regulatory goals and tools are increasingly unhelpful (Craig 2010; Garmestani et al. 2019). Climate change will require not only the *integration* of climate change risks into environmental law but also a transformation of goals, requirements, standards, and prohibitions in order to adapt to changing social-ecological systems and to allow these systems to exercise their own adaptive capacity (Garmestani et al. 2019; McDonald et al. 2018).

Legal scholarship has outlined some key features of new legal systems under climate change. These include expanded monitoring and research; reduction of non-climate pressures; increased coordination across media, sectors, interests, and governments; principled flexibility in substantive regulatory tools; procedural discretion; and shifts from front-end to back-end decision-making processes, including adaptive management (Craig 2010; Ruhl 2010; Garmestani et al. 2019). For ocean management in particular, the literature highlights several tools that can be used to improve the resilience of coastal and marine social-ecological systems. These include integrated coastal management, ecosystem-based Marine Spatial Planning (MSP), Marine Protected Area (MPA) networks, ecosystem restoration and protection, precautionary fisheries catch limits, and flexible and dynamic (real-time) management tools (IPCC 2019).

While Canada has not adopted a comprehensive oceans strategy or plan in the context of climate change,[2] the mitigation and adaptation priorities outlined in over-arching policy instruments have sectoral relevance. The *Pan-Canadian Framework on Clean Growth and Climate Change* (Government of Canada 2016) outlines the mitigation and adaptation commitments by the federal, provincial, and territorial governments. These are grouped around four pillars: pricing carbon pollution; complementary actions to reduce emissions; adaptation and climate resilience; and clean technology, innovation, and jobs. The policy documents put particular emphasis on vulnerable regions, including Canada's North, Indigenous communities, and coastal regions. The *Federal Adaptation Policy Framework* (Government of Canada 2011) further outlines key roles for the federal government in support of adaptation and resilience: generating and sharing knowledge; building adaptive capacity to respond and helping Canadians take action; and integrating adaptation into federal policy and planning (mainstreaming).

The mainstreaming mandate has found statutory recognition in the recently enacted *Impact Assessment Act,* which requires the Impact Assessment Agency of Canada or a review panel, as appropriate, to consider, *inter alia,* the "extent to which the effects of the designated project hinder or contribute to the Government of Canada's ability to meet its environmental obligations and its commitments in respect of climate change."[3] It was also clearly articulated in the Mandate Letters to the Minister of Fisheries, Oceans and the Canadian Coast Guard, which instructed the Minister to "use scientific evidence and the precautionary principle, *and take into account climate change,* when making decisions affecting fish stocks and ecosystem management" (emphasis added) (Prime Minister of Canada 2015, 2016, 2018).

This chapter focuses on the capacity of legal and policy frameworks for selected ocean sectors to integrate climate change considerations and respond to changing systems. In order to provide a selective but representative overview, we focus on different dimensions of oceans in the context of climate change. We consider ocean-based renewable energies as a potential contributor to mitigation efforts. We focus on the protection of aquatic species at risk, of particular importance for sustaining ecological resilience. Finally, we address resource-oriented activities sustaining Canadian livelihoods: fisheries and marine aquaculture. We close with some remarks on progressions and challenges.

We cannot address here the international climate change regime and regional efforts, while shipping (and particularly the potential for commercial shipping in the Canadian Arctic) has been covered extensively in the literature (Roach 2016; VanderZwaag 2015). Coastal infrastructure protection from the impacts of sea-level rise or extreme events is also outside the scope of this chapter, while community-based governance, adaptive management, and integrated management are addressed in detail in other chapters of this volume. Reconciliation with Indigenous Peoples and its cross-cutting influence in the conservation and sustainable use of coastal and marine resources are addressed in more detail in Chapters 2 and 14.

MARINE RENEWABLE ENERGY FOR CLIMATE CHANGE MITIGATION

The federal mitigation and adaptation policies discussed above all emphasize renewable energy as a means of emission reduction. The contribution of marine renewable energy (MRE) – including wind turbines in offshore wind farms (OWFs), tidal energy, wave energy, and offshore solar arrays – can be significant: a 2019 report found that up to 5.4% of required global mitigation actions could be met through MRE generation (Hoegh-Guldberg et al. 2019).

In 2018, industry estimates showed that although OWFs comprised only 4% of global capacity generated by wind power, they comprised almost 10% of the new installations that year (GWEC 2019). Forecasts indicate that the offshore component, driven by technological developments such as large-capacity turbines and public resistance to onshore facilities, will see significant growth, "with the total installed offshore wind capacity rising ... from just 23 GW in 2018 to 228 GW in 2030 and near 1 000 GW in 2050" (IRENA 2019). Tidal energy is much less well developed, with small numbers of facilities in Europe (including older lagoon and tidal barrage facilities), but the industry has great potential for growth given the development of technologies utilizing in-stream turbines on the seabed, on barges, or in semi-submersible arrays with reduced environmental impact and use conflicts (Doelle 2015).

Canada's potential for MRE development (MRC 2019; Cornett 2006) suggests that its growth in generating such energy could be above the global average and that pressure for MRE development sites will expand. Canada's long coastlines present advantages over onshore sites, including availability of open space, less impact on population centres, greater strength and lower variability of winds, and a technology extensively applied already in other regions (Cornett 2006; IRENA 2019; MRC 2019). Canada also has promising locations for in-stream tidal power development, particularly in Nova Scotia, British Columbia, and Nunavut (Cornett 2006).[4] Of particular interest in the Canadian context, tidal power and OWFs are adaptable both to larger projects feeding into the electrical grid and to small-scale facilities that can serve remote communities, including a number of First Nations that are now considering these potential sources (MRC 2019).

Despite all these, actual development of MRE capacity in Canada has been limited. Only a few OWF projects have been submitted, fewer have been approved, and none are yet under construction (MRC 2019; Henley, Stewart, and Waugh 2014). Project sites include the west coast of Newfoundland, the south shore of Nova Scotia, and Hecate Strait in northern British Columbia, in Haida Gwaii (MRC 2019; Henley, Stewart, and Waugh 2014). Previous interest in the tidal power energy potential of the Bay of Fundy led to the construction of a tidal barrage project, the Annapolis Generating Station in the 1980s, but cost and environmental concerns, along with the emergence of more effective in-stream technologies, have meant that a barrage or similar technologies are not going to be the future of this sector (Cornett 2006; Doelle 2015). The Annapolis Generating Station ceased operations in 2019 and, in recent decades, tidal power research and experimental and demonstration projects have focused on new stream technologies.[5]

The regulatory framework for MRE development involves both federal and provincial jurisdictions, depending on the location of the facility. For projects within the territorial boundaries of a province, provincial legislation related to environmental impact assessment, species at risk, and coastal zone management will all apply as they would to onshore projects (Doelle 2015). In addition, the province controls the nature and term of the tenurial rights required for projects utilizing Crown-owned seabed areas (Doelle at al. 2006). The federal government, in turn, while not controlling the *proprietary* interest in the seabed, still maintains *legislative* jurisdiction where relevant, including fisheries, shipping, and interprovincial energy transmission. Any project, even within a province, will require permits and authorizations under: the *Fisheries Act* for habitat alteration and disruption; the *Impact Assessment Act* for environmental impact assessments (EIAs); the *Canadian Navigable Waters Act* for interference with navigable waters; the *Species at Risk Act* (SARA) (see below); the *Oceans Act,* including MPA provisions; and the *Canada Shipping Act, 2001* (Doelle 2015).[6]

No federal legislation specifically addresses MRE projects in marine areas outside provincial territory, where the federal authorities would additionally be responsible for the grant of tenurial rights. The *Canadian Energy Regulator Act* does assert jurisdiction over "offshore renewable energy projects" in the "offshore area" (areas of internal waters, territorial sea, and continental shelf not within provinces).[7] Beyond this bare statement of jurisdiction, however, no criteria or processes are provided for the assignment of project development rights.

At the provincial level, British Columbia applies a MRE policy focused on permits and tenure over "aquatic Crown lands" (Government of British Columbia 2019). Nova Scotia, by contrast, has developed a legislative regime for MRE projects: in 2008–13, the province conducted a strategic environmental assessment (SEA) for the sector; in 2011, it commissioned a planning exercise on MRE legislation; and in 2012, it produced a strategy for the development of marine renewables (Henley, Stewart, and Waugh 2014; Doelle 2015). In 2015, it introduced the *Marine Renewable-energy Act* (MREA), which establishes two areas of priority for marine renewables development and a system of licensing and permits with extendable terms to a maximum of 18 years, allowing for demonstration projects.[8] Longer-term tenure arrangements would require Crown leases.[9]

Demand for renewable energy sources, the advance of MRE technology, and public resistance to onshore wind projects combine to make it likely that future development of MRE projects will accelerate (IRENA 2019; Hoegh-Guldberg et al. 2019). However, the technology for tidal power must be further refined and the environmental implications fully considered (McDonald and VanderZwaag 2015). There are also three main challenges to be addressed in creating an appropriate legal framework that enables the sector to progress sustainably.

First, the limits of Canada's jurisdictional entitlements under the *United Nations Convention on the Law of the Sea* (UNCLOS)[10] must inform any development of the industry. If projects move from internal waters to the territorial sea and the 200-nautical-mile exclusive economic zone (EEZ), significant entitlements of other states must be accommodated. Protection of navigational interests (Chircop and L'Esperance 2016) and of rights related to submarine cables and pipelines must also be incorporated in planning and approval of these projects. Further, global and regional agreements establishing environmental standards of relevance to MRE must be taken into account (McDonald and VanderZwaag 2015).

Second, the Canadian constitutional framework for control of MRE projects, from approval to decommissioning, is not fully addressed in existing legislation. While provincial governments clearly have *territorial* jurisdiction over significant areas of marine space, the federal government retains *legislative* jurisdiction for important functions (Doelle et al. 2006). Furthermore, First Nations entitlements to extensive consultation, if not outright control of particular spaces, will be highly relevant. If projects are developed outside provincial boundaries, any leasehold entitlements must be under

federal legislation that does not yet exist (Doelle et al. 2006). In 2011, the federal government initiated a policy and regulatory program (the Marine Renewable Energy Enabling Measures, or MREEM) to develop options and recommendations regarding the administration of marine renewable energy projects in the federal offshore, but no comprehensive federal legislative approach has yet emerged to bridge the overlapping jurisdictions.

Third, any legislative regime must deal with conflicting uses competing with MRE for ocean space. OWFs that have long-term tenure and large spatial footprints create conflicts with fishing, shipping, submarine cables, pipelines, offshore hydrocarbons, conservation, tourism, and traditional Indigenous uses. In the European Union, a Council Directive requires the broad application of maritime spatial planning to deal with use conflicts,[11] but Canada's efforts at integrated ocean management (mandated in Part II, Sections 31–34 of the *Oceans Act*) have focused on strategic plans for large marine areas, without the level of detail envisioned under a spatial planning system. The Nova Scotia legislation requires the identification of priority project areas, based on criteria including use conflicts, which is a positive step, but this cannot adequately plan for federally regulated uses. In any event, a prioritization located within one *sectoral* act will not substitute for a truly integrated approach that considers all marine user sectors and their individual and cumulative impacts on the marine systems.

PROTECTING NATURAL RESILIENCE THROUGH THE *SPECIES AT RISK ACT*

Canada's *Species at Risk Act* aims to provide various protections to listed species at risk that may be facing the threat of extinction. SARA prohibits killing, harming, harassing, or taking of listed endangered and threatened species, and requires the preparation of a recovery strategy and one or more action plan for each listed species (Sections 32, 37, 47). Critical habitat areas are required to be identified to the extent possible in recovery strategies and action plans, and no person is allowed to destroy any part of a critical habitat specified through a ministerial order (Sections 41[1][c], 49[1][a], 58). For listed species of special concern, SARA requires the preparation of a management plan for the species and its habitat (Section 65).

While SARA does not directly require climate change or ocean acidity threats to be addressed, with no mention of those terms in the legislation (Lemkow and Vander-Zwaag 2014), those threats may nevertheless be considered under various provisions. The Committee on the Status of Endangered Wildlife in Canada (COSEWIC), the expert body established to assess the status of wildlife species in Canada and to recommend appropriate listings, is required to carry out its functions on the basis of the best available information, including scientific knowledge, community knowledge, and Aboriginal Traditional Knowledge (Section 15[2]). This suggests the need to also take into account climate change information and projections. Recovery strategies are required to identify threats to the survival of species and threats to habitats (Section 41[1][b]), which may include changing environmental conditions. When preparing a recovery strategy, the competent minister may adopt an ecosystem approach (Section 41[3]). Action plans must include a statement of measures to be taken to implement the recovery strategy, including those that address the threats to the species and those that help achieve the population and distribution objectives (Section 49[1][d]).

Even without the complications of potential climate change and ocean acidity impacts, SARA's limitations in protecting marine species at risk are well documented (VanderZwaag and Hutchings 2005; Mooers et al. 2010). Listing of species for SARA protections is ultimately a political decision by the federal cabinet, and actually getting marine species listed has often been problematic (Findlay et al. 2009; Hutchings, Stephens, and VanderZwaag 2016). The finalization of recovery strategies has frequently lagged (McCune et al. 2013; CESD 2013), and the development of action plans has been exceedingly slow (Environment and Climate Change Canada 2018). For example, while the inner Bay of Fundy populations of Atlantic salmon were listed as endangered when SARA was proclaimed in June 2003, an action plan was finalized only in September 2019 (DFO 2019a). Identification of critical habitats and their protection through ministerial orders have also lagged (Taylor and Pinkus 2013; Broome 2010), while designation of unoccupied habitats that appear suitable for future occupation and recovery support has not occurred for most species (Camaclang et al. 2014).

"Mainstreaming" climate change considerations into the *Species at Risk Act* implementation has been problematic, as evidenced by the limited mentions in recovery strategies (Lemkow and VanderZwaag 2014; Hartman, VanderZwaag, and Fennel 2014). Climate change is almost always listed as a "potential" threat without further elaboration, which may be explained by the difficulty in documenting direct effects of climate change and by uncertainties in projecting future impacts (McCune et al. 2013). Furthermore, the procedural and substantive protections of the *Species at Risk Act* have lacked the flexibility needed to provide timely responses to emerging threats. An emblematic example is the recent plight of the North Atlantic right whale (NARW). The NARW population was estimated to be reduced to 411 animals by the end of 2017, with fewer than 100 breeding females (FOPO 2018). The reduction in numbers has become even more critical in recent years, with an estimated 336 animals in 2020 (DFO n.d.b).

A shift in the whale's primary food source, the copepod *Calanus finmarchicus,* has been linked to changing sea surface temperatures, with reduced *Calanus* abundance in the NARW's traditional feeding areas in the Bay of Fundy and Roseway Basin off southwestern Nova Scotia, but a greater biomass in the Gulf of St. Lawrence (DFO 2019f). In 2017, 12 NARWs were found dead in the southern Gulf of St. Lawrence waters while following prey; most of the mortalities were attributed to fishing gear entanglements and ship strikes (Davies and Brilliant 2019). Five right whale entanglements in the Gulf's waters were also reported in 2017 (Davies and Brilliant 2019).

Despite the fact that the NARW has been listed as endangered under SARA since January 2005, SARA has not played much of a role in responding to the climate-driven mortality crisis. Its recovery strategy, first published in June 2007 and revised in 2014 (DFO 2014), identifies the three main threats as fishing gear entrapment, vessel strikes, and contaminants and other forms of habitat degradation. The strategy paid little attention to climate change, merely noting that it could be affecting both the local spring and summer distribution of right whales and the calving rate of the North Atlantic population (DFO 2014, 26). The pledged action plan was not yet finalized at the time of writing. A critical habitat protection order was issued on November 30, 2017, but only the traditional feeding areas of Grand Manan and Roseway Basins were designated.[12] Furthermore, the protections granted were left vague in the Regulatory Impact Analysis Statement accompanying the order, which only provided examples of activities that may destroy critical habitat and emphasized that activities would be assessed on a case-by-case basis.

Numerous adaptation measures have been taken outside the SARA framework to address NARW mortalities. These have included early closures and gear restrictions for the snow crab fishery, static and dynamic closures to fixed-gear fisheries in areas where right whales are observed, and a system of static and dynamic speed reduction zones for ships in the Gulf of St. Lawrence (Koubrak, VanderZwaag, and Worm 2021).[13] No NARW mortalities were reported in Canadian waters in 2018, but additional measures were adopted after eight NARW mortalities were reported in June and July 2019. These included additional shipping measures,[14] increased aerial surveillance, additional funding for the Marine Mammal Response Program (Transport Canada 2019), and a three-day mission, Operation Ghost, by DFO fishery officers and members of the Canadian Coast Guard to recover 101 lost snow crab traps and over 9 km of rope from the Gulf of St. Lawrence in July 2019 (DFO 2019h). Other initiatives to support NARW recovery were funded under the Oceans Protection Plan (Transport Canada n.d.) and the 2018 federal budget (Government of Canada 2018). These included research on the impacts of ocean noise on endangered whales and the development of a real-time whale alert system for mariners.

Various avenues might be followed to make SARA more climate-ready and adaptive. It might be amended to specifically require climate change and ocean acidity to be considered in all key components of implementation, including listing, recovery planning, and designation and protection of critical habitats (Hartman, VanderZwaag, and Fennel 2014). SARA allows regulations to be passed to prescribe additional matters to be included in a recovery strategy (Section 41[4]), and a regulation could spell out the need to specifically address climate change and ocean acidity (Hartman, VanderZwaag, and Fennel 2014). A policy document on SARA and climate change might be developed to clarify how climate change and ocean acidity are expected to be considered in decision making.

Such a guidance document might draw from experience in the United States, where the National Marine Fisheries Service (NMFS) has developed guidance on how to treat climate change in *Endangered Species Act* decisions (NMFS 2016). That document identifies seven key climate change considerations needing attention, and emphasizes the importance of being proactive in designating unoccupied critical habitats in the face of species movements prompted by climate change. The need for adaptive management approaches is also highlighted, along with the need for adequate monitoring of climate and biological variables.

FISHERIES MANAGEMENT IN A CHANGING CLIMATE

DFO's main promotion of scientific research on climate change impacts on fisheries and marine ecosystems has been through the Aquatic Climate Change Adaptation Services Program (ACCASP). Initiated in 2011 – with $16.5 million invested over five years – to increase understanding of climate change impacts and to facilitate adaptation preparedness (Han, Lyon, and Mansour 2015), it was renewed through Budget 2016 with $5.6 million of funding over two years, and again in Budget 2017 with $21.6 million over four years (Madore and Nguyen 2017). Early deliverables were risk assessments on climate change impacts on biological systems and infrastructure within DFO's mandate for the Arctic (DFO 2013a), Pacific (DFO 2013c), and Atlantic (DFO 2013b). Over 60 research projects have also been supported (DFO n.d.c).

Translating scientific findings into fisheries management practice has been slow and challenging, however (Patterson et al. 2016). DFO has established an Ecosystem Approach to Fisheries Management Working Group to review how ecosystem advice has been applied in fisheries decision making and to provide guidance on how ecosystem changes may be better incorporated in the future. DFO has also developed a conceptual framework for including climate change considerations in fish stock assessments (DFO 2019e). Work to make such a framework operational is ongoing. Climate change and environmental variables are sometimes mentioned in fish stock assessments but without clear and explicit incorporation into management advice, as shown by 2019 assessments for Atlantic cod (DFO 2019g), snow crab (DFO 2019c), and northern shrimp (DFO 2019b).

Amendments to the *Fisheries Act* (receiving royal assent on June 21, 2019)[15] do not mention climate change but include several provisions that might support future adaptations. The precautionary and ecosystem approaches receive legal recognition as factors that may be considered by the Minister in making decisions under the act.[16] The amended act also emphasizes sustainability by providing that the Minister must "implement measures to maintain major fish stocks at or above the level necessary to promote the sustainability of the stock, taking into account the biology of the fish and the *environmental conditions* affecting the stock" (emphasis added).[17] Additionally, the Minister now has the explicit authority to *promptly* issue fisheries management orders to protect and conserve fish in any area of Canadian waters,[18] a provision that could be employed, *inter alia,* in case of climate-driven shifts in stock abundance or distribution.

Some existing fisheries policies under Canada's Sustainable Fisheries Framework also seem conducive to addressing climate change threats and uncertainties, particularly as they stress ecosystem, precautionary, and adaptive approaches. A policy on incorporating the precautionary approach into Canadian fisheries (DFO 2009a) pledges the development of precautionary reference points for key fished stocks and encourages consideration of changing ocean conditions and productivity in developing reference points. Canada has, however, struggled to set precautionary reference points for the 179 major fish stocks being managed (CESD 2016).[19]

The DFO Policy for Managing the Impacts of Fishing on Sensitive Benthic Areas adopts a precautionary and adaptive approach to proposed fishing in frontier areas without a history of fishing, specifically waters deeper than 2,000 m or areas of the Arctic where there is no history of fishing and little information available concerning benthic features (DFO 2009b). Fishing in frontier areas would be subject to an exploratory fishery protocol with only a limited area open to fishing and close monitoring of species caught. The Policy on New Fisheries for Forage Species urges a precautionary approach to opening new forage fisheries, such as krill, that are below the top of the aquatic food chain (DFO 2009c). One of the key objectives is to maintain target, bycatch, and ecologically dependent species within the bounds of natural fluctuations in abundance.

Canada has also adopted a policy framework to ensure that future commercial fisheries in the Beaufort Sea region will be subject to precautionary and adaptive approaches (Steiner et al. 2019). The Beaufort Sea Integrated Fisheries Management Framework for the Inuvialuit Settlement Region (Fisheries and Oceans Canada et al. 2014) establishes a multi-step decision process for any future commercial fisheries applications (Ayles, Porte, and Clark 2016). Key factors to be considered include possible adverse effects on Inuvialuit subsistence fisheries and preferential rights of the Inuvialuit to fish in the Inuvialuit Settlement Region. Decisions would also need to be founded on research relating to effects on target species, vulnerable ecosystem components, and the ecosystem (Fisheries and Oceans Canada et al. 2014).

MARINE AQUACULTURE AND A CHANGING CLIMATE

Aquaculture has a dual role to play in climate change adaptation. On the one hand, it is favoured as an adaptation strategy with mitigation co-benefits: it contributes to global food security, generates alternative sources of income and employment, replaces emission-intensive land-based sources of protein, and increases ocean carbon sinks capacity (Hoegh-Guldberg et al. 2019). On the other hand, as a marine-based industry, it will be affected by climate change and ocean acidification through multiple and complex pathways and with high regional and species variability.[20] Some of these impacts have already occurred.[21] Nevertheless, aquaculture has significant scope for adaptation through selective interventions in the rearing process (Bueno and Soto 2017; Reid and Gurney-Smith 2016) and therefore offers significant potential for maintaining and expanding economic opportunities under changing climate. Realizing this potential, however, requires a clear policy vision and an institutional and legal framework that enables and stimulates adaptation (Bueno and Soto 2017; Biesbroek et al. 2013).

Although Canada, particularly British Columbia and the Atlantic provinces, has a significant marine aquaculture sector, the federal and provincial governments have given little attention to its challenges and opportunities under climate change. The most comprehensive assessment stems from two partnership workshops held in 2015 (Reid and Gurney-Smith 2016). There is no federal climate change adaptation plan or vulnerability assessment addressing aquaculture in particular.[22] Policy documents, including the federal *Aquaculture Development Strategy 2016–2019* and the *Sustainable Aquaculture Program,* do not refer to climate change, although some initiatives may indirectly improve the sector's adaptation capacity. Provincial responses, in turn, either do not consider the needs of the aquaculture sector in the context of climate change or contain only aspirational statements.[23]

The lack of consideration of the impacts of climate change on the aquaculture sector extends to data and research (Reid and Gurney-Smith 2016). Research addressing climate change and Canadian aquaculture remains limited (Rice et al. 2018, 439), and needs have not been prioritized (Office of the Chief Science Advisor of Canada 2018; Reid and Gurney-Smith 2016). Systematic and real-time monitoring at relevant scales is also lagging (Clements and Chopin 2016, 334), but some government and collaborative initiatives offer promising developments. These include, for example, the Canadian Integrated Ocean Observing System, a web-based real-time and standardized fish health data management system (Fish-iTrends, currently only for salmon lice data), and the installation of real-time monitoring buoys.

The current legal and regulatory framework is a further roadblock to seizing new aquaculture opportunities and realizing the adaptive capacity of the sector. Official assessments (CESD 2018; DFO 2018; Standing Senate Committee on Fisheries and Oceans 2016), academic literature (Doelle and Saunders 2016; Young et al. 2019), and judicial decisions[24] have questioned the ability of the regulatory framework to govern an economically, ecologically, and socially sustainable marine aquaculture. Climate change will put these shortcomings in sharper relief. Key challenges arise from the lack of clear roles and responsibilities, insufficient development of zoning tools and siting guidance, and gaps in regulatory standards.

Clear roles and responsibilities are a prerequisite for an enabling adaptive framework (Van Putten et al. 2018; Biesbrock et al. 2013). Canada's institutional and regulatory landscape for aquaculture is notoriously complex and fragmented, however. The constitutional status of aquaculture regulation remains uncertain, with significant overlap of federal and provincial powers and responsibilities (Doelle and Saunders 2016). A pragmatic and cooperative approach emerged through province-specific

memoranda of understanding, but a recent judicial decision[25] added complexity to the legal landscape by ruling that fish-rearing activities on the coasts of British Columbia fall under federal jurisdiction (Doelle and Saunders 2016). As a result, Canada's aquaculture sector is managed under three different regimes: federal-led regulation in British Columbia, a unique regime in Prince Edward Island, and province-led regulation in the rest of Canada. Further differentiation resulted from the recent federal commitment to develop a "plan to transition from open net-pen salmon farming in coastal British Columbia waters by 2025" (Prime Minister of Canada 2019).

In all provinces, the federal government retains important responsibilities under several legislative instruments.[26] DFO has been designated the lead or coordinating department at the federal level, in addition to its own regulatory mandate (Doelle and Saunders 2016). Nevertheless, there are numerous federal agencies that have responsibility for some aspects of aquaculture operations or that are consulted in licensing processes. Agencies' concurrent and overlapping responsibilities also occur at the provincial level.

This fragmentation has several implications for the capacity of the industry to adapt to the challenges of climate change. For example, the Commissioner of the Environment and Sustainable Development has recently noted that there is a need to clearly outline the roles and responsibilities for emerging diseases between DFO, under the *Fisheries Act* and the *Fishery (General) Regulations,*[27] and the Canadian Food Inspection Agency under the *Health of Animals Act* (CESD 2018). This is a critical issue since distribution of disease-causing agents is likely to shift in warming waters (Reid et al. 2019). In turn, the industry has voiced concern about regulatory red tape for adaptive measures. For example, the lengthy processes to approve new feed ingredients or for emergency drug approval are seen as barriers to timely responses to threats that are predicted to increase in frequency and severity (Reid and Gurney-Smith 2016).

Appropriate siting is considered a key management tool for mitigating environmental impact and reducing social conflict, as well as a determinant of adaptation. Zoning has therefore been emphasized as one adaptation strategy (Bueno and Soto 2017; Brander et al. 2018). Canada, however, does not have consistent policies for MSP

(Young et al. 2019). Siting decisions are made on a case-by-case basis in lengthy, costly, and often controversial processes involving provincial and federal authorities, public consultation, and engagement with Indigenous communities. Furthermore, the involvement of the federal government in the siting decisions in provincial-led processes has been criticized both for procedural uncertainty and for lack of consistent guidelines (DFO 2018). All these features represent practical obstacles to temporary and permanent site relocation. Effective use of area-based planning tools could also guide the development of new aquaculture opportunities, including offshore aquaculture and aquaculture that contributes to coastal protection and carbon capture.

A further area of regulatory concern is gaps in regulatory standards or inconsistent standards. Containment provides a key example, considering the high risk of more frequent and stronger extreme future weather events that can lead to infrastructure damage. Canada does not have a national standard for the design and maintenance of aquaculture infrastructure, a matter that according to the federal-provincial memoranda of agreement falls mostly under provincial responsibility (with the exception of British Columbia and Prince Edward Island). The provinces in turn adopt different approaches and requirements,[28] and only Newfoundland and Labrador has recently begun to require International Organization for Standardization (ISO) or certified third-party engineering standard for sea-cage systems, components, and installation (Fisheries and Land Resources 2019, 9).

Some key developments may turn the tide for aquaculture management. Although they have not been set up to address climate change, they offer opportunities for mainstreaming climate change in the regulatory system and improving adaptive capacity. The External Advisory Committee on Aquaculture Science[29] may lay out a prioritized data collection and research plan. A framework for aquaculture risk management, including the risks of climate change, is currently under elaboration as an overarching framework for future policies addressing aquaculture sustainability. Additionally, DFO has committed to pilot projects for area-based aquaculture management (DFO n.d.a) and for MSP (DFO 2019d), although the prospects of implementing such strategies in provincial-led jurisdictions are not clear.

Crucial are two recent legal and regulatory commitments: the development of general aquaculture regulations (DFO 2019d),[30] and the preparation of a federal aquaculture act as agreed by the Canadian Council of Fisheries and Aquaculture Ministers in 2018.[31] These developments provide an unparalleled opportunity to construct a solid and forward-looking framework that ensures clear roles and responsibilities, consistent objectives and standards, flexible management at the regional and local scale, incentives for innovation, and transparent and accountable decision making.

CONCLUDING REMARKS

Are Canadian ocean laws and policies climate-ready? This chapter has offered a nuanced answer to this question in terms of progressions and challenges.

Key progressions include: a clear commitment by federal, provincial, and territorial governments to promote adaptation and climate resilience, with emphasis on mainstreaming climate change into decision making; investments in monitoring networks and scientific research; regional risk-based assessments of climate change impacts and risks on biological systems and infrastructure within DFO's mandate; ongoing work to establish a framework for incorporating climate change considerations into fisheries stock assessments; and the elaboration of a framework for aquaculture risk management. Amendments to the *Oceans Act* to strengthen marine protection,[32] and pilot projects for MSP and area-based aquaculture management, also contribute to ocean resilience and adaptive capacity. The endorsement of adaptive management in the *Impact Assessment Act* and in sectoral fisheries policies are also positive developments, although lack of guidance on the content, process, and accountability of adaptive management plans and measures remains a concern.

Sectoral legislation, in turn, has significant adaptive and transformative capacity. The Minister of Fisheries and Oceans has broad discretion under the *Fisheries Act*. This discretion allows for regulatory objectives and tools that respond to new social-ecological realities. Furthermore, recent amendments to the *Fisheries Act* strengthened principled fisheries management with emphasis on sustainability, precautionary, and ecosystem approaches. The *Species at Risk Act*, in turn, includes several programmatic provisions that can be developed through regulations, recovery strategies, and action plans to accommodate the risks, variability, and uncertainties of climate change. Ministerial discretion, however, does not ensure that climate change will be considered, or by which means. In the absence of a legal mandate in SARA or the *Fisheries Act* to incorporate climate change and ocean acidification considerations into decision making, a business-as-usual model may prevail. Both aspects of regulatory discretion are clearly exemplified in the case of the North Atlantic right whale.

This chapter has also identified several challenges, including the lack of an explicit and comprehensive vision of the role of oceans and ocean-based activities in a low-emission future. Rather than react to potential or actual impacts to existing ecosystems and activities, this vision could shape ocean-based economies and ocean protection to support climate action and to adapt to unprecedented realities. Several avenues could be pursued. A stand-alone ocean and climate change policy or plan that includes provincial and Aboriginal governments could be developed. The Oceans Strategy under the *Oceans Act* could be renewed and modernized to address the vital relationship between oceans and climate change. The announcement of work toward a comprehensive blue economy strategy (Prime Minister of Canada 2019) may represent a turning point in this respect.

Further challenges result from insufficient use of strategic planning tools, including MSP or other area-based management tools. This hinders horizontal and vertical coordination and prioritization of risks and responses. While the uncertainty and time horizon of climate change impacts makes long-term strategic planning difficult, iterative MSP processes (Craig 2019) and other area-based tools are a non-regret approach to address climate change and other stressors.

In addition to medium- and long-term strategic planning, regulatory frameworks need to enable timely and flexible responses to factors already affecting ocean resources, including extreme events and distribution shifts. This has proven challenging under current legislative frameworks. Fragmented, tangled, and unclear roles and responsibilities between jurisdictions, agencies, and programs, as well as overly structured decision-making processes (Garmestani et al. 2019), can delay responses

or create regulatory gaps and red tape, as demonstrated in the reviews of legal responses to marine aquaculture, marine renewable energy, and species at risk set out in this chapter.

Perhaps the most important challenge is the rate of progress. More than 20 years after the adoption of Canada's leading *Oceans Act,* its implementation is, by all accounts, incomplete at best. The same can be said of the *Species at Risk Act.* While Canada's progressions toward effective ocean management under climate change can be commended, there is a dire need to pick up the pace. Climate change can no longer be considered a future concern; it is a palpable and costly reality.

ACKNOWLEDGMENTS
The authors acknowledge the research support provided by the Canada First Research Excellence Fund through the Ocean Frontier Institute based at Dalhousie University. This chapter attempts to be accurate as of March 31, 2020.

NOTES

1 *Oceans Act,* SC 1996, c 31.

2 Canada's Oceans Strategy (Government of Canada 2002) under the *Oceans Act* only recognizes the need to better understand ecosystem dynamics, including climate, variability, and the impact of change on living marine resources. The federal government has not developed climate change adaptation plans for the oceans or for specific ocean sectors.

3 *Impact Assessment Act,* SC 2019, c 28, s 22.1(i). A designated project is required to identify not only its GHG emissions but also its impacts on carbon sinks and how the project is resilient to both the current and future impacts and risks of a changing climate (Government of Canada 2019). These factors are also relevant for the Minister's determination of whether the project's adverse effects are in the public interest, a provision that could lead to debatable but perhaps necessary trade-offs between environmental protection and climate action (*Impact Assessment Act,* s 63[e]). It should be noted, however, that the relevance of the new act for ocean-based projects is limited by the restricted scope of projects that require federal impact assessment.

4 Cornett (2006) identified 191 sites with "mean potential" of 1 MW or more. A more recent estimate placed

"practicable" capacity at 35,700 MW – enough to displace over 113 million tonnes of carbon dioxide (MRC 2019).

5 The first demonstration in-stream turbine project operated between 2006 and 2011 at Race Rocks in British Columbia, generating 65 kW (and successfully replacing diesel generation at the research station during the project) (MRC 2019). In Nova Scotia, the Fundy Ocean Research Centre for Energy (FORCE) has carried out five demonstration projects, involving both large in-stream turbines and floating arrays (Doelle 2015; MRC 2019). In 2019, a 280 kW floating in-stream turbine was installed in Grand Passage near Digby, outside the FORCE site (SME 2019).

6 *Fisheries Act*, RSC 1985, c F-14; *Impact Assessment Act*, SC 2019, c 28, s 1; *Canadian Navigable Waters Act*, RSC 1985, c N-22; *Species at Risk Act*, SC 2002, c 23; *Oceans Act*, SC 1996, c 31; *Canada Shipping Act, 2001*, SC 2001, c 26

7 *Canadian Energy Regulator Act*, SC 2019, c 28, s 10, Part 5.

8 *Marine Renewable-energy Act*, Chapter 32 of the Acts of Nova Scotia 2015, Chapter 32, as amended by SNS 2017, c 12, in force 2018 (MREA).

9 The essential scheme of the MREA is as follows: (1) two areas of priority for marine renewables development are established in the Bras d'Or Lake and portions of the Bay of Fundy (Schedules A and B); (2) within these areas, the province may, after consultation, designate smaller "Marine Renewable Electricity Areas" where projects may be permitted; (3) all federally required permits are to be in place; (4) a system of licensing and permitting is established, with limited but extendable terms (to a maximum of 18 years), allowing for demonstration projects, both connected and unconnected to the grid (s 14); (5) longer-term tenure arrangements would require Crown Lands leases, but no details are yet prescribed.

10 *United Nations Convention on the Law of the Sea*, December 10, 1982, 1833 UNTS (UNCLOS).

11 Directive 2014/89/EU of the European Parliament and of the Council of July 23, 2014, establishing a framework for maritime spatial planning (MSP Directive). Maritime spatial planning is defined as "a process by which the relevant Member State's authorities analyse and organise human activities in marine areas to achieve ecological, economic and social objectives."

12 Critical Habitat for the North Atlantic Right Whale (*Eubalaena glacialis*) Order, SOR/2017-262.

13 Beginning in 2017, a mandatory 10-knot slower speed zone was implemented for ships of 20 m or more in length (Meyer-Gutbrod, Greene, and Davies 2018).

14 The slow-down speed restriction was extended to vessels 13 m in length and over, and the speed reduction zones were expanded (Koubrak, VanderZwaag, and Worm 2021).

15 *An Act to amend the Fisheries Act and other Acts in consequence*, SC 2019, c 14.

16 Ibid., s 2.5.

17 Ibid., s 6.1(1). Nevertheless, s 6.1(2) allows the Minister to set a limit reference point and implement measures to maintain the fish stock above that point, if in his or her opinion it is not feasible or appropriate, for cultural reasons or because of adverse socio-economic impacts, to implement measures referred to in subsection 1.

18 Ibid., s 9.1(1): "The Minister may, if he or she is of the opinion that prompt measures are required to address a threat to the proper management and control of fisheries and the conservation and protection of fish, make a fisheries management order with respect to any aspect of fisheries in any area of Canadian fisheries waters specified in the order: (a) prohibiting fishing of one or more species, populations, assemblages or stocks of fish; (b) prohibiting any type of fishing gear or equipment or fishing vessel from being used; (c) limiting the fishing of any specified size, weight or quantity of any species, populations, assemblages or stocks of fish; and (d) imposing any requirements with respect to fishing."

19 For 2017, only 80 of the 179 stocks (45%) had upper stock references, while 105 of 179 (59%) had limit reference points (DFO 2019i).

20 Climate-induced physical, chemical, and biological hazards are projected to affect the suitability of marine areas in both negative and positive ways. Warmer waters may lengthen the growing season and accelerate growth rates, but also carry risks such as exceeding thermal maxima, increased variation in production success, introduction of invasive species or pathogens, or increased potential for hypoxia. Aquaculture depending on wild marine resources (seed, fishmeal, and fish oil) may be indirectly affected by projected impacts to their productivity. Extreme events (storms, bouts of warming or hypoxia, harmful algal blooms) may increase mass fish mortality, seafood contamination, and infrastructure damage. Environmental changes can also affect the susceptibility of host species to disease-causing agents; the distribution, abundance, or virulence of disease-causing agents; and the effectiveness of therapeutants

(Reid et al. 2019; Froehlich, Gentry, and Halpern 2018; Bueno and Soto 2017; Reid and Gurney-Smith 2016).

21 Aquaculture losses resulting from warming waters have been suggested in several jurisdictions (Reid et al. 2019, 574). Ocean acidification, in turn, has already caused catastrophic losses of oyster larvae in Washington State and in British Columbia (Clements and Chopin 2016, 330–31; Reid et al. 2019, 577).

22 DFO's regional risk-based assessments of climate change impacts and risks on the biological systems and infrastructure within its mandate (DFO 2013a, 2013b, 2013c) addressed threats and opportunities for aquaculture only in general terms. They did not include an assessment of risks to fisheries and aquaculture management systems ("soft infrastructure") (DFO 2013c, 8). Neither did they include an assessment of risks of damage to aquaculture infrastructure since "DFO's responsibilities do not extend to infrastructure at aquaculture sites" (DFO 2013c, 9).

23 For example, the action plan of the Government of Newfoundland and Labrador (2019) titled *The Way Forward on Climate Change in Newfoundland and Labrador* addresses the need to support the aquaculture industry to increase food production in a manner that takes into consideration GHG emissions. It also states its intention to work with the industry to increase knowledge and build resilience. Similarly, the BC Agrifood and Seafood Strategic Growth Plan (Government of British Columbia 2015) acknowledges climate change as one key challenge to its growth agenda, and includes adaptation measures such as diversifying aquaculture species and working with BC farmers to increase their adaptation capacity.

24 *Morton v Canada (Minister of Fisheries and Oceans) et al.,* [2015] FCJ No 566, 2015 FC 575; *Morton v Canada (Minister of Fisheries and Oceans),* [2019] FCJ No 178, 2019 FC 143.

25 *Morton v British Columbia (Agriculture and Lands),* [2009] BCJ No 193, 2009 BCSC 136.

26 Key relevant legislative instruments include: the *Fisheries Act,* the *Species at Risk Act,* the *Oceans Act,* the *Health of Animals Act* (SC 1990, c 21), the *Fish Inspection Act* (RSC 1985, c F-12), and the *Food and Drugs Act* (RSC 1985, c F-27).

27 *Fishery (General) Regulations,* SOR/93-53.

28 Nova Scotia requires an engineer's approval of the design of the structures in place for containment management as well as third-party audit of the containment management section of the Farm Management Plan (*Aquaculture Management Regulations,* NS Reg 348/2015, amended by NS Reg 118/2019, ss 15, 34, 35). New Brunswick producers follow an industry Code of Containment, while provincial regulations require notifications of breach of containment, which should include information on inspections performed on mooring systems, cage system components, and net structures (New Brunswick Salmon Growers' Association 2008; New Brunswick Regulation 91-158 under the *Aquaculture Act* [OC 91-806], s 14.1). In British Columbia, DFO requires, as a condition of the aquaculture licence, that licence holders take all reasonable measures to prevent escapes, including a general obligation of proper maintenance of cages and nets (DFO 2016).

29 The committee was recommended by Canada's Chief Science Advisor to ensure that DFO's science program is designed to meet the emerging needs of decision makers.

30 The goal of the general aquaculture regulations is to "consolidate all aquaculture-related regulatory content under the Fisheries Act into one comprehensive set of aquaculture regulations," to enhance the aquaculture regulatory framework, and to support business competitiveness (DFO 2019d).

31 The content of the proposed federal aquaculture legislation is currently in consultation process (see also Prime Minister of Canada 2019).

32 The amendment to the *Oceans Act* strengthened the designation process of MPAs by allowing their interim designation through ministerial orders rather than regulations (ss 35.1–35.3). It also strengthens the enforcement of MPAs (s 39). Canada has also adopted an ambitious protection goal: conserving 25% of its oceans by 2025 and working to increase protection to 30% by 2030 (Prime Minister of Canada 2019).

REFERENCES

Ayles, B., L. Porte, and R.M. Clark. 2016. "Development of an Integrated Fisheries Co-Management Framework for New and Emerging Commercial Fisheries in the Canadian Beaufort Sea." *Marine Policy* 72: 246–54.

Biesbroek, G.R., J.M. Klostermann, C.A.M. Termeer, and P. Kabat. 2013. "On the Nature of Barriers to Climate Change Adaptation." *Regional Environmental Change* 13 (5): 1119–29.

Brander, K., K. Cochrane, M. Barange, and D. Soto. 2018. "Climate Change Implications for Fisheries and Aquaculture." In *Climate Change Impacts on Fisheries and Aquaculture: A Global Analysis,* edited by B.F. Phillips and M. Pérez-Ramírez, 1: 45–62. Newark, NJ: John Wiley and Sons.

Broome, K. 2010. "SARA's Wagging Finger: Canada's Courts Are Not Allowing the Government to Shirk Its Responsibility to Protect Biodiversity." *Alternatives Journal* 36 (6): 18–19.

Bueno, P.B., and D. Soto. 2017. *Adaptation Strategies of the Aquaculture Sector to the Impacts of Climate Change.* FAO Fisheries and Aquaculture Circular No. 1142. Rome: Food and Agriculture Organization of the United Nations.

Camaclang, A.E., M. Maron, T.G. Martin, and H.P. Possingham. 2014. "Current Practices in the Identification of Critical Habitat for Threatened Species." *Conservation Biology* 29 (2): 482–92.

Chircop, A., and P. L'Esperance. 2016. "Functional Interactions and Maritime Regulation: The Mutual Accommodation of Offshore Wind Farms and International Navigation and Shipping." *Ocean Yearbook* 30: 439–87.

Clements, J.C., and T. Chopin. 2016. "Ocean Acidification and Marine Aquaculture in North America: Potential Impacts and Mitigation Strategies." *Reviews in Aquaculture* 9: 326–41.

CESD (Commissioner of the Environment and Sustainable Development). 2013. "Chapter 6 – Recovery Planning for Species at Risk." In *Report of the Commissioner of the Environment and Sustainable Development.* Ottawa: Office of the Auditor General of Canada. https://publications. gc.ca/collections/collection_2013/bvg-oag/FA1-2-2013 -1-6-eng.pdf.

–. 2016. *Report 2 – Sustaining Canada's Major Fish Stocks – Fisheries and Oceans Canada.* 2016 Fall Reports of the Commissioner of the Environment and Sustainable Development. Ottawa: Office of the Auditor General of Canada. https://www.oag-bvg.gc.ca/internet/English/ parl_cesd_201610_02_e_41672.html.

–. 2018. *Report 1 – Salmon Farming.* 2018 Spring Reports of the Commissioner of the Environment and Sustainable Development. Ottawa: Office of the Auditor General of Canada. https://www.oag-bvg.gc.ca/internet/English/ parl_cesd_201804_01_e_42992.html.

Cornett, A. 2006. *Inventory of Canada's Marine Renewable Energy Resources.* Technical Report CHC-TR-041. Ottawa: Canadian Hydraulics Centre, National Research Council Canada. https://ressources-naturelles.canada.ca/sites/www. nrcan.gc.ca/files/canmetenergy/files/pubs/CHC-TR-041.pdf.

Craig, R.K. 2010. "Stationarity Is Dead, Long Live Transformation: Five Principles for Climate Change Adaptation Law." *Harvard Environmental Law Review* 34: 9–73.

–. 2019. "Fostering Adaptive Marine Aquaculture through Procedural Innovation in Marine Spatial Planning." *Marine Policy* 110: 103555.

Davies, K.T.A., and S.W. Brilliant. 2019. "Mass Human-Caused Mortality Spurs Federal Action to Protect Endangered North Atlantic Right Whales in Canada." *Marine Policy* 104: 157–62.

DFO (Fisheries and Oceans Canada). 2009a. "A Fishery Decision-Making Framework Incorporating the Precautionary Approach." https://www.dfo-mpo.gc.ca/reports -rapports/regs/sff-cpd/precaution-eng.htm.

–. 2009b. "Policy for Managing the Impacts of Fishing on Sensitive Benthic Areas." https://www.dfo-mpo.gc.ca/ reports-rapports/regs/sff-cpd/benthi-eng.htm.

–. 2009c. "Policy on New Fisheries for Forage Species." https://www.dfo-mpo.gc.ca/reports-rapports/regs/sff-cpd/ forage-eng.htm.

–. 2013a. *Risk-Based Assessment of Climate Change Impacts and Risks on Biological Systems and Infrastructure within Fisheries and Oceans Canada's Mandate Arctic Large Aquatic Basin.* DFO Canadian Science Advisory Secretariat Science Response 2012/042. Ottawa: Fisheries and Oceans Canada.

–. 2013b. *Risk-Based Assessment of Climate Change Impacts and Risks on the Biological Systems and Infrastructure within Fisheries and Oceans Canada's Mandate Atlantic Large Aquatic Basin.* DFO Canadian Science Advisory Secretariat Science Response 2012/044. Ottawa: Fisheries and Oceans Canada.

–. 2013c. *Risk-Based Assessment of Climate Change Impacts and Risks on the Biological Systems and Infrastructure within Fisheries and Oceans Canada's Mandate Pacific Large Aquatic Basin.* DFO Canadian Science Advisory Secretariat Science Response 2013/016. Ottawa: Fisheries and Oceans Canada.

–. 2014. *Recovery Strategy for the North Atlantic Right Whale (*Eubalaena glacialis*) in Atlantic Canadian Waters.* Species at Risk Act Recovery Strategy Series. Ottawa: Fisheries and Oceans Canada. https://www.sararegistry. gc.ca/virtual_sara/files/plans/rs_bnan_narw_am_0414 _e.pdf.

–. 2016. "Pacific Region Marine Finfish Integrated Management of Aquaculture Plan: July 2016, Version 2.1." https:// www.pac.dfo-mpo.gc.ca/aquaculture/management -gestion/marine-marin/index-eng.html.

–. 2018. *Final Report: Evaluation of the Sustainable Aquaculture Program.* Project Number 96031, Evaluation Directorate, Chief Financial Officer Sector, Fisheries and Oceans Canada. https://waves-vagues.dfo-mpo.gc.ca/ library-bibliotheque/40761976_full.pdf.

–. 2019a. *Action Plan for the Atlantic Salmon (*Salmo salar*), Inner Bay of Fundy Population in Canada.* Species at Risk

Act Action Plan Series. Ottawa: Fisheries and Oceans Canada.

–. 2019b. *Assessment of Northern Shrimp on the Eastern Scotian Shelf (SFAs 13–15)*. DFO Canadian Science Advisory Secretariat Science Advisory Report 2019/013. Ottawa: Fisheries and Oceans Canada.

–. 2019c. *Assessment of Snow Crab* (Chionoecetes opilio) *in the Southern Gulf of St. Lawrence (Areas 12, 19, 12E and 12F) to 2018 and Advice for the 2019 Fishery*. DFO Canadian Science Advisory Secretariat Science Advisory Report 2019/010. Ottawa: Fisheries and Oceans Canada.

–. 2019d. "Departmental Plan 2019–2020." https://www.dfo-mpo.gc.ca/rpp/2019-20/dp-eng.html.

–. 2019e. *Framework for Incorporating Climate-Change Considerations into Fisheries Stock Assessments*. DFO Canadian Science Advisory Secretariat Science Advisory Report 2019/029. Ottawa: Fisheries and Oceans Canada.

–. 2019f. *Review of North Atlantic Right Whale Occurrence and Risk of Entanglements in Fishing Gear and Vessel Strikes in Canadian Waters*. DFO Canadian Science Advisory Secretariat Science Advisory Report 2019/028. Ottawa: Fisheries and Oceans Canada.

–. 2019g. *Stock Assessment of Atlantic Cod* (Gadus morhua) *in NAFO Divisions 4x54*. DFO Canadian Science Advisory Secretariat Science Advisory Report 2019/015. Ottawa: Fisheries and Oceans Canada.

–. 2019h. "Successful Three-Day Operation Ghost Recovers 101 Traps and Nine Kilometres of Rope from the Water in the Gulf of St. Lawrence." Press release, July 24. https://www.canada.ca/en/fisheries-oceans/news/2019/07/successful-three-day-operation-ghost-recovers-101-traps-and-nine-kilometres-of-rope-from-the-water-in-the-gulf-of-st-lawrence.html.

–. 2019i. "Summary of 2017 Sustainability Survey for Fisheries." https://www.dfo-mpo.gc.ca/reports-rapports/regs/sff-cpd/survey-sondage/results-resultats-s-2017-en.html.

–. n.d.a. "Aquaculture Initiatives and Management Measures." http://www.dfo-mpo.gc.ca/campaign-campagne/aquaculture/initiatives-eng.html.

–. n.d.b. "North Atlantic Right Whale." https://www.dfo-mpo.gc.ca/species-especes/mammals-mammiferes/whales-baleines/narw-bnan/index-eng.html.

–. n.d.c. "Research Funded by the Aquatic Climate Change Services Program (ACCASP)." http://www.dfo-mpo.gc.ca/science/rp-pr/accasp-psaccma/index-eng.html.

Doelle, M. 2015. "Offshore Renewable Energy Governance in Nova Scotia: A Case Study of Tidal Energy in the Bay of Fundy." *Ocean Yearbook* 29: 271–98.

Doelle, M., D. Russell, P. Saunders, D. VanderZwaag, and D. Wright. 2006. "The Regulation of Tidal Energy Development Off Nova Scotia: Navigating Foggy Waters." *University of New Brunswick Law Journal* 55: 27–70.

Doelle, M., and P. Saunders. 2016. "Aquaculture Governance in Canada: A Patchwork of Approaches." In *Aquaculture Law and Policy: Global, Regional and National Perspectives*, edited by N. Bankes, I. Dahl, and D. Vander-Zwaag. Cheltenham, UK: Edward Elgar.

Environment and Climate Change Canada. 2018. "Horizontal Evaluation at a Glance: Species at Risk Program." https://www.canada.ca/en/environment-climate-change/corporate/transparency/priorities-management/evaluations/species-at-risk.html.

Findlay, C.S., S. Elgie, B. Giles, and L. Burr. 2009. "Species Listing under Canada's Species at Risk Act." *Conservation Biology* 23 (6): 1609–17.

Fisheries and Land Resources. 2019. *Aquaculture Policy and Procedures Manual*. St. John's: Government of Newfoundland and Labrador, Fisheries and Land Resources. https://www.gov.nl.ca/ffa/files/licensing-pdf-aquaculture-policy-procedures-manual.pdf.

Fisheries and Oceans Canada, Fisheries Joint Management Committee, Inuvialuit Game Council, and Inuvialuit Regional Corporation. 2014. "Beaufort Sea Integrated Fisheries Management Framework for the Inuvialuit Settlement Region." https://fjmc.ca/wp-content/uploads/2016/08Beaufort-Sea-Integrated-Fisheries-Management-Framework-2014-FINAL-version.pdf.

FOPO (House of Commons Standing Committee on Fisheries and Oceans). 2018. *Protection and Recovery of Endangered Whales: The Way Forward*. Report of the Standing Committee on Fisheries and Oceans. Ottawa: House of Commons.

Froehlich, H.E., R.R. Gentry, and B.S. Halpern. 2018. "Global Change in Marine Aquaculture Production Potential under Climate Change." *Nature Ecology and Evolution* 2: 1745–50.

Garmestani, A., J.B. Ruhl, B.C. Chaffin, R.K. Craig, H.F.M.W. van Rijswick, D.G. Angeler, C. Folke, L. Gunderson, D. Twidwell, and C.R. Allen. 2019. "Untapped Capacity for Resilience in Environmental Law." *Proceedings of the National Academy of Sciences* 16 (40): 19899–904.

Gattuso, J.-P., A.K. Magnan, L. Bopp, W.W.L. Cheung, C.M. Duarte, J. Hinkel, E. Mcleod, et al. 2018. "Ocean Solutions to Address Climate Change and Its Effects on Marine Ecosystems." *Frontiers in Marine Science* 5 (337).

Government of British Columbia. 2015. "Roadmap to 2020 – B.C. Agrifood and Seafood Strategic Growth Plan

Launched." News release, December 2. https://news.gov.bc.ca/releases/2015AGRI0071-002008.

–. 2019. "Land Use Operational Policy: Ocean Energy Projects." https://www2.gov.bc.ca/assets/gov/farming-natural-resources-and-industry/natural-resource-use/land-water-use/crown-land/ocean_energy.pdf.

Government of Canada. 2002. *Canada's Oceans Strategy: Our Oceans, Our Future. Policy and Operational Framework for Integrated Management of Estuarine, Coastal and Marine Environments in Canada.* Ottawa: Fisheries and Oceans Canada. https://waves-vagues.dfo-mpo.gc.ca/library-bibliotheque/264678.pdf.

–. 2011. *Federal Adaptation Policy Framework.* Gatineau, QC: Environment and Climate Change Canada.

–. 2016. *Pan-Canadian Framework on Clean Growth and Climate Change: Canada's Plan to Address Climate Change and Grow the Economy.* Gatineau, QC: Environment and Climate Change Canada. https://publications.gc.ca/site/eng/9.828774/publication.html.

–. 2018. "Budget 2018: Equality and Growth for a Strong Middle Class." Press release, February 27. https://www.canada.ca/en/department-finance/news/2018/02/budget-2018-equality-and-growth-for-a-strong-middle-class.html.

–. 2019. "Draft Strategic Assessment of Climate Change." https://www.canada.ca/en/services/environment/conservation/assessments/environmental-reviews/get-involved/draft-strategic-assessment-climate-change.html.

Government of Newfoundland and Labrador. 2019. "The Way Forward on Climate Change in Newfoundland and Labrador." https://www.gov.nl.ca/ecc/occ/action-plans/.

GWEC (Global Wind Energy Council). 2019. *Global Wind Report 2018.* Brussels: Global Wind Energy Council. https://indianwindpower.com/pdf/GWEC_Global_Wind_2018.pdf.

Han, G., P. Lyon, and A. Mansour. 2015. "Introduction to the Special Issue on the Aquatic Climate Change Adaptation Services Program." *Atmosphere-Ocean* 53 (5): 447–51.

Hartman, W., D.L. VanderZwaag, and K. Fennel. 2014. "Recovery Planning for Pacific Marine Species at Risk in the Wake of Climate Change and Ocean Acidification: Canadian Practice, Future Courses." *Journal of Environmental Law and Practice* 27: 23–56.

Henley, D., C. Stewart, and J. Waugh. 2014. "Regulation of Alternative Energy Projects in Atlantic Canada." *Dalhousie Law Journal* 37 (1): 175–204.

Hoegh-Guldberg, O., K. Caldeira, T. Chopin, S. Gaines, P. Haugan, M. Hemer, J. Howard, et al. 2019. *The Ocean as a Solution to Climate Change: Five Opportunities for Action.* Washington, DC: World Resources Institute. http://www.oceanpanel.org/climate.

Hutchings, J.A., T. Stephens, and D.L. VanderZwaag. 2016. "Marine Species at Risk Protection in Australia and Canada: Paper Promises, Paltry Progressions." *Ocean Development and International Law* 47 (3): 233–54.

IPCC (Intergovernmental Panel on Climate Change). 2019. "Summary for Policymakers." In *IPCC Special Report on the Ocean and Cryosphere in a Changing Climate.* Bremen, Germany: IPCC.

IRENA (International Renewable Energy Agency). 2019. *Future of Wind: Deployment, Investment, Technology, Grid Integration and Socio-Economic Aspects.* Abu Dhabi: International Renewable Energy Agency. https://www.irena.org/publications/2019/Oct/Future-of-wind.

Koubrak, O., D.L. VanderZwaag, and B. Worm. 2021. "Saving the North Atlantic Right Whale in a Changing Ocean: Gauging Scientific and Law and Policy Responses." *Ocean and Coastal Management* 200: 105109. https://doi.org/10.1016/j.ocecoaman.2020.105109.

Lemkow, A., and D.L. VanderZwaag. 2014. "Recovery Planning under Canada's *Species at Risk Act* in a Changing Ocean: Gauging the Tides, Charting Future Coordinates." *Journal of Environmental Law and Practice* 26: 121–56.

Madore, O., and T. Nguyen. 2017. "Update Climate Change: Implications for Canadian Marine Fisheries and Aquaculture." *Hillnotes,* June 8, 1–7.

Mageau, C., D. VanderZwaag, K. Huffman, and S. Farlinger. 2010. "Ocean Policy: A Canadian Case Study." In *Routledge Handbook of National and Regional Ocean Policies,* edited by B. Cicin-Sain, D.L. VanderZwaag, and M. Balgos. London: Routledge.

McCune, J.L., W.L. Harrower, S. Avery-Gomm, J.M. Brogan, A.M. Csergo, L.N.K. Davidson, A. Garani, et al. 2013. "Threats to Canadian Species at Risk: An Analysis of Finalized Recovery Strategies." *Biological Conservation* 166: 254–65.

McDonald, J., P.C. McCormack, M. Dunlop, D. Farrier, J. Feehely, L. Gilfedder, A.J. Hobday, and A.E. Reside. 2018. "Adaptation Pathways for Conservation Law and Policy." *WIREs Climate Change* 10 (1): e555. https://doi.org/10.1002/wcc.555.

McDonald, S,. and D. VanderZwaag. 2015. "Renewable Ocean Energy and the International Law and Policy Seascape: Global Currents, Regional Surges." *Ocean Yearbook* 29: 299–326.

Meyer-Gutbrod, E.L., C.H. Greene, and K.T.A. Davies. 2018. "Marine Species Range Shifts Necessitate Advanced Policy

Planning: The Case of the North Atlantic Right Whale." *Oceanography* 31 (2): 19–23.

Mooers, A.O., D.F. Doak, C.S. Findlay, D.M. Green, C. Grouios, L.L. Manne, A. Rashvand, M.A. Rudd, and J. Whitton. 2010. "Science, Policy, and Species at Risk in Canada." *BioScience* 60 (10): 843–49.

MRC (Marine Renewables Canada). 2019. *2018 State of the Sector Report: Marine Renewable Energy in Canada.* Halifax: Marine Renewables Canada. https://marine renewables.ca/wp-content/uploads/2018/06/MRC-State -of-the-Sector-2018.pdf.

New Brunswick Salmon Growers' Association. 2008. "Code of Containment for Culture of Atlantic Salmon in Marine Net Pens in New Brunswick." https://static1.squarespace. com/static/56e827cb22482efe36420c65/t/570ed80db 09f950e801cde72/1460590605851/2008_NBSGA_Code_ of_Containment_June_2008.pdf.

NMFS (National Marine Fisheries Service). 2016. "Guidance for Treatment of Climate Change in NMFS Endangered Species Act Decisions. Procedural Instruction 02-110-18." https://www.fisheries.noaa.gov/national/endangered -species-conservation/endangered-species-act-guidance -policies-and-regulations.

Office of the Chief Science Advisor of Canada. 2018. *Report of the Independent Expert Panel on Aquaculture Science: December 2018.* Ottawa: Office of the Chief Science Advisor of Canada. https://www.ic.gc.ca/eic/site/052.nsf/eng/ 00011.html.

Patterson, D.A., S.J. Cooke, S.G. Hinch, K.A. Robinson, N. Young, A.P. Farrell, and K.M. Miller. 2016. "A Perspective on Physiological Studies Supporting the Provision of Scientific Advice for the Management of Fraser River Sockeye Salmon (*Oncorhynchus nerka*)." *Conservation Physiology* 4: 1–15.

Prime Minister of Canada. 2015. "ARCHIVED – Minister of Fisheries, Oceans and the Canadian Coast Guard Mandate Letter." https://pm.gc.ca/en/mandate-letters/ 2015/11/12/archived-minister-fisheries-oceans-and -canadian-coast-guard-mandate.

–. 2016. "ARCHIVED – Minister of Fisheries, Oceans and the Canadian Coast Guard Mandate Letter." https://pm. gc.ca/en/mandate-letters/2016/08/19/archived-minister -fisheries-oceans-and-canadian-coast-guard-mandate.

–. 2018. "ARCHIVED – Minister of Fisheries, Oceans and the Canadian Coast Guard Mandate Letter." https://pm.gc. ca/en/mandate-letters/2018/08/28/archived-minister -fisheries-oceans-and-canadian-coast-guard-mandate.

–. 2019. "ARCHIVED – Minister of Fisheries, Oceans and the Canadian Coast Guard Mandate Letter." https://pm.gc.

ca/en/mandate-letters/2019/12/13/archived-minister -fisheries-oceans-and-canadian-coast-guard-mandate.

Reid, G.K., and H.J. Gurney-Smith, eds. 2016. "Proceedings of the Atlantic and Pacific Climate Change and Aquaculture Workshops." *Bulletin of the Aquaculture Association of Canada 2015–2.* https://aquacultureassociation.ca/ wp-content/uploads/2017/01/Climate-Change-and -Aquaculture-Workshop-Proceedings-2016.pdf.

Reid, G.K., H.J. Gurney-Smith, D.J. Marcogliese, D. Knowler, T. Benfey, A.F. Garber, I. Forster, et al. 2019. "Climate Change and Aquaculture: Considering Biological Response and Resources." *Aquaculture Environment Interactions* 11: 569–602.

Rice, J., R. Beamish, D. Duplisea, G. Reid, and H. Gurley-Smith. 2018. "Canadian Fisheries and Aquaculture: Prospects under a Changing Climate." In *Climate Change Impacts on Fisheries and Aquaculture: A Global Analysis,* edited by B.F. Phillips and M. Pérez-Ramírez, 1: 45–62. Newark, NJ: John Wiley and Sons.

Roach, J.A. 2016. "Arctic Navigation: Recent Developments." In *Challenges of the Changing Arctic: Continental Shelf, Navigation and Fisheries,* edited by Myron H. Nordquist, John Norton Moore, and Ronan Long, 173–285. Leiden: Brill Nijhoff.

Ruhl, J.B. 2010. "Climate Change Adaptation and the Structural Transformation of Environmental Law." *Environmental Law* 40: 363–435.

SME (Sustainable Marine Energy). 2019. "Results of First Phase PLAT-I Testing at Grand Passage." June 6. https:// www.inerjys.com/press-releases/2019/6/6/results-of-first -phase-plat-i-testing-at-grand-passage.

Standing Senate Committee on Fisheries and Oceans. 2016. *Volume 3 – An Ocean of Opportunities: Aquaculture in Canada.* Report of the Standing Senate Committee on Fisheries and Oceans. Ottawa: Senate of Canada.

Steiner, N.S., W.W.L Cheung, A.M. Cisneros-Montemayor, H. Drost, H. Hayashida, C. Hoover, J. Lam, et al. 2019. "Impacts of the Changing Ocean–Sea Ice System on the Key Forage Fish Arctic Cod (*Boreogadus saida*) and Subsistence Fisheries in the Western Canadian Arctic Evaluating Linked Climate, Ecosystem and Economic (CEE) Models." *Frontiers in Marine Science* 6: Article 179.

Taylor, E.B., and S. Pinkus. 2013. "The Effects of Lead Agency, Nongovernmental Organizations and Recovery Team Membership on the Identification of Critical Habitat for Species at Risk: Insights from the Canadian Experience." *Environmental Reviews* 21: 93–122.

Transport Canada. 2019. "News Release: Government of Canada Introduces New, Additional Measures to Protect the North Atlantic Right Whale." News release, July 8. https://www.canada.ca/en/transport-canada/news/2019/ 07/government-of-canada-introduces-new-additional -measures-to-protect-the-north-atlantic-right-whale.html.

–. n.d. "Report to Canadians: Investing in Our Coasts through the Oceans Protection Plan." https://tc.canada.ca/ en/initiatives/oceans-protection-plan/report-canadians -investing-our-coasts-through-oceans-protection-plan.

Van Putten, I., A. Breckwoldt, A. Bundy, P. Guillotreau, P.K. Nayak, H. Österblom, and R.I. Perry. 2018. "Conclusion: Lessons from Global Change Response to Advance Governance and Sustainable Use of Marine Systems." In *Global Change in Marine Systems Integrating Natural, Social and Governing Responses,* edited by P. Guillotreau, A. Bundy, and R.I. Perry, 295–313. Abingdon, UK: Routledge.

VanderZwaag, D.L. 2015. "Climate Change and the Shifting International Law and Policy Seascape for Arctic Shipping." In *Climate Change Impacts on Ocean and Coastal Law: U.S. and International Perspectives,* edited by Randall S. Abate, 299–314. Oxford: Oxford University Press.

VanderZwaag, D.L., and J.A. Hutchings. 2005. "Canada's Marine Species at Risk. Science and Law at the Helm, but a Sea of Uncertainties." *Ocean Development and International Law* 36: 219–59.

Young, N., C. Brattland, C. Digiovanni, B. Hersoug, J.P. Johnsen, K.M. Karlsen, I. Kvalvik, et al. 2019. "Limitations to Growth: Social-Ecological Challenges to Aquaculture Development in Five Wealthy Nations." *Marine Policy* 104: 216–24.

13

Coastlines, Communities, and Cameras: How Participatory Video Can Enhance Ocean Research

Vincent L'Hérault, Natalie Baird, Hillary Beattie, Ian Mauro, and Eric Solomon

New approaches to ocean management and conservation integrate social-ecological considerations (Long, Charles, and Stephenson 2015; Charles 2012; Ommer 2007). Given the complexity of social-ecological systems, putting theory into practice is challenging for an effective integration of human dimensions within management processes (Bennett et al. 2016). Social science methodologies in conservation studies have refined participatory approaches and have substantially strengthened community engagement and communication (Reed 2008), especially management where Indigenous communities and their knowledge and territories are a priority (Augustine and Dearden 2014; Jones, Rigg, and Lee 2010) (see also Chapter 2). This approach is more inclusive and able to bridge knowledge and epistemological gaps between local communities and research, and has demonstrated its ability to contribute to meaningful, trust-based relationships that lead to genuine collaboration (Augustine and Dearden 2014).

OCEANCANADA'S COMMITMENT TO COMMUNITY ENGAGEMENT AND KNOWLEDGE MOBILIZATION

The OceanCanada Partnership (OCP) sees both community engagement and communication as tightly linked and essential components of meaningful ocean research. At the onset of the partnership, a knowledge mobilization working group (KMWG) was convened with the explicit goal of developing participatory projects on all three of Canada's coastlines that meaningfully connected community knowledge with an effective communications strategy co-designed and co-developed with community members to ensure that they had leadership over research and decision making. Our approach to knowledge mobilization was multi-dimensional, developed iteratively through respectful dialogue between researchers and community members (Figure 13.1). To avoid common institutionalized barriers that prevent meaningful community participation, OCP's knowledge mobilization team had a philosophy and practice that contextualized our approach, including recognition that it is critical that the work be done in an equitable way; early participation and relationship building; team identification overseen by the community; co-design of objectives and methods; and co-creation of knowledge, synthesis, and associated outreach strategy (reviewed in Reed 2008) (Figure 13.1).

Community benefits were a critical outcome of the participatory process if it were to be deemed effective. Our team was composed of experienced community-based researchers, facilitators, filmmakers, and students who, by approaching research through community and communication-centred objectives, helped ensure a "people-first" approach within OCP.

This chapter explores:

- the use of participatory video to promote engagement, collective capacity building, and enhanced communication

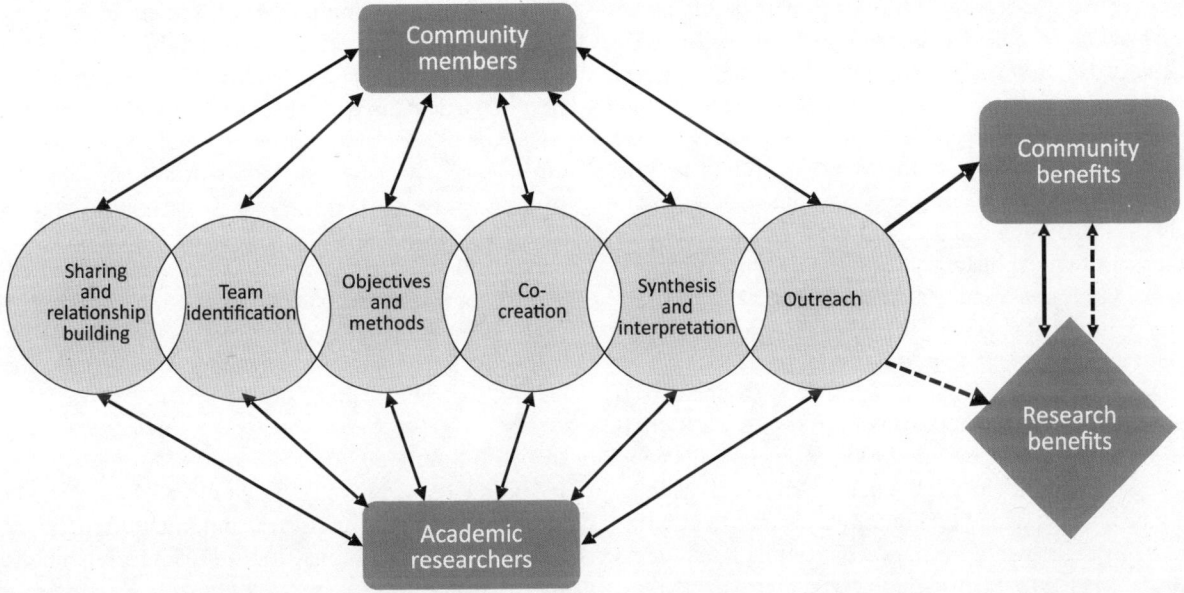

Figure 13.1 Conceptual framework illustrating OceanCanada's engagement and knowledge mobilization model based on community-based participatory research methods.

- how to maximize benefits of research so that it is equitable for both scholars and communities
- strategies that increase the ability of participatory knowledge mobilization processes to inform decision making.

We present a brief review of community-based participatory research (CBPR) and participatory video (PV) methods and a series of case studies developed through the KMWG with meaningful participation of communities across Canada's three oceans. With each case, we also generally examined some of the themes addressed by the larger OCP initiative: Changing Oceans, Access to Ocean Resources, and Ocean Governance (see Chapters 3–12). Some case studies address Indigenous communities and their holistic relationships with the ocean, acknowledging the long and challenging history of colonial approaches in scientific research. By combining community-based research and communications, these efforts helped amplify new voices to new audiences, and create value for partners beyond traditional academic outcomes.

COMMUNITY-BASED PARTICIPATORY RESEARCH

Background

Community-based participatory research is a collaborative approach to research that values and supports partners having equal participation in decision making (Israel et al. 1998), which generates knowledge co-production and co-ownership for positive change that yields more direct benefits for communities (Halseth et al. 2016; Minkler 2005; Castleden, Morgan, and Lamb 2012). This destabilizes traditional power hierarchies between the "researched" and the "researcher" (Rice and Mündel 2018). It includes local and traditional knowledge that is action-oriented, with the ultimate goal of supporting positive change and community empowerment (Tremblay and Jayme 2015). These approaches are particularly effective when combined with video.

Participatory Video

Participatory video is an action-oriented research approach that uses video and collective filmmaking as a means of identifying and addressing participants' needs and provoking collective action (Milne, Mitchell, and De Lange 2012). It allows participants to express themselves in their own words, and retains an "extended language" of participants' subtle gestures, allowing them to "speak for themselves" in a way that other mediums do not. Combined with local and Indigenous knowledge, video has rich visual information and is an effective tool for transmitting oral history, lived experience, and social and environmental context (Borish et al. 2021). From a community standpoint, it is also an effective way to document, archive, and share local and Indigenous knowledges, and contribute to bridging knowledge across generations, further supporting local resiliency. Participatory video allows local and Indigenous knowledge to be represented, not in text but in the rich visual medium of video, which allows communities to represent themselves in their own words and environmental contexts (Borish et al. 2021). This is a powerful approach that allows other communities – including scientists, decision makers, and other, often external, partners – to experience the lived realities and priorities of people embedded within their place.

CBPR and Indigenous Research Methodologies

Our work aligns with several aspects of Indigenous research methodologies (Smith 2013; Wilson 2008) that have been formulated by Indigenous scholars to provide a conceptual framework for engaging with Indigenous epistemology. It is first and foremost a relational approach – focused on building and reinforcing relationships based on respect, relevance, reciprocity, and responsibility – and thus associated research processes and products are grounded deeply in community priorities and knowledge (Wilson 2008). It does not search for "objective" reality, but rather recognizes that knowledge is *subjectively situated* (Wilson 2008).

Although this approach may account for mutual respect, trust, and self-governance within research, it is crucial to note that it is not a panacea for Indigenous research (Stanton 2014). Limited relationships and timelines can lead to superficial exchange and engagement where the project reverts to mainstream methods and control (Stanton 2014). Challenges remain, including researcher position, outsider status, and the dissemination

of results in a meaningful, ethical, and culturally appropriate way.

Another important challenge lies in the implementation of participatory methods and tools (Wilson 2008). As technological changes lower barriers to access, scholars are increasingly using visual community-based methods to document and share research. These methods include photo-voice, digital storytelling, digital mapping, and participatory video (Gubrium and Harper 2016). Packard (2008) argues that such approaches are more accessible to community members than nonvisual methods, which helps alter the uneven power dynamics between researchers and community members.

OUR APPROACH

We used a PV approach centred on local and Indigenous epistemologies, inquiries, and knowledge through the application of a robust ethic of relationship building and respect for diverse ways of knowing. Communities were involved in every aspect of the research/production process (planning, production, editing, and dissemination), and were allowed to determine what types of end products would suit their purpose (Figure 13.2). Research questions differed from one case study to another (climate change impacts on oceans, fisheries, and local economy, cultural revitalization, resource access, governance, etc.), and were addressed through the lens of PV. Success was determined by the total audience reached by the end products and, most importantly, by the positive impacts that PV generated in the communities.

We carried out four case studies covering Canada's three coastlines: (1) climate change across coastal British Columbia; (2) the *glwa* canoe resurgence in Bella Bella, BC; (3) youth perspectives on ocean changes in Pangnirtung, Nunavut; and (4) the sustainable lobster fishery in the Magdalen Islands, Quebec. Put together, they demonstrated social-ecological diversity among Canada's three oceans.

Case 1 – *Beyond Climate*

Across the Pacific coast, ocean ecosystems and Indigenous and non-Indigenous communities are increasingly under threat due to complex factors, including climate change, industrial development, and numerous other anthropogenic stressors. These interconnected impacts –

Figure 13.2 Our research approach, engaging communities in a modified participatory video method, adapted to work with local and Indigenous communities and knowledges. The whole process revolves around building and maintaining meaningful and equitable relationships that sustain an iterative research cycle encompassing (1) trust and exchange; (2) co-creation; (3) analysis, synthesis, and consensus; and (4) sharing and celebration.

and associated human narratives across the landscape – are often difficult to capture, synthesize, and communicate. Given the importance of holistically understanding the coastline (e.g., Ommer 2007), PV was used to assess research objectives related to local-level climate change impacts and experiences across the Pacific region, and how this linked with climate communications.

As the OceanCanada Partnership initiative commenced, Dr. Ian Mauro was planning his third feature film about climate change in British Columbia: *Beyond Climate* (Mauro and Suzuki 2018), a collaboration with scientist, broadcaster, and environmentalist David Suzuki and numerous community partners. It was designed to document the impacts and opportunities that climate change presented for British Columbia as a coastal province and to increase public awareness and political

support for climate action. Filmed over several years, the research documentary included interviews with 50 people – from First Nations, farmers, fishers, and foresters to scientists, policy-makers, and industry leaders – enabling the filmmakers to establish trust-based relationships with various individuals and communities (Figure 13.1). The film weaves their human stories in an iteratively developed visual narrative that documented climate change unfolding in real time and with real consequences.

Mauro and Suzuki brought a preliminary draft of the film to 12 coastal communities across British Columbia on a "Listening Tour" in 2015, which was designed to further hear from and document additional community perspectives and to allow communities to provide specific feedback on the film in a participatory fashion. More than 3,500 people attended these events in Nanaimo, Port Alberni, Comox, Campbell River, Alert Bay, Port Hardy, Bella Bella, Smithers, Kitimat, Prince Rupert, Masset, and Skidegate, and a report about the tour and its findings was published and presented to the United Nations (Grames 2016). In some communities, additional video interviews were conducted, and feedback on the draft film iteratively informed the final documentary. The film was structured using climate communication theory and following similar methodologies published elsewhere (Borish et al. 2021).

Indigenous communities throughout the Pacific coast were seeing substantial changes to oceans as well as the rivers that support salmon species and their migration throughout the region. For example, Peter Lantin, former president of the Haida Nation, spoke about historic drought conditions in his people's island homelands. "Haida Gwaii is a rainforest, so that has a huge impact on us. All the rivers that we've relied on historically, we're seeing the levels being low, it's affecting fishing, it's affecting everything." Given the cultural and nutritional importance of salmon for many Indigenous communities, these communities remained extremely concerned about climate change and its impact on food security, cultural harvesting rights, and traditions.

The film also features a case study on Island Scallops, a business based in Qualicum Beach, Vancouver Island. CEO Rob Saunders discussed how increased acidity of the oceans caused the company millions of dollars in financial losses, and how this has affected his ability to

Figure 13.3 Participants from the *Beyond Climate* film project – Rob Saunders, CEO of Island Scallops (left) and Peter Lantin, former president of the Haida Nation (right) – speaking about the impacts of climate change on the Pacific Ocean and marine species such as shellfish and salmon. These place-based narratives were geographically located within the film using animations.

employ people within his community. Reflecting on 2013, a particularly challenging year, Saunders noted: "Ten million animals gone in such a short period of time. No company can withstand those kinds of losses. There's no question that atmospheric CO_2 is increasing. Global warming, I don't think there's any doubt."

Beyond Climate generated significant outreach opportunities for BC communities, the province, and the country, generating news and awareness regarding climate change for millions of people in Canada and internationally. Since its release at the Planet in Focus International Environmental Film Festival in Toronto in the fall of 2018, the film has been screened at over 25 film festivals around the world, and has won numerous awards, including multiple Best Documentary Feature and People's Choice Awards. Mauro and Suzuki continued to tour the film, engaging audiences across Canada in

discussions regarding climate change, and they eventually released the film online (www.beyondclimate.ca) to support communities navigating the climate crisis. Lasting partnerships, specifically with First Nations communities, were established through this filmmaking process, and ongoing co-developed research continues to be generated through this work (see Case 2 below).

Beyond Climate is an example of "landscape-level listening," with research conducted across a broad geography with the explicit goal of seeking diverse perspectives within and participation across the region as a whole. This combination of social sciences and filmmaking increases the communicative potential of research, creating new audiences for scholarship, and mobilizes knowledge in a manner that is accessible, high-impact, and designed to cultivate and promote both community knowledge and real-world action.

Case 2 – Heiltsuk Cultural Resurgence in Bella Bella

Tribal Canoe Journeys is an annual Indigenous gathering where several "canoe families" from different Indigenous Nations along the Northwest Pacific coast of Canada and the United States paddle to a host community to share cultural experiences over several days. In 1986, a traditional dugout canoe, known as a *glwa* in the Heiltsuk language of Hailhzaqvl, had been carved for the first time in over a hundred years in the community of Bella Bella and paddled down to Vancouver by Hereditary Chief Frank Brown and other members of the Heiltsuk Nation for Expo 86. For many members of the Nation, this was a critical moment for Heiltsuk cultural resurgence, which we have further reflected upon and co-published elsewhere based on the findings from this PV research project (Brown et al. 2021).

In the summer of 2014, the Heiltsuk community of Bella Bella hosted Tribal Canoe Journeys. During the filming of *Beyond Climate*, Mauro attended the gathering and met Frank Brown, his daughter, Vina, and other Heiltsuk community members and leaders. Through conversations about the importance of Indigenous leadership in research and how participatory processes help facilitate this, they quickly established a partnership. They conducted interviews with Heiltsuk community members about the history and impact of Tribal Canoe Journeys, while filming cultural songs, dances, and stories that were shared during the gathering.

Following the production of a 13-minute documentary about the gathering called *Qatuwas 2014*, Brown invited Mauro and master's student Hillary Beattie to collaborate on a larger research project about Tribal Canoe Journeys. The main research objective of this initiative was to systematically document and understand the impact of Tribal Canoe Journeys on the resurgence of the Heiltsuk Nation, using participatory video.

After obtaining approval from the Heiltsuk Integrated Resource Management Department (HIRMD) and from university research ethics boards, Hillary Beattie and Vina Brown conducted video-based interviews with an additional 40 community members and leaders from Bella Bella over a two-year period. They also travelled down the Northwest Pacific coast in the summer of 2016 with a group of Heiltsuk youth as they paddled to a Tribal Canoe Journeys gathering in Nisqually, Washington (Figure 13.4). As part of the process, they taught participatory video techniques to the youth, who helped record the journey.

Beattie and Mauro edited drafts of the film that were regularly sent to Bella Bella for feedback, and the team collectively made the film using collaborative editing software and extensive dialogue. This process was overseen

Figure 13.4 Heiltsuk youth paddling to Nisqually, Washington, for Tribal Canoe Journeys in 2016.

by a larger Qatuwas committee, made up of local elders and community members, ensuring community oversight and control over the initiative. Vina Brown visited Winnipeg multiple times to continue the co-creation process. Early in 2017, the team finalized a 46-minute documentary titled *Glwa: Resurgence of the Ocean-Going Canoe* (Beattie and Brown 2017). Prior to public release, the team hosted a screening in Bella Bella in September 2017 to receive any final feedback from community members, and ultimately received approval for release. The film premiered at the imagineNATIVE Film + Media Arts Festival in Toronto and subsequently screened at multiple events and film festivals across Canada. The video trailer also reached almost a quarter of a million people online through social media.

The team documented oral narratives regarding the importance of Tribal Canoe Journeys and its impact on the community, especially the fact that travelling by canoe through traditional territories helped people reconnect with the natural environment and their ancestors. "It's different than being out on the water in any other way because you see it differently ... You're closer, you're in relation to it," one young man said of canoe journeys. Participants also said the journeys helped them heal from intergenerational trauma caused by the residential school system and from other impacts of colonial policy. It also helped them develop confidence and self-esteem, especially for youth participants, who indicated that the journeys taught them about their traditional language, cultural practices, and laws, which had been outlawed by the potlatch ban between 1885 and 1951. Tribal Canoe Journeys also helped participants develop strong relationships between their "canoe families," across generations within their community, and between different Indigenous Nations attending the gatherings.

Overall, the film has helped support the Heiltsuk Nation in sharing its story of cultural revitalization and the important intergenerational work it is doing – linking elders and youth – to reclaim Heiltsuk traditional knowledge and practices and ensure continued connection with and stewardship of the lands and oceans. The PV approach helped the community to document its oral history, and use video to communicate with and reach a much larger audience with its stories and knowledge. Since the release of the film, the documentary and interviews have contributed to the Sacred Journeys exhibit (https://sacredjourneyexhibit.com/), which has toured major museums across the Pacific coast and demonstrates the enduring value of PV projects when overseen and controlled by Indigenous communities.

Case 3 – Youth Perspective on Ocean Change and Inuit Culture in Pangnirtung

Changes in land use and access, community activities, and animal populations related to climate change are a source of concern for Arctic residents. Such observations are documented in *Qapirangajuq: Inuit Knowledge and Climate Change,* the first feature-length film about climate change in Inuktitut, which was collaboratively developed with communities across Nunavut (Mauro and Kunuk 2010). In this film, Elders and hunters were particularly vocal about the threat of climate change to their traditional and modern lifestyles.

In 2016, building on a long-term relationship with the community of Pangnirtung, Nunavut, Mauro was invited back to the community with master's student Natalie Baird to develop a new community-based research project with the main objective of visualizing changing oceans through video. Many community members highlighted the need to support youth skills and development as well as intergenerational knowledge exchange as key components of adapting and responding to social-ecological change. Local fisher and community leader Johnny Mike shared: "As Inuit living in the Arctic, we are at the frontlines to witness the change. Younger generations will be facing much more climate change than we can see today." In partnership with local youth organizations, the second research objective of the project was to explore youth perspectives on land, ocean, identity, and environmental change using video and storytelling approaches.

Baird partnered with local filmmaker David Poisey, the Pangnirtung Youth Centre, and Attagoyuk Ilisavik (local high school) to facilitate two video and photography workshops for youth and Elders, in the community and on the land (Figure 13.5). In the first workshop, she engaged youth and Elders in planning, producing, and editing short films in English and Inuktitut. In the second workshop, youth were introduced to "pinhole cameras" – made from coffee tins, recycled material, and

Figure 13.5 Pangnirtung film-maker David Poisey with grand-daughter Tatiana Mike. Poisey was born on the land at Kivituq, an outpost camp near Qikiqtarjuaq. An award-winning cinematographer and director, he is recognized as an important mentor to Indigenous filmmakers.

black-and-white photo paper – a very simplified camera technology. Youth built their cameras and took to the shoreline to create photos based on the question: Why are *imaq* (sea water) and *siku* (sea ice) so important to you?

The team collectively produced five short films that offer glimpses into the multiple identities and expressions of youth, including their strengths, the challenges they face, and their hopes and dreams for the future. The short film *Tariuq Takujannik – The Ocean from My Eye* (Baird and Poisey 2018) documented the process and photography results of the pinhole camera workshop. As the youth voices share in the film, "the sea holds animals and passage to our ancestors' lands," revealing the importance of ocean and sea ice as continued pathways to youth individual and collective identity and well-being. All five short films highlighted the fact that the youth feel a sense of personal responsibility – as well as hope – for their future, which is tied to healthy ocean environments. These films are highly mobile, travelling to audiences across Canada and internationally through film festivals, online screenings, and academic conferences.

Pangnirtung youth have gained momentum in climate action. Their multi-sensory depictions of identity, land, and changes break away from the standard narrative of

Inuit youth as passive victims of climate change, instead sharing the hope of Inuit youth and highlighting their vitality, persistence, and adaptation in the face of change. The project bridged generations and supported youth in further developing self-confidence, communication skills, and inspiration for the future. The ocean remains an important pathway to identity as well as physical and mental well-being.

Case 4 – Sustainability and Resilience in the Magdalen Islands Lobster Fishery

The Magdalen Islands are situated in the middle of the Gulf of St. Lawrence. The region is rich in fish and shellfish species that have attracted Indigenous and European fishers for centuries (Figure 13.6) (personal communication by Claude Painchaud). For over 125 years, the lobster fishery has been a central pillar of the socioeconomic and cultural landscape in its remote communities (Gendron, Camirand, and Archambault 2000). Over the last decades, however, a number of new stressors, including ocean warming (Cetina-Heredia et al. 2015) and offshore oil development (Bourgault et al. 2014), have emerged in the Maritime provinces, with the potential to impact the lobster fishery. Given the region's importance from both economic and cultural standpoints, our main

Figure 13.6 Two generations of lobster fishers in the Magdalen Islands (*left to right:* Francis, Denis, and Simon Deraspe).

research objective was to explore the contemporary challenges faced by the lobster fishery of the Magdalen Islands, and the fishery's resiliency.

As with the cases presented above, the Magdalen Islands case study was developed through community dialogue and partnerships from Vincent L'Hérault's pre-established relationships with the community. Between summer 2018 and spring 2019, he teamed up with a local author, Annie Landry, to initiate a PV project about the lobster fishery. Typical of many participatory initiatives, the project started out broad in scope and progressively took shape based on community input regarding content shared through interviews and as opportunities unfolded with participants. The film team conducted approximately 50 video-based interviews with key actors involved in the lobster fishery, and also filmed major components of the industry: fishing, processing, transport, marketing, business, and other activities (Table 13.1). During this process, given the diversity of the testimonies documented, we chose to assemble the information according to the access framework developed by OCP team members (Bennett et al. 2018). Access is defined as the ability to use and benefit from this resource.

Interviews showed that access to lobster has greatly evolved in the past and still does today, as determined by a variety of issues and drivers (Table 13.1A). The wealth of the stock itself has greatly fluctuated over time (personal communication by L. Gendron) due to fishing pressure – resulting from changes in technical capacity and competencies – management, and environmental

changes. Changes in rights, allocation methods, and market development have also influenced access to the lobster (DFO 2010). Interviews also highlighted a number of mechanisms contributing to local resilience within the fishery (Table 13.1B).

The Magdalen Islands lobster fishery is a complex business that requires the sound and timely contribution of many actors operating across different sectors and a continuum of competencies that bring the product from sea to plate. One of the main contributions of PV was provision of a fertile ground for a diverse range of stakeholders involved in the fishery to reflect on their industry and the future of the Magdalen Islands. Although the film has yet to be released, given its participatory nature it has generated substantial interest in the community and, like the previously described case studies, has significant potential to communicate the outcomes of the project and the perspectives of fishers to a larger audience. Our approach – focused on fishers but inclusive of the larger industry – also generated considerable interest across the entire fisheries sector, and now various businesses and government actors have expressed interest in supporting its continuation and potential expansion.

DISCUSSION AND INTEGRATION

We approached each of our four case studies with involvement of community partners from the start to ensure co-design of objectives and methods, co-creation and co-interpretation of results, and widespread dissemination of the findings (Figure 13.1). The case studies were distinct from each other in significant ways (Table 13.2). To understand these nuances – and the larger lessons learned – we reflect on the results while also considering the original objectives of this chapter.

Participatory Video Promoting Engagement and Communication

Our work succeeded in bringing about community engagement and associated knowledge mobilization. In total, 15 communities and nearly 200 participants were engaged across Canada's three coastlines, and numerous digital media outputs were developed that foregrounded community knowledge, perspectives, and priorities within and beyond the places where projects were developed. These research outcomes enabled communities

Table 13.1

Summary of 50 interviews conducted in the Magdalen Islands and abroad on the topics of (A) potential challenges to the lobster fishery and impacts on resource access, and (B) components of resiliency in the lobster fishery and benefits for resource access.

Topic	*Access*
A – Challenges	**Impacts on access**
Environmental change; ocean warming and the associated shift in ecosystems	Short- to medium-term increase in lobster stocks, uncertain future linked to ocean patterns and trophic interactions
Competing uses: (1) offshore development; (2) North Atlantic right whale protection measures	(1) Increased risk to resources; (2) physically reduced access to fishing zones
Globalization and market evolution that result in obsolete marketing models	Decrease in competitive advantage and benefits to local fishers/industries
Shortage of labour force in processing plants	Losses and/or decreased volume of production that affects market for lobster
B – Components of resiliency	**Benefits for access**
Federal policy under the *Fisheries Act:* Policy for Preserving the Independence of Inshore Fleet in Canada's Atlantic Fisheries (DFO 2010)	Puts benefits in the hands of the fishers, promotes financial capacity, promotes local lifestyle and well-being, and brings indirect benefits to communities
Provincial funding programs to fishers and the industry	Promote capacity: technical innovation and enrolment in the industry
Professionalization of the industry: Fishing School – professional post-secondary diploma	Promotes capacity: knowledge transfer; technical capacity
Consultation committee made up of managers and fishers	Promotes understanding and respect for a smoother, consensus-based application of management measures
Development of resource wealth indicators	Promotes responsiveness to changes in the resource and effectiveness of management practices
Commercialization agreement between fishers and industry	Promotes consistency in ex-vessel price and benefits to fishers
Marine Stewardship Council certification (first in North American Atlantic – 2013)	Opens niche market opportunities and provides competitive advantage

to speak their truths to diverse audiences and stimulate collective action (Milne, Mitchell, and De Lange 2012).

We operated at various scales and levels of participation, depending on local specific objectives. For example, the *Beyond Climate* film was co-developed with numerous communities and partners along the entire coast of British Columbia. In local case studies – in Bella Bella, Pangnirtung, and the Magdalen Islands – PV projects focused on single communities and their priorities. *Beyond Climate* has the power of being reflective of the whole province and much of its coastline, through the practice of "landscape-level listening" with iterative cycles of engagement, and a significant media and outreach impact regionally, nationally, and globally. Given its provincial focus and scale, arguably this project had more diffuse participation, capacity building, and co-creation compared with other case studies. These types of trade-offs – on this "continuum of participation" – are important to consider early on in project co-design with communities. This kind of research requires significant technical

Table 13.2

Summary of the four case studies, including the scale of the social-ecological system, nature of partners, nature of participants, type of participatory video (PV), OCP cross-cutting themes addressed, and main contributions

	Beyond Climate	Glwa (Bella Bella)	Qapirangajuq (Pangnirtung)	Magdalen Islands lobster fishery
Scale	Provincial, especially BC coast	Local	Local	Local
Partners	David Suzuki Foundation First Nations leaders and knowledge keepers Shellfish industry	First Nations leaders and community members	Inuit community members High school teachers Professional cinematographer	Local artist Local fishers
Participants	Community members, youth and Elders, business owners, municipal authorities, non-profit organizations, academic researchers ($n = 50$)	Community members, youth and Elders ($n = 50$)	Community members, youth and Elders ($n = 19$)	Fishers, youth and Elders, business owners, government representatives ($n = 50$)
PV	Feature film	Feature film	Short film	Feature film
Audience	>1 million	>100,000	>30,000	TBD
OCP theme	Changing Ocean Access to Ocean Resources Ocean Governance	Access to Ocean Resources Ocean Governance	Changing Ocean Access to Ocean Resources	Access to Ocean Resources Ocean Governance
Main contribution	Large-scale engagement Iterative process with community validation Knowledge synthesis and translation across local and provincial scales	Community empowerment through cultural revitalization	Youth empowerment through skills and voice expression	Community empowerment through "gatherer" effect Linking people in adversity Working together

expertise, funding, and time for completion. Projects must be carefully planned, balancing the need to potentially create training opportunities for students and post-doctoral fellows while ensuring community leadership, capacity building, and ultimately benefits.

Reinforced Participation

Throughout each of the projects, we co-developed, in collaboration with communities, relationships and processes that reinforced participation so that there were meaningful, consistent, and collaborative interactions between academic and community-based partners. Figures 13.1

and 13.2 show how relationship building and trust support co-creation, which can take time. Indeed, some have suggested that scholars might spend "the first year drinking tea" (Castleden, Morgan, and Lamb 2012) before commencing formal research. For the most part, we benefited from having pre-existing relationships with individuals and/or communities, enabling us to quickly build rapport and co-design projects with partners, which ensured meaningful consent to proceed with research that was directed by, with, and for community. Mauro and L'Hérault have long-term relationships and ongoing research with various Arctic communities, which enabled opportunities in Pangnirtung to unfold relatively easily. However, a long-term relationship is not an absolute requirement for effective partnership, as the Tribal Canoe Journey research shows. Mauro was invited by the community to attend the Tribal Canoe Journey gathering, and was instructed by Hereditary Chief Brown on the protocols and obligations regarding filming within Heiltsuk Territory. He agreed to and followed these protocols, and the collaboration quickly flourished and continues to the present day. The Magdalen Islands research stemmed from long-term relationships between L'Hérault and a number of individuals in the community. The film created common ground and messaging among the community and has helped raise interest and interaction among local, provincial, and international business players. Since the work in Pangnirtung, Bella Bella, and the Magdalen Islands was highly localized and highly relational, the linkages between academic and community partners was, and continues to be, very rich.

Equitable Research

Before starting this kind of research, academic partners must critically reflect on their privileged positions and the power imbalances this may bring to research, especially when this might involve researching historical relationships between settlers and Indigenous Peoples. Important questions must be asked: How are community members participating in the research process? Is the project challenging unequal relations between researchers and participants? Will the research actually benefit the community, and who decides this?

In all the case studies presented, our team made significant efforts to make sure that these critical questions were being asked, and that communities were the ones who ultimately decided the direction, scope, and outcomes of the research. In particular, training, capacity building, meaningful co-creation, synthesis, and interpretation of results were central to the work and helped ensure equitable power dynamics between researchers and community members. Given that we explored challenges facing communities – involving aspects of both environmental and cultural change – the training was particularly important for community-based youth and university graduate students as it provided pragmatic ways to create and communicate community-based solutions.

The Pangnirtung and Bella Bella case studies were particularly salient in developing community learning centred on youth engagement through research and video skills development. Here, youth opportunities and empowerment had been specifically identified as a priority by the communities, and young people collaborated with graduate students to meaningfully contribute to original research and associated video making. Similarly, the *Beyond Climate* project also engaged youth in various communities through the Listening Tour, and schools in each community were encouraged to create art projects linked to the themes of the film. These participatory and art-based approaches position research so that it is overseen by communities and conducted in a manner that respects and represents their views. Given the video-based outputs, the co-created knowledge can be mobilized widely, and demonstrates what equitable research looks like in practice.

Strategies to Inform Decision Making

Unlike standard approaches to interview-based research, research videos are easier to share with policy-makers and the general public than academic texts (Anderson and McLachlan 2016), and video outputs can contribute to public dialogue and support political change. Since participatory video processes are explicitly designed to identify and address participants' needs and stimulate collective action (Milne, Mitchell, and De Lange 2012), they are ideal outputs for informing decision making. *Beyond Climate* has been screened across British Columbia, Canada, and the world to mayors, Indigenous leaders, academics, and communities, and has been extensively reported on by the media in discussions of issues ranging from greenhouse gas emissions to oil tankers and pipelines. The videos of Bella Bella and Pangnirtung show that community-based and intergenerational knowledge in the videos was shared with local leaders, and helps ground the community in its history as a baseline for navigating the future. In the Magdalen Islands, interviews have been conducted with locals, policy-makers, and a range of other actors, bringing about a cross-sectoral dialogue that is critical for the future of the lobster fishery. Our approach has helped strengthen the linkages between social science and conservation policy, which in turn may help clearly identify management or policy implications of research (Game, Schwartz, and Knight 2015).

CONCLUSION

It is critical to have communities not only involved in research but actually driving the process, which ensures that it benefits local people and results in more relevant and durable management and policy. We have argued that participatory approaches, when tightly linked with communication techniques, lead to effective and ethical ocean research. These participatory approaches rebalance power dynamics between researchers and knowledge holders and contribute to:

- portraying local voices, opinions, perspectives, and Indigenous knowledges in deeper, more accurate, and more authentic ways
- giving local communities a stronger voice and increased recognition for their value in conservation and management decisions
- producing research results and associated policy that are more relevant, robust, and durable
- enabling effective, relevant, and meaningful communication of research results across a broad spectrum of key audiences, from the participating communities to resource managers, policy-makers, and the general public.

Our participatory and community-led approach – working with cameras across coastlines and communities – produced high-impact research that transmitted knowledge about Canada's oceans to audiences regionally, nationally, and internationally. Throughout this work, the voices and priorities of communities have always been front and centre, which we believe is one of the best ways to understand and conserve oceans now and into the future.

REFERENCES

Anderson, C.R., and S.M. McLachlan. 2016. "Transformative Research as Knowledge Mobilization: Transmedia, Bridges, and Layers." *Action Research* 14 (3): 295–317.

Augustine, S., and P. Dearden. 2014. "Changing Paradigms in Marine and Coastal Conservation: A Case Study of Clam Gardens in the Southern Gulf Islands, Canada." *Canadian Geographer/Le Géographe canadien* 58 (3): 305–14.

Baird, N., and D. Poisey. 2018. *Tariuq Takujannik – The Ocean from My Eye*. Video. https:/www.youtube.com/watch?v=HTYQEeCiC4M.

Beattie, H., and V. Brown. 2017. *Glwa: Resurgence of the Ocean-Going Canoe*. Feature film. Winnipeg: Prairie Climate Centre, https://www.glwafilm.com/.

Bennett, N.J., M. Kaplan-Hallam, G. Augustine, N. Ban, D. Belhabib, I. Brueckner-Irwin, A. Charles, et al. 2018. "Coastal and Indigenous Community Access to Marine Resources and the Ocean: A Policy Imperative for Canada." *Marine Policy* 87: 186–93.

Bennett, N.J., R. Roth, S.C. Klain, K.M.A. Chan, D.A. Clark, G. Cullman, G. Epstein, et al. 2016. "Mainstreaming the Social Sciences in Conservation." *Conservation Biology* 31 (1): 56–66.

Borish, D., A. Cunsolo, I. Mauro, C. Dewey, and S.L. Harper. 2021. "Moving Images, Moving Methods: Advancing Documentary Film for Qualitative Research." *International Journal of Qualitative Methods* 20. https://doi.org/10.1177/16094069211013646.

Bourgault, D., F. Cyr, D. Dumont, and A. Carter. 2014. "Numerical Simulations of the Spread of Floating Passive Tracer Released at the Old Harry Prospect." *Environmental Research Letters* 9 (5): 054001.

Brown, F., H. Beattie, V. Brown, and I. Mauro. 2021. "Tribal Canoe Journeys and Indigenous Cultural Resurgence: A Story from the Heiltsuk Nation." In *The Politics of the Canoe,* edited by B. Erickson and S. Kotz. Winnipeg: University of Manitoba Press.

Castleden, H.C., V.S. Morgan, and C. Lamb. 2012. "'I Spent the First Year Drinking Tea': Exploring Canadian University Researchers' Perspectives on Community-Based Participatory Research Involving Indigenous Peoples." *Canadian Geographer/Le Géographe canadien* 56 (2): 160–79.

Cetina-Heredia, P., M. Roughan, E. van Sebille, M. Feng, and M.A. Coleman. 2015. "Strengthened Currents Override the Effect of Warming on Lobster Larval Dispersal and Survival." *Global Change Biology* 21 (12): 4377–86.

Charles, A. 2012. "People, Oceans and Scale: Governance, Livelihoods and Climate Change Adaptation in Marine Social-Ecological Systems." *Current Opinion in Environmental Sustainability* 4 (3): 351–57.

DFO (Fisheries and Ocean Canada). 2010. "Policy for Preserving the Independence of Inshore Fleet in Canada's Atlantic Fisheries." http://www.dfo-mpo.gc.ca/reports-rapports/regs/piifcaf-policy-politique-pifpcca-eng.htm.

Game, E.T., M.W. Schwartz, and A.T. Knight. 2015. "Policy Relevant Conservation Science." *Conservation Letters* 8 (5): 309–11.

Gendron L., R. Camirand, and J. Archambault. 2000. "Knowledge Sharing between Fishers and Scientists: Toward a Better Understanding of the Status of Lobster Stocks in the Magdalen Islands." In *Finding Our Sea Legs: Linking Fishery People and Their Knowledge with Science and Management.* St. John's: ISER Press.

Grames, P. 2016. *Charting Coastal Currents: Canada's Pacific Communities Talk Climate, Culture, Oceans and the Future.* Vancouver: David Suzuki Foundation. https://davidsuzuki.org/science-learning-centre-article/charting-coastal-currents/.

Gubrium, H., and K. Harper. 2016. *Participatory Visual and Digital Methods.* London: Routledge.

Halseth, G., S.P. Markey, D. Manson, and L. Ryser. 2016. *Doing Community-Based Research: Perspectives from the Field.* Montreal and Kingston: McGill-Queen's University Press.

Israel, B.A., A.J. Schulz, E.A. Parker, and A.B. Becker. 1998. "Review of Community-Based Research: Assessing Partnership Approaches to Improve Public Health." *Annual Review of Public Health* 19: 173–202.

Jones, R., C. Rigg, and L. Lee. 2010. "Haida Marine Planning: First Nations as a Partner in Marine Conservation." *Ecology and Society* 15 (1): 12. http://www.jstor.org/stable/26268116.

Kral, M.J. 2014. "The Relational Motif in Participatory Qualitative Research." *Qualitative Inquiry* 20 (2): 144–50.

Long, R.D., A. Charles, and R.L. Stephenson. 2015. "Key Principles of Marine Ecosystem-Based Management." *Marine Policy* 57: 53–60.

Mauro, I., and Z. Kunuk. 2010. *Qapirangajuq: Inuit Knowledge and Climate Change.* Short film. Igloolik Isuma Productions.

Mauro, I., and D. Suzuki. 2018. *Beyond Climate.* Feature film. Winnipeg: Prairie Climate Centre. https://www.beyondclimate.ca.

Milne, E.J., C. Mitchell, and N. De Lange. 2012. *Handbook of Participatory Video.* Lanham, MD: AltaMira Press.

Minkler, M. 2005. "Community-Based Research Partnerships: Challenges and Opportunities." *Journal of Urban Health* 82: ii3–ii12.

Ommer, R. 2007. *Coasts under Stress. Restructuring and Social-Ecological Health.* Montreal and Kingston: McGill-Queen's University Press.

Packard, J. 2008. "I'm Gonna Show You What It's Really Like Out Here: The Power and Limitation of Participatory Visual Methods." *Visual Studies* 23 (1): 63–77.

Reed, M.S. 2008. "Stakeholder Participation for Environmental Management: A Literature Review." *Biological Conservation* 141 (10): 2417–31. https://doi.org/10.1016/j.biocon.2008.07.014.

Rice, C., and I. Mündel. 2018. "Story-Making as Methodology: Disrupting Dominant Stories through Multimedia Storytelling." *Canadian Review of Sociology/Revue canadienne de sociologie* 55 (2): 211–31.

Smith, L.T. 2013. *Decolonizing Methodologies: Research and Indigenous Peoples.* London: Zed Books.

Stanton, C.R. 2014. "Crossing Methodological Borders: Decolonizing Community-Based Participatory Research." *Qualitative Inquiry* 20 (5): 573–83.

Tremblay, C., and B. Jayme. 2015. "Community Knowledge Co-Creation through Participatory Video." *Action Research* 13 (3): 298–314. https://doi.org/10.1177/1476750315572158.

Wilson, S. 2008. *Research Is Ceremony: Indigenous Research Methods.* Winnipeg: Fernwood Publishing.

14

Policy Direction for Reconciliation and Indigenous Ocean Management in Canada

Russ Jones, Nancy Doubleday, Megan Bailey, Ken Paul, Fraser Taylor, and Peter Pulsifer

This chapter examines the best practices for reconciliation identified in Chapter 2, and provides recommendations on the future direction of reconciliation in Canada given the current trajectory. Reconciliation is a "wicked problem," meaning one that is complex, presenting dilemmas that are resistant to solutions and are usually not solved once and for all but tend to reappear (Rittel and Webber 1973; Zurba et al. 2019).[1] It is difficult but possible to solve if we build on best practices and continue policy changes based on core principles of truth, justice, historical responsibility, and restructuring of the social and political relationships between Canada and Indigenous Nations (e.g., Rouhana 2011, 297). To this end, we build on the framework developed in Chapter 2 (Figure 14.1) to further examine the reconciliation criteria in Table 2.3 and we refer frequently to the examples provided there.[2] The discussion below focuses on changes underway in governance, resource access, and protection of culture and values that are having mixed success in transforming relationships. The policy recommendations focus on changes needed to establish a just and equitable reconciliation framework, and measures to advance shared management, planning, and governance of ocean spaces.

A PRINCIPLED ANALYSIS

Canada's approach to reconciliation moved in 2015 from a policy requiring surrender of lands and extinguishment of rights to one of recognition of rights.[3] Recognition of rights requires a shift to governance agreements that promote equality and power sharing. Canada has committed to reconciliation between Indigenous Peoples and Canadian society, but it has not made commitments typical of transitional justice (as cited in Hughes 2012[4]) despite such commitments having broad political and societal support. As a result, results are mixed in terms of achieving justice in the domains of political domination, loss of territory, and cultural imposition, which are the main impacts of colonization (Moore 2016).

Policy changes have occurred unevenly over decades, as discussed in Chapter 2, leaving reconciliation of many Indigenous ocean issues still unaccomplished. The effect of Canada's new policy of recognition of Indigenous rights, and adoption of international standards requiring Indigenous consent for resource development, could have far-reaching effects, but success depends on political will, and progress may require court direction.

GOVERNANCE: PROGRESS AND CHALLENGES

Twenty years ago, the Royal Commission on Aboriginal Peoples (RCAP) identified governance as a key area requiring transformative change in Canada (Chartrand 2004), and proposed comprehensive changes to Indigenous governance that were largely ignored.[5] Current Indigenous governance systems range from modern structures, including those recognized or established through modern treaties, to those established through colonial systems, such as *Indian Act* Band Councils.[6] Canadian government policies today, including those to implement the *United Nations Declaration on the Rights of Indigenous Peoples*

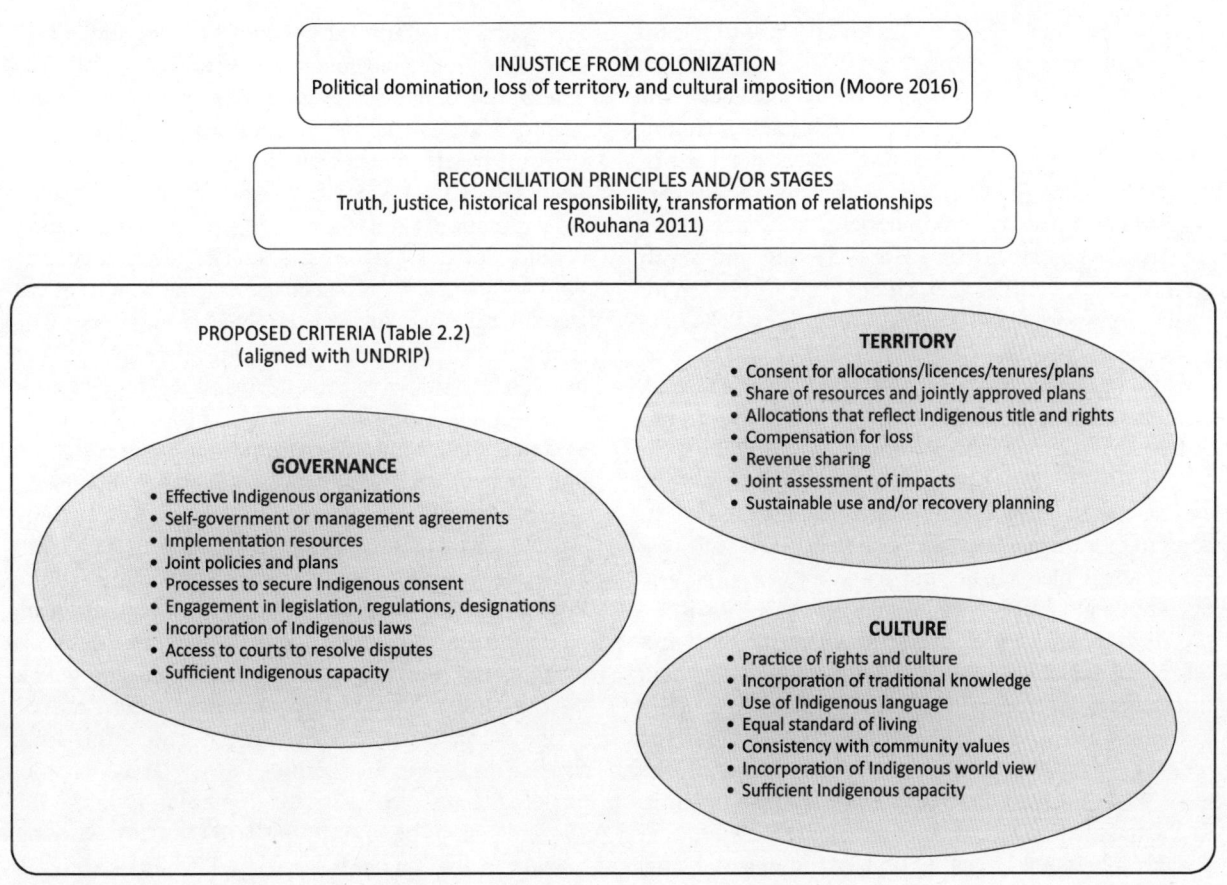

Figure 14.1 Framework for analysis of Indigenous reconciliation (based on Chapter 2).

(UNDRIP 2007) and Canada's Ten Principles (Department of Justice Canada 2018) state that Indigenous peoples have a right to self-determination and support self-government.[7] As indicated in Table 2.3 and the case studies in Chapter 2, governance systems are continuing to evolve and adapt in response to issues and challenges, including pressures on oceans and ocean resources and the need to address some of the issues identified by the RCAP. Polycentric approaches to governance are required to give standing to formal (i.e., governmental) and informal (extra-governmental, community, citizen initiatives) processes to support reconciliation.

Ocean governance structures in the Arctic, Pacific, and Atlantic regions differ significantly in role and mandate. In the Arctic, the scope of land claims agreements (LCAs)

or treaties often did not include ocean issues or were not at the scale required for Indigenous groups to work together on ocean issues. Dialogue on most Arctic issues, including oil and gas and commercial fisheries, has been possible through existing LCA structures (Case 3 in Chapter 2). Inuit have continued to engage internationally in dialogue about polar issues through the Inuit Circumpolar Council and the Arctic Council to advance common issues such as the international Pikialasorsuaq Commission (Case 3). In the Pacific, new governance structures have emerged among Indigenous groups and with federal or provincial agencies in advance of treaties to support integrated marine use planning (Case 1), oceans protection, fisheries, and treaty negotiation issues (e.g., Cases 4 and 5). British Columbia–wide bodies have played a coordinating role but lack a mandate to address Indigenous title and rights.[8] New structures that are interim to treaty settlements are largely consistent with reconciliation criteria identified in Table 2.2, including effective Indigenous organizations, presence of management agreements, joint policies, and plans, and the capacity and resources to support implementation. Issues in the Atlantic continue to be addressed at the community level (Case 2), but aggregate bodies such as the Atlantic Policy Congress of First Nations Chiefs (APCFNC) play a role in regional issues.[9]

Although Canada committed to UNDRIP, there are uncertainties about the degree of political commitment to free, prior, and informed consent (Patzer 2019). The current UNDRIP wording was a compromise but did envisage that "*some circumstances* should require the consent of Indigenous groups – namely, large-scale projects that will have a major impact on the lands and lives of Indigenous peoples" (Patzer 2019, 225). This interpretation is consistent with interpretation of Aboriginal title by Canadian courts that concludes that Indigenous title holders have the right to exclusive use and "the right to determine the uses to which the land is put."[10] New policies of recognition of rights and the seeking of free, prior, and informed consent (FPIC) have the potential to change the power dynamics of negotiations and may provide a basis for revisiting understandings of former treaties and agreements, but there is much work still to be done on this front at the federal level.

Co-management, a form of power sharing, is an evolving governance process that has the potential to contribute to reconciliation. Effective co-management requires meaningful power sharing, but also building of capacity and capability to ensure that the parties' intent can be implemented and fully realized. As outlined in Table 2.3 and the case studies in Chapter 2, self-government is strongest at community levels but has been recognized at broader scales in some modern LCAs.[11] It requires formal avenues and resources for Indigenous agency and perspectives, including traditional knowledge, to contribute to the co-management decision-making process (Snook et al. 2019). It can provide a model for reconciliation consistent with reconciliation principles by supporting Indigenous self-determination, but decisions and recommendations based on consensus also meet require-

ments for consultation and accommodation under Canadian law. Co-management is generally high for issues where Indigenous rights have been court-determined, such as fisheries, marine mammals, and protected areas. Governance structures need to be developed on a practical scale (i.e., site, territory, region) that supports self-determination; and where they pre-exist, Indigenous organizations should form the basis for governance structures. One challenge is that co-management requires capacity that is often developed only after major issues are resolved. While multiple co-management boards have been created under various LCAs in the Arctic, it is unclear whether the spirit and intent of true co-management as conceptualized in the agreements is being realized in practice, particularly for fisheries (Snook et al. 2019).

Commitments to recognize Indigenous law and governance are incomplete, as evidenced by the recent federal initiative to develop a framework for recognition and implementation of rights, including "new legislation and policy that will make the recognition and implementation of rights the basis for all relations between Indigenous Peoples and the federal government going forward" (Prime Minister of Canada 2018). Since 2015, this initiative has resulted in the creation of over 80 reconciliation of rights tables across Canada, representing more than 390 Indigenous communities (CIRNA 2020). Development of legislation has been delayed by the need to gain support from Indigenous organizations and legislation is sometimes passed with limited consultation (Youds 2020). For some, this means nothing less than full partnership in drafting the legislation (Assembly of First Nations 2018). Recognition of Indigenous law constitutes an important gap despite acknowledgment in Canada's Ten Principles, remaining a challenge due to the diversity of Indigenous Peoples and Crown denial of Indigenous jurisdiction, thus becoming the subject of Indigenous activism.[12]

ACCESS TO RESOURCES: PROGRESS, OPPORTUNITIES, AND CHALLENGES

Treaty and reconciliation processes throughout Canada are attempting to address dispossession, or taking of lands, resources, and territories,[13] a key injustice arising from colonization and settlement (Moore 2016, 455; Chartrand 1999, 96). Moore (2016, 458) recognizes that corrective justice measures to address past taking of land are limited to compensation, resource and revenue transfer, return of public lands, use rights for Indigenous People over land that is now occupied by other people, and symbolic apology. She also recognizes that a full measure of justice is unavailable since these measures cannot truly compensate.

Most development in Canada has occurred without Indigenous consent, meaning that a paradigm shift is required to make policies consistent with current direction under UNDRIP.[14] Canada's Ten Principles are silent on redress or compensation, and commit only to "engagement with the aim to achieve FPIC."

Interim transfers of fisheries access to Indigenous groups on both the Atlantic and Pacific coast in advance of treaty settlements (Case 2 in Chapter 2, and Table 2.3) constitute a major federal policy shift in the past two decades, but many issues about resource access and governance remain to be resolved despite court decisions and negotiation attempts. A recent policy decision in British Columbia that salmon farm tenures require consent (Table 2.3) is consistent with criteria identified in Table 2.2. Federal actions regarding FPIC have been mixed, with decisions not to develop an oil pipeline or ship oil from northern BC ports consistent with the support of a majority of First Nations while at the same time the federal government has promoted a pipeline in southern British Columbia over objections by some First Nations, but with support of others (Case 4).

Modern treaties typically include a fiscal component consisting of federal or provincial transfers, corporate earnings, Crown resource revenue sharing, and other revenue sources.[15] Settlements have typically focused on Crown assets, or assets that can be transferred or purchased for transfer to Indigenous groups.[16] Distributive justice is necessary but requires the powerful to share or relinquish power.[17] However, an equitable or fair amount of land or share of resources such as fish are difficult to determine.[18] As a result, settlements are often driven by federal policies or mandates that may be subject to or guided by LCAs or court decisions, but are often met by institutional inertia and require further negotiation, or have led to a return to court in some cases.

In its *Delgamuukw* and *Tsilhqot'in* decisions,[19] the Supreme Court of Canada determined that Aboriginal title exists in Canada and includes a right to determine the use of land. Negotiations recently resulted in an agreement on a six-year pathway to Tsilhqot'in governance involving a vision including "consensus decisions utilizing FPIC."[20] The Nuu-chah-nulth fishing rights decision and a subsequent appeal recognized a "commercial rights fishery that had priority" over regular commercial and recreational fisheries; the appeal followed unsuccessful negotiation of an alternative fisheries regulatory scheme, and found that some infringements were justified whereas some were not on a case-by-case basis. Despite the court decision and new federal reconciliation policies, the regulatory framework continues to be subject to negotiation.[21] The Supreme Court of Canada's *Marshall* decision affirmed a right to a commercial fishery that would provide a "moderate livelihood," and is the subject of ongoing negotiation and assertive actions.[22] Lack of clear direction from the courts contributes to uncertainty in negotiations and typically plays into the length of time to resolve issues, e.g., a decade or more.[23] As described in Chapter 10, Arctic fisheries allocations are viewed by some as inequitable and inconsistent between regions, and federal allocation decisions are at times not in line with recommendations from co-management boards formed under LCAs. Major issues such as seabed ownership and oil and gas development in the Arctic were not addressed during land claims negotiations and are being addressed by seeking consent and further negotiating on specific issues.

Reconciliation of resource access from Table 2.3 and the case studies has included revenue sharing and impact/benefit agreements (e.g., Case 5 in Chapter 2). Policies that attempt to address development issues include compensation, revenue sharing, and requirements to partner with Indigenous groups before tenures are issued. Compensation has not commonly been included in modern treaty settlements, but treaties usually include a cash or land component that could be interpreted as compensation. Revenue sharing has been utilized for mining, forestry, and alternative energy projects, but could be applied to marine development such as oil and gas (e.g., Case 5), aquaculture and marine-based energy, such as tidal or offshore wind farms (see, e.g., Coates 2015; Fiser and Pendakur 2017). Funding for Indigenous initiatives has also been provided by means of carbon credits related to preservation of old-growth forests (see, e.g., Coastal

First Nations n.d.). Canada's new policy of recognizing Indigenous rights is relevant to access and decision making in ocean areas. The law is less clear than for land, with no definitive court decisions about Aboriginal title to ocean spaces as yet.

Economic access, including access to resources, is a major factor affecting human well-being, and timely settlement is a key factor affecting justice. Reconciliation through modern treaties should provide remedies for past injustices affecting past, as well as present and future, generations. Delays in settlements typically benefit the Crown and disadvantage Indigenous People in the present. Tools such as revenue sharing are applicable once consent is obtained, which in government parlance amounts to accommodation for impacts of a development.

PROTECTING CULTURE AND VALUES

We distinguish between natural cultural change and imposed change resulting from political and economic subordination (Moore 2016, 452),[24] and recognize that wise ocean policies can support culture and nature. Culture includes roles in society, expectations about behaviour, beliefs, and the sense of well-being from belonging to a group (Fryberg, Covarrubias, and Burack 2016, 2).[25] For example, the Indigenous world view is a belief system that commonly includes understandings regarding the natural world and humans' relationship with it, and values such as respect and reciprocity (Kinnear 2007, 12). However, cultural well-being also involves self-efficacy and agency – and needs recognition that Indigenous Peoples take a proactive approach to culture.

As described earlier, Indigenous cultures have undergone significant change since the time of contact, including drastic population declines due to infectious diseases and displacements. Policies during the assimilation era, such as the removal of children from families (Pyne and Taylor 2019), were particularly damaging, leading to systematic language loss and hindering cultural transmission. Resource depletion and climate change can also significantly affect cultural practices (Eckert et al. 2018). Cultural change resulting from participation in the wage economy could be seen as voluntary (e.g., Knight 1996), but is closely connected to dispossession from territories and resources.[26] Canada has taken modest steps to address components of cultural imposition resulting from col-

onization, but policies such as UNDRIP and Canada's Ten Principles fall short of reconciling cultural changes.[27]

While Indigenous practices such as fishing and hunting are protected as rights under Canada's Constitution, and courts have recognized the evolving nature of Indigenous rights and rights to fish using preferred means, requirements are not prescriptive and engagement varies widely between and among regions (Table 2.3).[28] Similarly, Canadian policies and legislation currently encourage Indigenous engagement in integrated ocean, Marine Protected Area, and species-at-risk planning and management, but do not require it.

However, frameworks for approval of developments such as aquaculture, oil and gas, and shipping that can have major effects on environment and culture, depending on the location and scale of development, operate under a system of project approvals that is largely political. While decisions affecting industry will be balanced against potential court challenges based on Aboriginal rights, and mitigative measures may provide some relief, seldom is the no-development option adopted (despite requirements for FPIC in UNDRIP), even though it may be the best protection for culture.

Structural changes such as co-management systems and ecosystem-based management (EBM) approaches are well positioned to adapt management processes to external changes and support transformative change (Barnes et al. 2017). As well, collaborative or Indigenous-led planning processes and a shift toward integrated management as required under the *Oceans Act* provide opportunities to incorporate the Indigenous world view, values, and knowledge into policies and plans (e.g., Paul 2018).

EBM approaches are less common but can provide more resilience and ability to adapt to changes than single-species fishery science approaches that do not include key species, community, or economic aspects.[29] An example of a holistic approach to management is the conceptual model of the herring social-ecological system (Figure 14.2). Importantly, this holistic view also situates culture in social-ecological systems as a source of agency and as a key driver for adaptation (Doubleday 2019). The Maritimes Region of Fisheries and Oceans Canada (DFO) is currently developing an EBM framework based on four pillars of sustainability (governance, ecological, social, and economic), but the extent to which this framework can take on the necessity of the challenge of reconciliation and power sharing is unclear.

Indigenous traditional knowledge[30] about territories and species can improve understanding of ecosystem relationships and management. Many species are culturally important, and traditional knowledge can be influential in decision making or critical to conservation outcomes, e.g., Committee on the Status of Endangered Wildlife in Canada (COSEWIC) and species-at-risk assessments of polar bears, Columbia River chinook, and northern abalone (Table 2.3). Traditional knowledge holders need to be engaged directly, as researchers may lack understanding of the potential benefits of knowledge co-production or have implicit bias (see, e.g., Ulicsni et al. 2018). Indigenous knowledge transmission is at risk due to cultural changes, and processes for sharing knowledge can be challenging due to past legacies and the need to build trust; despite recognition in policy and legislation,[31] there has been limited utilization in planning and management (e.g., Hill, Schuster, and Bennett 2019).

Isolation is not enough to preserve distinct languages and cultures, given the distribution of TV, internet, broadband communications, and social media. While reconciliation outcomes such as self-government, self-determination, and economic self-sufficiency offer positive tools for reinforcing culture and maintaining identity, broader action will be needed to support Indigenous connections to the land and threatened and endangered languages and culture.[32]

SUCCESS AT TRANSFORMING RELATIONSHIPS

Our review of the status of reconciliation of ocean issues across Canada's three coasts, here and in Chapter 2, shows uneven progress. We acknowledge broad similarities, and serious and nuanced differences. For instance, Inuit have a vastly different history in relation to self-organizing and self-governance as a result of their much more recent experience of colonialism, with much of the greatest impact occurring within living memory (Doubleday 1989).[33] Economic history accounts for Inuit dependency on the fur trade that emerged, and the relocations that ensued. Again, the fundamental issue is loss of resource access and control, recognizable in numerous cases across Canada.

In 2015, the federal Liberals campaigned on prioritizing Indigenous reconciliation; Prime Minister Justin Trudeau

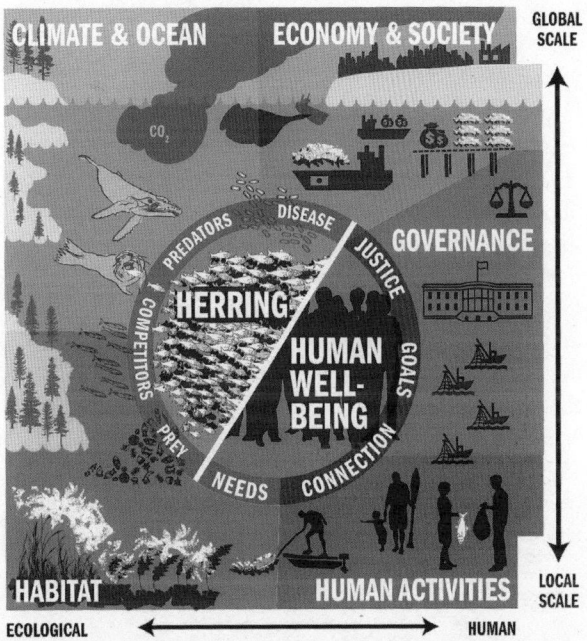

Figure 14.2 Conceptual model of the herring social-ecological system (from Levin, Francis, and Taylor 2016, used with permission from Taylor and Francis Journals).

provided direction in mandate letters to federal ministers upon taking office (Prime Minister of Canada 2018). Federal policy has been revamped and concrete steps toward reconciliation have been taken, such as the establishment of more than 80 discussion tables across Canada to explore new ways of working together to advance recognition of Indigenous rights and self-determination (CIRNA 2020). As described earlier, Canada adopted UNDRIP in 2016 and published Ten Principles for reconciliation in 2018. The BC Treaty Process has been revamped, and loans incurred by First Nations since the process started in 1990 were recently forgiven. A number of new reconciliation agreements have been reached as a result of the new tables (CIRNA 2020).

Contradictions remain, such as approval of the Trans Mountain Pipeline (Case 4) despite objections and legal challenges by some Indigenous Nations. In fall 2019, British Columbia adopted a legislative approach toward UNDRIP implementation.[34] The federal government subsequently passed legislation providing a framework for implementation of UNDRIP after a private member's bill failed to pass the Senate in 2019.[35] Adoption of UNDRIP with enabling federal legislation and a policy of recognition of rights should result in changes in Canada's court pleadings. At the same time, recent court decisions regarding title and rights continue to define Canada's obligations. Canada's approach has been more progressive than that of Australia, where there was a state apology for the Stolen Generations (the removal of Indigenous children from their families) but internal debate is continuing on whether or not to negotiate a treaty with Indigenous Peoples there (Short 2016). And it may be less advanced than that of New Zealand, where the 1840 Treaty of Waitangi, and a Waitangi Tribunal that has been in place since 1975, has guided progress toward self-determination of Maori *iwi* and development of the Maori economy, including fisheries settlements (Sullivan 2016).

Canada's Ten Principles recognize that transformative change is needed in relationships but leaves the details for negotiation. Transformative change is necessary because current systems are unworkable or have shown little capacity for change. Some sharing of power has resulted from negotiation of treaties.

A just reconciliation framework must begin with acknowledgment of and by the parties of historical injustices, including their relative positionality. It must be consistent with an approach based on justice, truth, responsibility, and transformation of relationships as identified by Rouhana (2011). Key elements include balanced negotiations, tailored approaches that meaningfully address issues, and timeliness. Inquiries and truth commissions have started the process of recognition and healing but have fallen short of addressing broader problems, including ocean issues. Timeliness has been a major issue, with most treaties requiring decades to negotiate.[36] The new incremental approach sees sequential reconciliation agreements as building blocks to a final treaty or agreement.

Major barriers have been a limited federal mandate for negotiations, particularly on issues such as land and fisheries, and the need for governance structures capable of adaptation and transformation (Barnes et al. 2017). The mix of federal, provincial, and territorial jurisdictions over ocean issues has also been a challenge to reconciliation, as outlined in Table 2.3. Progress requires a paradigm shift to governance agreements that are more equal and

that share power. Major gaps are areas in the Pacific and Atlantic without treaties or that have treaties but do not address territory and resources (e.g., Peace and Friendship Treaties in the Atlantic). New policies on recognition of rights have not yet become part of agreements or been reflected in positions during litigation. Attention also needs to be paid to ensuring adequate capacity for long-term implementation of agreements in order to deliver concrete outcomes (Paul 2018).

THE PATHWAY TO A JUST RECONCILIATION

What role will Indigenous people play in ocean management in Canada in the future? Based on criteria developed according to UNDRIP,[37] we expect that the Indigenous role will be enhanced as reconciliation processes mature, given recent commitments to recognition of Indigenous rights and requirements for FPIC prior to developments.

Canadian policies relating to Indigenous Peoples have moved gradually over the past five decades from pursuing assimilation toward seeking reconciliation. Positive steps toward reconciliation are occurring on all three coasts, but there are gaps and shortcomings from the viewpoint of true reconciliation and the criteria in this chapter. As case studies in Chapter 2 show, best practices exist but are not necessarily mobilized to inform policy and management practices in other regions and communities. Reconciliation is a long-term ongoing process that requires political commitment and resources in order to be successful. Canada has identified that implementation of UNDRIP requires transformative change (Department of Justice Canada 2018). Based on the discussion above, we make five policy recommendations to move in that direction, each followed by a brief rationale.

Adopt a just and equitable reconciliation framework

Revise Canada's reconciliation framework based on fundamental requirements for justice, truth, responsibility, and transformation of social and political relationships.[38] UNDRIP and, to a lesser extent, Canada's Ten Principles, provide a platform for making these changes. Key changes must include:

- *Power sharing.* This requires an understanding of current injustices and the relative positionality of the parties and a transition to meaningful power sharing.

Sharing of power and resources will require political will and further government compromise.

- *Balanced negotiations.* True reconciliation requires an asymmetric approach to issues in which the less powerful consent to processes and solutions without undue pressure and power holders commit to just outcomes.
- *Tailored approaches.* Reconciliation of marine issues requires regionally specific approaches.

  - *Historic treaties.*[39] Acknowledge past injustices and seek lasting solutions building on recent experiences and best practices from other regions. Support Indigenous governance and self-determination, encourage collaboration among Indigenous partners on ocean issues, and take positive steps toward management and sharing of ocean resources.
  - *Modern treaties.* Utilize agreed-upon processes to build on existing governance systems and adapt them to emerging issues. Support Indigenous leadership, provide appropriate and sufficient resources for treaty implementation, and establish a legislated commission to regularly monitor and report on progress (Land Claims Agreements Coalition 2017). Unlike some regional governance initiatives (e.g., Cases 2 and 4 in Chapter 2), there is no single Inuit voice because the implementation falls to the Inuit beneficiary organizations responsible for land claims in their regions.
  - *No treaties.* Negotiate co-governance structures that recognize Indigenous laws and decisions at scales to seek FPIC and support formation of aggregates to address policy issues at larger scales. Develop new negotiating mandates for economic issues based on Aboriginal title and cultural practices. Establish incremental steps to address specific ocean issues, including co-governance of fisheries and other living marine resources, and allocations among users, privileging Indigenous economic, cultural, social, and nutritional requirements.[40]

- *Timeliness.* Set a goal of achieving justice for living generations within a reasonable time frame.[41] Accelerate negotiations by rapidly implementing Canada's policy of recognition of Indigenous rights.

- *Recognition and implementation of rights legislation.* Work with Indigenous partners to cooperatively develop legislation to align Crown policy and legislation across departments and recognize Indigenous law with recognition principles as promised by Prime Minister Trudeau in 2018 (Prime Minister of Canada 2018).[42]

Broadly engage Indigenous communities in visioning and planning of ocean spaces to reinforce connections to place and better incorporate Indigenous traditional knowledge, including Inuit *Quajimajatuqangit,* and world views

Fulfill requirements/promises to meaningfully involve Indigenous Peoples in policy and strategic decisions and incorporate traditional knowledge in policies, plans, and decisions through culturally appropriate processes. Develop and follow best practices and provide space and resources to engage with all Indigenous Nations across Canada's coasts. Engage early on issues, build on existing co-governance structures or develop new ones, and provide adequate resources. Include all major issues within the scope of planning initiatives.

Cooperatively pursue holistic approaches to management, such as integrated management and ecosystem-based management

Enhance existing or develop new strategic partnerships between Indigenous, federal, provincial, and/or territorial governments to support integrated ocean management and ecosystem-based management at local, subregional, and regional scales. Support holistic approaches to management and planning by aligning plans with Indigenous values and world view and incorporating Indigenous teachings and knowledge. Broaden the current ecosystem approach to include social systems and human well-being.[43]

Support and establish co-governance or co-management arrangements that integrate knowledge systems, share power and responsibility, and redefine relationships

Build on successful co-governance and co-management partnerships consistent with government-to-government relationships and recognition of Indigenous rights and title. Enable co-governance through policy directives and mandates for federal ministers, provide sufficient resources, and overcome structural resistance. Develop new co-governance structures across Canada's coasts based on best practices. Support development of co-governance institutions at appropriate scales. Respect the traditions of co-governance that exist.

Support Indigenous self-determination and development of resource management capacity

Support nation building led by Indigenous Nations through long-term investments in governance structures and human resource capacity. This includes support for community planning and development of governance structures and institutions at local, subregional, and regional scales. Develop co-governance capacity using a long-term capacity development approach and/or capacity-sharing perspectives.[44] Adopt principles of successful polycentric/biocentric models of co-governance, including the principle of subsidiarity, meaning that the responsibility aligns with decision-making capacity and resides with those most affected.

CONCLUSION

Reconciliation in Canada is an ongoing process to confront and address injustices from colonization, including political domination, loss of territory, and cultural imposition. Progress across Canada's three coasts has been slow and uneven, but best practices are evident in the areas of governance, resource access, and cultural support and renewal. Indigenous groups both exercise their own jurisdiction and laws and participate in joint management and stewardship of ocean initiatives. Meaningful power sharing and resource access have been stumbling blocks in negotiation of modern treaties as well as decisions on resource developments. There are examples of management systems aligning with Indigenous world views and embracing concepts such as integrated and ecosystem-based management based on Indigenous leadership and knowledge. Adequate resources are essential for negotiation, planning, and implementation, including capacity to support Indigenous governance and co-management.

The reconciliation criteria described in Chapter 2 based on UNDRIP align with Canada's stated policy direction of recognition of Indigenous rights. Current practices

fall short of this objective, and Indigenous leadership will be important in developing a credible framework and legislation. Canada has not made commitments to reconciliation typical of transitional justice, which leaves many difficult issues to be addressed through negotiation, political activism, or the courts. More progress has been made in situations where Indigenous rights (e.g., rights to fish or hunt) have been proven in Canadian courts and activities such as Marine Protected Areas or species-at-risk recovery plans may infringe these rights. Less progress has been seen in decisions about major developments such as oil and gas pipelines or marine shipping. Even in these areas, however, some degree of reconciliation has been possible through measures such as collaborative governance structures, joint participation in environmental assessments, or revenue-sharing agreements. A greater challenge exists where Indigenous groups oppose development. Although the stated policy is to seek FPIC, decisions often fail to recognize Indigenous rights, particularly Indigenous title, with the result that disputes often end up in the courts.

ACKNOWLEDGMENTS

Our appreciation goes to Louise Mandell, QC, who reviewed an earlier version of this chapter, and to Steve Diggon and Miles Richardson for their helpful review and comments.

NOTES

1 Zurba and colleagues (2019) identify aspects of protected areas that present wicked problems. Fisheries allocation has also been identified as a wicked problem. See, e.g., Jones 2010.

2 See Chapter 2, Table 2.3 and the case studies reported.

3 Modern land claims agreements and treaties provide a comprehensive approach to ocean issues that were negotiated under a policy of extinguishment of Indigenous rights within a symmetric framework. As identified by Rouhana (2011, 295), a symmetric framework "attempts to impose 'balance' by treating all parties to a conflict in parallel fashion. Consequently, in the service of the symmetrical analysis of conflict that is one of their (conflict settlement's) fundamental tenets, such methods purposely and willfully avoid core issues such as historical responsibility, justice, truth, and other transitional justice-related issues." Early Arctic experience with negotiations was not

always observed to apply this framework to the disadvantage of Inuit, who held a much stronger position as co-equals with respect to sovereignty.

4 Hughes (2012) cites the following regarding the Canadian Truth and Reconciliation Commission: "Transitional justice is normally employed in cases of regime change, usually as a transition from a form of authoritarian rule to a democracy. By contrast, Canada is applying a transitional-justice framework in the context of a more politically stable nation-state"; "the problem with the Canadian TRC is its conceptualization by the Canadian state as an instrument that draws a line through history, in effect finalizing or perfecting the colonial project rather than being part of a transformation and decolonization."

5 The RCAP recommended a new nation-to-nation relationship governed by historical and new treaty agreements, and a Canada-wide process of building or rebuilding Indigenous Nations as a third order of government within Canada that would include Métis and Non-Status Indians (Chartrand 2004, 122). The new relationship would be brought about by four basic elements: a royal proclamation and legislation to implement the renewed relationship, efforts to rebuild Indigenous Nations through development of constitutions and citizenship codes recognized through a new law; a Canada-wide framework agreement that sets the stage for an Indigenous order of government within the Canadian federation; and negotiation of new or renewed treaties between Indigenous Nations and other Canadian governments (122). This included reconciling different understandings and implementing the spirit and intent of treaties through negotiation (123).

6 Chartrand (2004, 130) notes that the BC Treaty Process is conducted "with bands that vary in population size from 136 to 7517 people, with an average of 1782 and a median of 800 people. Even the largest of these will have a relatively small physical capacity to govern effectively on the RCAP standards."

7 UNDRIP limits the right of autonomy and self-government to internal and local affairs (Article 4). Canada identifies an inherent right of self-government (Principle 1) that exists within an evolving system of cooperative federalism and distinct orders of government (Principle 4).

8 In the Pacific, British Columbia–wide organizations such as the First Nations Summit coordinate responses on treaty-related issues but are not the holders of Indigenous title and rights. Similarly, a First Nations Fisheries Council disseminates information and coordinates responses to fisheries issues focused largely on salmon.

9 The APCFNC is the Policy Research and Advocacy Secretariat for 31 Chiefs (of Mi'kmaq, Maliseet, Innu, and Passamaquoddy), Nations, and Communities in Atlantic Canada, Quebec, and Maine (https://www.apcfnc.ca/).

10 See *Delgamuukw v British Columbia,* 1997 SCC 3, 1111–12; *Tsilhqot'in Nation v British Columbia* 2014 SCC 44, 121.

11 Inuit never entered into treaties with the British Crown before Confederation, or the Crown in right of Canada following Confederation. This makes an examination of self-organization and self-government by Inuit critical to recognition of their interests in marine and coastal environments, resources, seabed, and sea ice.

12 See, for example, the conflicts over the Pacific North West LNG and Trans Mountain Pipelines (Case 5).

13 Canada's Ten Principles (Department of Justice Canada 2018, 6) refers to "the reality that Indigenous peoples' ancestors owned and governed the lands which now constitute Canada prior to the Crown's assertion of sovereignty" and that "all of Canada's relationships with Indigenous people are based on recognition of this fact and supported by the recognition of Indigenous title and rights, as well as the negotiation and implementation of pre-Confederation, historic, and modern treaties."

14 UNDRIP identifies an Indigenous right to redress or compensation for lands, territories, and resources that are taken or damaged without free, prior, and informed consent (Article 28[1]). As well, states are required to obtain FPIC for projects, lands, territories, and resources before development (Article 32[1–3]).

15 The Conference Board of Canada (Fiser and Pendakur 2017) estimated revenue sources for an average BC First Nation with a revenue-sharing agreement to consist of federal/provincial transfers (55%), corporate earnings (12%), Crown resource revenue sharing (10%), and other sources (23%).

16 For example, various fisheries licence retirement programs (Allocation Transfer Program, Atlantic Integrated Commercial Fisheries Initiative, Marshall Response Initiative, Pacific Integrated Commercial Fisheries Initiative) or forest tenures that may entail compensation to third parties.

17 "We know from the history of peoples everywhere that those who have power and those who have resources do not give them up easily; they need good reasons to do so" (Chartrand 1999, 96). This is one reason why allocation

issues often appear in the courts, where a long history of settler domination exists.

18 This is a common problem in other jurisdictions. As described by Cohen (1986), the Boldt decision on fisheries allocation in the US Pacific Northwest was necessary before acceptance of a federal/state/tribal government-to-government relationship and fisheries co-management. In British Columbia, the implementation of fishing rights continues to be negotiated and litigated. McRae and Pearse (2004) estimated that 30% of Fraser River sockeye would be transferred to Indigenous groups based on recent treaties; the First Nation Panel on Fisheries (2004) identified transfer of 50% of fisheries to Indigenous Nations as a starting point with a rationale that the underlying title was Indigenous.

19 See note 10 above.

20 *Gwets'en Nilt'i Pathway Agreement,* July 25, 2019, https://www2.gov.bc.ca/assets/gov/environment/natural-resource-stewardship/consulting-with-first-nations/agreements/gwetsen_nilti_pathway_agreement_signed_15august2019.pdf.

21 *Ahousaht Indian Band and Nation v Canada (Attorney General),* 2009 BCSC 1494; *Ahousaht Indian Band and Nation v Canada,* 2018 BCSC 633; *Ahousaht Indian Band and Nation v Canada (Attorney General),* 2021 BCCA 155; *Ahousaht First Nation v Her Majesty the Queen in Right of Canada, as represented by the Minister of Indian Affairs and Northern Development,* 2021 FCA 135, leave to appeal to SCC dismissed, 2022 CanLII 14378 (SCC).

22 *R v Marshall,* [1999] 3 SCR 456; see, e.g., Case 2 in Chapter 2 and Assembly of Nova Scotia Mi'kmaq Chiefs 2019.

23 For example, the first phase of the Nuu-chah-nulth fishing rights decision took seven years (2006–13) to move through the courts. The second phase, which focused on potential justification for infringements of Aboriginal rights, began in 2013 and concluded in 2022. See note 21 above.

24 Moore (2016, 456) suggests as a baseline for corrective justice whether individuals are worse off now than they would have been "if they had lived in a more just, reciprocal and non-exploitative, and non-dominating political order."

25 "Culture refers to the implicit and explicit patterns of historically and socially derived ideas and images, and their manifestation in the institutions and interactions (i.e., policies and practices) that constitute society" (Fryberg, Covarrubias, and Burack 2016).

26 Knight (1996) provides an informal Indigenous history of logging, transport, construction, commercial fishing, and canning in British Columbia.

27 Cultural values are identified throughout UNDRIP (Table 2.2). UNDRIP provides protection for cultural values but does not extend these to cultural rights. According to Engle (2011, 149), the United Nations was unable to reach consensus on collective cultural rights due to concerns that this might undermine individual rights. Canada's Ten Principles do not directly address culture, but the preamble does acknowledge that strong Indigenous cultural traditions and customs, including languages, are fundamental to rebuilding Indigenous Nations (Department of Justice Canada 2018, 4).

28 For example, Inuit have been involved in marine planning initiatives of government since the Lancaster Sound hearings, and the scale and nature of this continues to increase.

29 See analysis of New Zealand and Canadian fisheries by Le Heron and colleagues (2006) and Jones (2010), respectively.

30 Berkes (1999, 8) defines Traditional Ecological Knowledge as "a cumulative body of knowledge, practice, and belief, evolving by adaptive processes and handed down through generations by cultural transmission, about the relations of living beings (including humans) with one another and with their environment."

31 Required in legislation for species-at-risk assessment and National Marine Conservation Area management.

32 Positive messaging in the media and education has been identified as an intervention to alleviate colonization (Fryberg, Covarrubias, and Burack 2016).

33 The last Inuit family living on the land moved to government settlements in 1970. Between 1953 and 1955, 87 Inuit were relocated to Resolute, Grise Fiord, and Arctic Bay to support Canadian sovereignty claims. Prior to this, smaller groups had been moved to more northerly areas in support of the fur trade. Although earlier interactions with churches, government agents, and traders seriously impacted Inuit lifeways, sometimes tragically, Inuit did not experience the loss of land and sea access in the same way as other Indigenous Peoples in Canada living further south. However, what they did experience was overwhelming competition for marine resources from European whalers and sealers that undermined the sustainability of their livelihoods, from the 1700s to 1964, when commercial whaling ceased in the Eastern Arctic. In the Western Arctic, Inuvialuit whalers had

limited participation on commercial whaling vessels, enabling them to maintain access to whale meat and *muktak,* a traditional food consisting of whale skin and blubber (Raddi and Weeks 1985; Doubleday 1994).

34 *Declaration on the Rights of Indigenous Peoples Act,* SBC 2019, c 44. See Chapter 2, note 24 for details.

35 *United Nations Declaration on the Rights of Indigenous Peoples Act,* SC 2021, c 14. See Chapter 2, note 24 for details.

36 For example, about 30 years for the Nisga'a treaty, 20–30 years for Arctic land claims agreements, and only eight agreements under the BC Treaty Process after 27 years. The current treaty and reconciliation process is based on defined steps, including identification of issues, negotiation of a framework agreement, and negotiation of a detailed treaty or reconciliation agreement. Interim agreements, although possible according to the BC Treaty Process, have been rare. Some gradual progress is being made under the new federal policy of an incremental approach to reconciliation agreements.

37 See Chapter 2, Table 2.1.

38 Fisheries and Oceans Canada and the Canadian Coast Guard developed a reconciliation strategy (DFO 2019) soon after release of Canada's Ten Principles. The strategy is intended as a living document and was developed internally by DFO "based on federal policy directions and feedback from engagement activities with Indigenous peoples and departmental stakeholders." One of the indicators for the strategy is the number of agreements with Indigenous groups. A major limitation is the narrow involvement of Indigenous Peoples in the strategy, and reliance on existing policies, processes, and decision-making structures that may not be consistent with UNDRIP.

39 Some Atlantic Nations are engaged in community-driven negotiating tables for recognition of Indigenous rights and self-determination (CIRNA 2015).

40 Efforts to rejuvenate the BC Treaty Process are ongoing. Marine issues, including fisheries, have been major stumbling blocks. Recent changes are described in a Principals' Accord that rejects extinguishment, recognizes Aboriginal title and rights in negotiating mandates, and incorporates an incremental approach (i.e., a series of agreements). See https://www2.gov.bc.ca/assets/gov/environment/natural-resource-stewardship/consulting-with-first-nations/agreements/principals_accord_signed_dec_1_2018.pdf.

41 For example, short term (6–12 months), medium term (6 months–2 years), and long term (2–5 years).

42 The Assembly of First Nations (2018) has emphasized the need for this work to proceed in full partnership with Indigenous Peoples.

43 Doubleday (2019) identifies the need to shift to a social-ecological system (SES) model that privileges culture. See also DFO 2007.

44 Many communities need additional capacity for planning, management, and implementation, e.g., governance, science, business, administration, and international diplomacy expertise.

REFERENCES

Assembly of First Nations. 2018. *Affirming First Nations Rights, Title and Jurisdiction.* https://www.afn.ca/wp-content/uploads/2018/09/Affirming-FN-Rights-Title-and-Jurisdiction_EN.pdf.

Assembly of Nova Scotia Mi'kmaq Chiefs. 2019. "Brief to Standing Senate Committee on Fisheries and Oceans." April 17. https://sencanada.ca/content/sen/committee/421/POFO/Briefs/AssemblyofNSMi'kmaqChiefs_e.pdf.

Barnes, M.L., Ö. Bodin, A.M. Guerrero, R.J. McAllister, S.M. Alexander, and G. Robins. 2017. "The Social Structural Foundations of Adaptation and Transformation in Social-Ecological Systems." *Ecology and Society* 22 (4): 16. https://doi.org/10.5751/ES-09769-220416.

Berkes, F. 1999. *Sacred Ecology: Traditional Ecological Knowledge and Resource Management.* Philadelphia: Taylor and Francis.

Chartrand, P. 1999. "Aboriginal Peoples in Canada, Aspirations for Distributive Justice as Distinct Peoples: An Interview with Paul Chartrand by Murray Dobbin." In *Indigenous Peoples' Rights in Australia, Canada, and New Zealand,* edited by P. Havemann, 88–107. Auckland, NZ: Oxford University Press.

–. 2004. "Towards Justice and Reconciliation: Treaty Recommendations of Canada's Royal Commission on Aboriginal Peoples 1996." In *Honour Among Nations? Treaties and Agreements with Indigenous People,* edited by M. Langton, L. Palmer, M. Tehan, and K. Shain, 120–32. Carlton, Victoria: Melbourne University Press.

CIRNA (Crown-Indigenous Relations and Northern Affairs Canada). 2015. "Peace and Friendship Treaties." https://www.rcaanc-cirnac.gc.ca/eng/1100100028589/1539608999656#wb-cont.

–. 2020. "Recognition of Rights Discussion Tables." https://www.rcaanc-cirnac.gc.ca/eng/1511969222951/1529103469169.

Coastal First Nations. n.d. "Carbon Credits." https://coastal firstnations.ca/our-land/carbon-credits/.

Coates, K. 2015. *Sharing the Wealth: How Resource Revenue Agreements Can Honour Treaties, Improve Communities, and Facilitate Canadian Development*. Ottawa: Macdonald-Laurier Institute. https://www.macdonaldlaurier.ca/files/pdf/MLIresourcerevenuesharingweb.pdf.

Cohen, F. 1986. *Treaties on Trial: The Continuing Controversy over Northwest Indian Fishing Rights*. Seattle: University of Washington Press.

Department of Justice Canada. 2018. *Principles Respecting the Government of Canada's Relationship with Indigenous Peoples*. Ottawa: Department of Justice Canada. https://www.justice.gc.ca/eng/csj-sjc/principles.pdf.

DFO (Fisheries and Oceans Canada). 2007. "A New Ecosystem Science Framework in Support of Integrated Management." https://www.dfo-mpo.gc.ca/science/publications/ecosystem/index-eng.htm.

–. 2019. "DFO–Coast Guard Reconciliation Strategy." September. https://waves-vagues.dfo-mpo.gc.ca/library-bibliotheque/40947208.pdf.

Doubleday, N. 1989. "Aboriginal Subsistence Whaling: The Right of Inuit to Hunt Whales and Implications for International Environmental Law." *Denver Journal of International Law and Policy* 17 (2): 353–93.

. 1994. "Arctic Whales: Sustaining Indigenous Peoples." In *Elephants and Whales Resources for Whom?* edited by M.M.R. Freeman and U. Kreuter, 241–61. Basel, Switzerland: Gordon and Breach Science Publishers.

–. 2019. "Culture as Vector: (Re)Locating Agency in Social-Ecological Systems Change." In *Agency and Time in the Environmental Humanities*, edited by R. Boschman and M. Trono, 327–47. Waterloo, ON: Wilfrid Laurier University Press.

Eckert, L.E., N.C. Ban, S.-C. Tallio, and N. Turner. 2018. "Linking Marine Conservation and Indigenous Cultural Revitalization: First Nations Free Themselves from Externally Imposed Social-Ecological Traps." *Ecology and Society* 23 (4): 23. https://doi.org/10.5751/ES-10417-230423.

Engle, K. 2011. "On Fragile Architecture: The UN Declaration on the Rights of Indigenous Peoples in the Context of Human Rights." *European Journal of International Law* 22: 141–63.

First Nation Panel on Fisheries. 2004. *Our Place at the Table: First Nations in the B.C. Fishery*. http://fns.bc.ca/wp-content/uploads/2016/10/FNFishPanelReport0604.pdf.

Fiser, A., and K. Pendakur. 2017. *Options and Opportunities: Resource Revenue Sharing between the Crown and Indigenous Groups*. Ottawa: Conference Board of Canada. https://www.researchgate.net/publication/322198112_Options_And_Opportunities_Resource_Revenue_Sharing_Between_The_Crown_And_Indigenous_Groups_In_Canada.

Fryberg, S., R. Covarrubias, and J.A. Burack. 2016. "The Ongoing Psychological Colonization of North American Indigenous People: Using Social Psychological Theories to Promote Social Justice." In *The Oxford Handbook of Social Psychology and Social Justice,* edited by P.L. Hammack Jr., 113–28. https://dx.doi.org/10.1093/oxfordhb/9780199938735.013.35.

Hill, C.J., R. Schuster, and J.R. Bennett. 2019. "Indigenous Involvement in the Canadian Species at Risk Recovery Process." *Environmental Science and Policy* 94: 220–26.

Hughes, J. 2012. "Instructive Past: Lessons from the Royal Commission on Aboriginal Peoples for the Canadian Truth and Reconciliation Commission on Indian Residential Schools." *Canadian Journal of Law and Society* 27 (1): 101–27.

Jones, R. 2010. "The Trajectory of Canada's Pacific Coast Fisheries: Are Current Fisheries Policies Adequate to Cope with Environmental, Social and Economic Change?" Unpublished report prepared for Evaluation Directorate, Fisheries and Oceans Canada, Ottawa.

Kinnear, L. 2007. "Contemporary Mi'kmaq Relationships between Humans and Animals: A Case Study of the Bear River First Nation Reserve in Nova Scotia." MES thesis, Dalhousie University. https://www.collectionscanada.gc.ca/obj/thesescanada/vol2/002/MR39166.PDF.

Knight, R. 1996. *Indians at Work: An Informal History of Native Labour in British Columbia 1858–1930*. Vancouver: New Star Books.

Land Claims Agreements Coalition. 2017. *A Modern Treaties Implementation Review Commission: A Proposal to the Government of Canada by the Land Claims Agreements Coalition*. Ottawa: Land Claims Agreements Coalition. http://landclaimscoalition.ca/wp-content/uploads/2018/02/MTIRC-Doc-and-Letter-to-Trudeau-1.pdf.

Le Heron, R., E. Rees, E., Massey, M. Bruges, and S. Thrush. 2006. "Improving Fisheries Management in New Zealand: Developing Dialogue between Fisheries Science and Management (FSM) and Ecosystem Science and Management (ESM)." *Geoforum* 39: 48–61.

Levin, P.S., T.B. Francis, and N.G. Taylor. 2016. "Thirty-Two Essential Questions for Understanding the Social-Ecological System of Forage Fish: The Case of Pacific Herring." *Ecosystem Health and Sustainability* 2 (4): e01213. https://doi.org/10.1002/ehs2.1213.

McRae, D.M., and P.H. Pearse. 2004. *Treaties and Transition: Towards a Sustainable Fishery on Canada's Pacific Coast.* Vancouver: Fisheries and Oceans Canada. https://waves-vagues.dfo-mpo.gc.ca/Library/280188.pdf.

Moore, M. 2016. "Justice and Colonialism." *Philosophy Compass* 11 (8): 447–61. https://doi.org/10.1111/phc3.12337.

Patzer, J. 2019. "Indigenous Rights and the Legal Politics of Canadian Coloniality: What Is Happening to Free, Prior and Informed Consent in Canada?" *International Journal of Human Rights* 23 (1–2): 214–33. https://doi.org/10.1080/13642987.2018.1562915.

Paul, K. 2018. "First Nations, Oceans Governance and Indigenous Knowledge Systems." In *The Future of Ocean Governance and Capacity Development: Essays in Honor of Elizabeth Mann Borgese (1918–2002),* edited by International Ocean Institute, 46–53. Leiden, Netherlands: Koninklijke Brill NV.

Prime Minister of Canada. 2018. "Remarks by the Prime Minister in the House of Commons on the Recognition and Implementation of Rights Framework." February 14. https://pm.gc.ca/en/news/speeches/2018/02/14/remarks-prime-minister-house-commons-recognition-and-implementation-rights.

Pyne S.A., and D.R.F. Taylor, eds. 2019. *Cybercartography in a Reconciliation Community: Engaging Intersecting Perspectives.* Elsevier.

Raddi, S., and N. Weeks. 1985. *The Prehistoric and Historic Utilization of Bowhead Whales in the Canadian Western Arctic: A Community-Based Study.* Toronto: World Wildlife Fund.

Rittel, H.W.J., and M.M. Webber. 1973. "Dilemmas in a General Theory of Planning." *Policy Sciences* 4: 2155–69.

Rouhana, N. 2011. "Key Issues in Reconciliation Challenging Traditional Assumptions on Conflict Resolution and Power Dynamics." In *Intergroup Conflicts and Their Resolution: A Social Psychological Perspective,* edited by Daniel Bar-Tal, 291–314. New York: Psychology Press. https://doi.org/10.4324/9780203834091.

Short, D. (2008) 2016. *Reconciliation and Colonial Power: Indigenous Rights in Australia.* Abingdon, UK: Routledge.

Snook J., J. Akearok, T. Palliser, A. Cunsolo, C. Hoover, and M. Bailey. 2019. "Enhancing Fisheries Co-Management in the Eastern Arctic." *Northern Public Affairs* 6 (2): 70–74. https://www.researchgate.net/publication/341193947_Enhancing_fisheries_co-management_in_the_Eastern_Arctic.

Sullivan, A. 2016. "The Politics of Reconciliation in New Zealand." *Political Science* 68 (2): 124–42.

Ulicsni, V., D. Babai, C. Vadász, V. Vadász-Besnyői, A. Báldi, and Z. Molnár. 2019. "Bridging Conservation Science and Traditional Knowledge of Wild Animals: The Need for Expert Guidance and Inclusion of Local Knowledge Holders." *Ambio* 48: 769–78. https://doi.org/10.1007/s13280-018-1106-z.

UNDRIP (*United Nations Declaration on the Rights of Indigenous Peoples*). 2007. https://www.un.org/development/desa/indigenouspeoples/declaration-on-the-rights-of-indigenous-peoples.html.

Youds, M. 2020. "Consultation Lacking as New UNDRIP Bill Tabled." *Ha-Shilth-Sa,* December 8. https://hashilthsa.com/news/2020-12-08/consultation-lacking-new-undrip-bill-tabled.

Zurba, M., K.F. Beazley, E. English, and J. Buchmann-Duck. 2019. "Indigenous Protected and Conserved Areas (IPCAs), Aichi Target 11 and Canada's Pathway to Target 1: Focusing Conservation on Reconciliation." *Land* 8 (1): 1–20. https://doi.org/10.3390/land8010010.

15

Toward a Healthy Future for Canada's Oceans and Coasts

U. Rashid Sumaila and the OCP Team

The social, cultural, and economic well-being of all Canadians is inextricably linked to the future of our oceans and coasts (Chapter 1). Yet, as the chapters in this volume emphasize, the future of our oceans and coasts is uncertain, rendering our future uncertain as well. Additionally, this volume highlights how this uncertainty is experienced and governed differently in communities across the Arctic, Atlantic, and Pacific regions of Canada. Each of these ocean and coastal regions is unique in terms of policy contexts, human history, relationships to ocean and coastal resources, and experiences with biophysical and ecological change. Given these differences, and while also recognizing our commonalities, we ask: How can and should we navigate pathways forward to foster viable and desirable ocean and coastal futures?

We have two objectives in this final chapter. First, we summarize the main findings reported in this book and draw attention to some of the core themes that have emerged from our collective efforts and the Ocean-Canada Partnership: reconciliation; changing oceans; access to ocean resources; ocean governance; and the relationship among law, policy, and knowledge mobilization. Second, we suggest practical pathways and recommendations for achieving a healthy ocean and supporting thriving coastal communities in Canada. Our aim is to synthesize insights in ways that resonate with all Canadians concerned about the long-term sustainability of our oceans and coasts and the social, cultural, and economic activities that depend on them (Bennett et al. 2018).

RECONCILIATION AND INDIGENOUS OCEAN MANAGEMENT IN CANADA

Reconciliation and Indigenous governance is a central concern in this volume (Chapters 2, 6, 8, 10, and 14). In particular, Chapters 2 and 14 articulate the history and context for Canada's reconciliation with Indigenous Peoples and how doing so is foundational to ensuring the sustainability of our oceans. Chapter 2 highlights the role of Indigenous Peoples as traditional knowledge holders about Canada's ocean and as rights holders with respect to ocean management and governance, and the effects of colonization on undermining Indigenous governance principles. Ongoing work is needed to understand and follow through on Canada's commitments to reconciliation with Indigenous Peoples, including via the adoption of the 2007 *United Nations Declaration on the Rights of Indigenous Peoples* (UNDRIP). Drawing on the case studies and analysis of ocean issues in this volume, Chapter 14 highlights and examines best practices and challenges in governance, resource access, and protection of Indigenous culture and values. The analysis was informed by Canada's commitment to recognition of Indigenous rights, including a commitment to secure free, prior, and informed consent (FPIC) for development by UNDRIP, and by Ten Principles that Canada identified for reconciliation based on UNDRIP (Department of Justice Canada 2018).

Reconciliation is a long-term and ongoing process that requires political commitment and resources to address three forms of injustices resulting from colonization:

political domination, loss of territory, and cultural imposition. As reflected in this volume, Canada has made mixed progress toward reconciliation in the context of our oceans and coasts. We identified the following challenges to reconciliation: the mix of federal, provincial, and territorial jurisdictions; diversity of Indigenous Peoples; and political and structural resistance to power sharing. Meaningful power sharing and access to resources have been stumbling blocks in negotiation of modern treaties as well as decisions on resource developments. However, best practices can be seen in management systems aligning with Indigenous world views and embracing concepts such as integrated and ecosystem-based management based on Indigenous leadership and knowledge.

Reconciliation is a driver for sustainable ocean and coastal futures in Canada, but for it to be fully achieved requires transformative change in social and political relationships, including governance, access to resources, and cultural renewal. Progress with ocean reconciliation in Canada has been slow and uneven, and greater efforts are necessary across geographic scales and jurisdictions. In this regard, *key policy recommendations* emerging from this volume as it relates to the ocean include: (1) adopting a just and equitable reconciliation framework; (2) engaging Indigenous communities in visioning and planning of ocean spaces; (3) cooperatively pursuing holistic approaches to management, such as integrated management (IM) and ecosystem-based management (EBM); (4) supporting and establishing co-governance or co-management arrangements that engage plural knowledge systems, share power and responsibility, and redefine relationships; and (5) supporting Indigenous self-determination and development of resource management capacity.

CHANGING OCEANS

Canada's marine environments are undergoing rapid (Chapter 3) and large-scale (Chapter 4) changes defined by shifts and trends in their mean physical or chemical states and their variabilities from interannual to decadal time scales and local to large spatial scales (ocean basin–wide). These changes affect the ecosystem services provided by Canada's three oceans, which include provisioning and cultural services for all, but especially Indigenous Peoples because of their centuries of inter-

actions with the ocean. The authors set out to understand the magnitude and pace of change and the impacts of changing oceans on ecosystem services and the people, as well as to explore scenarios that depict the potential futures and pathways of these changes in biophysical and human systems.

To achieve the set goals, the chapters explored some of the rapid environmental changes and their impacts across the three bordering oceans of Canada: the Arctic, Atlantic, and Pacific. They reviewed past and present observed rapid changes, followed by potential futures of rapid change. They also provided an interdisciplinary approach to highlight the cultural/phenomenological, scientific, and societal impacts that these rapid changes will have on the human societies that depend on the ocean not only for food security and wealth but also for their way of life. Furthermore, Chapter 4 described how large changes in Canada's oceans can affect ecosystems and fisheries through a three-lens perspective: (1) local/Indigenous views of large changes through quotes and storytelling; (2) scientific perspectives through evaluation of observed and modelled trends and projections; and (3) impacts on human communities that depend on marine resources. Three narratives describing different socio-economic development pathways for the future of Canadian oceans were developed: (1) sustainability; (2) business as usual; and (3) decline. National-scale scenarios were developed by integrating global- and national-scale models and data (i.e., earth system models, biological models of fish distribution, demographic and economic trend data) to generate projections of social-ecological outcomes at the national level. Chapter 4 used fisheries as a lens to investigate these outcomes, which were evaluated based on quantitative biological and economic indicators. Finally, Chapter 4 used local-level case studies to add depth and perspectives to the projected national outcomes. Each case study was linked to the appropriate national-scale scenario through scenario elements, outcomes, and/or the development process.

Both the national and local case study scenarios indicate that the projected outlook for Canadian fisheries is one of less catch and fewer fishers, since marine ecosystems under climate change are projected to have reduced species abundance, except in the Arctic, where abundance would likely increase (Chapter 5). Canada's

Arctic, Atlantic, and Pacific Oceans have all been recently impacted by rapid changes, including but not limited to marine heatwaves, with repercussions for fisheries; harmful algal bloom events affecting biodiversity and fisheries; extreme storms threatening coastal structures, and non-climatic events such as oil spills (leading to mortality of organisms) and sediment runoffs (leading to eutrophication and hypoxic zones). Naturally, rapid changes in oceanic conditions will have direct impacts on human communities along Canada's coasts, from losses in fisheries revenues due to marine heatwaves, to potential damage to tourism from oil spills, to emerging fisheries as a result of localized species invasions. Unfortunately, regardless of mitigation efforts, climate change will continue to intensify ocean warming, increasing the frequency and severity of heatwaves on land and in the ocean in the next few decades. This is because of the huge amount of carbon dioxide that has already accumulated; it takes time for the ocean to be restored to low CO_2 levels even with aggressive global greenhouse gas mitigation.

Chapter 4 showed that large changes in environmental conditions and in marine species composition, distribution, and abundance are taking place, based on information from research scientists and local and Indigenous traditional knowledge, and are projected to continue and potentially intensify in the future. These large changes are affecting ecosystem function and fisheries through multiple pathways, and corresponding impacts can already be seen and felt in coastal (especially northern) communities. Changes and impacts vary regionally due to differences in environmental conditions, the pace of warming and CO_2 uptake, and human uses in Canada's Pacific, Arctic, and Atlantic regions. Most changes are accelerated and intensified in the Arctic, but some changes may be more prominent or have a greater impact in other regions, such as oxygen decline and hypoxia in the Pacific. Furthermore, superimposing potentially intensifying (in magnitude and frequency) rapid changes on longer-term trends (i.e., large changes) can lead to significantly larger extreme events (heatwaves, storm surges, flooding, loss of sea ice, permafrost thaw, etc.) than those experienced previously in any particular region. This may necessitate a much higher level of preparedness in affected communities. It is worth stating that while modelling approaches are advancing rapidly, the complexity of

physical and biological dynamics and their interactions with human activities in Canada's marine ecosystems can confound projections of ecological outcomes. However, Chapter 5 describes how the use of multiple lines of knowledge (Indigenous traditional knowledge and Western science) can help communities understand the scale and urgency of changing oceans and help improve their adaptation to them.

The following are *key policy recommendations* stemming from our study of changing oceans:

1 Scientific research alone will be insufficient to account for, and respond to, rapid changes in Canada's oceans. Addressing climate change adaptation therefore requires not only that the best science be available but also that the diverse and cumulative challenges facing Indigenous Peoples be recognized so that proactive strategies to support their ways of life can be developed. Only with the integration of different types of knowledge, such as Indigenous ecological knowledge, will society achieve sustainable use of the ocean.

2 Policies have to be informed and based on the best available science and must be inclusive at all levels of society. Only a truly equitable process that respects the values of different groups can lead to sustainable ocean development.

3 Collaborative frameworks and implementation of participatory approaches can help ensure that local authorities prioritize solutions for managing degradation of coastal and marine ecosystems during local and provincial development planning. Within science, joint development of research projects with Indigenous or local communities are key to building effective collaborations.

4 The path to sustainability requires the enhancement of human, technical, and financial capacity; adoption of best fisheries management practices that integrate measures to promote conservation financing and corporate social responsibility; development of legal and binding instruments; and implementation and enforcement of global agreements for responsible fisheries and ocean management.

Connecting scenarios across scales allows for insights to emerge that may not have been apparent at only one scale. These insights can be used to inform the development of locally relevant solutions for addressing priorities that apply across scales, including climate change adaptation strategies, balancing marine protection with economic development, food security, and integration of traditional, local, and ecological knowledge into the assessment of current and future impacts of global change. The multiscale linkages can help elucidate regionally and locally relevant focal points and priorities for governance. For example, the Pacific emphasizes stronger co-management relationships, while the case of Port Mouton Bay highlights the urgency of addressing an apparent lack of government responsiveness.

ACCESS TO OCEAN RESOURCES

Part 3 of this book is primarily concerned with access to fisheries resources, and ways in which such access is impacted by institutional and historical realities. The contributions here are driven by several motivations, all linked deeply to our own experiences and participation in studying access. Chapters 6, 7, and 8 draw motivation from what their authors perceive, or others have argued, are suboptimal, and often unjust, outcomes of the ways in which fisheries access has been granted across Canada. First, in the fall of 2020, Mi'kmaq in Atlantic Canada exercised their treaty right through the launching of several livelihood lobster fisheries. Opposition to these new fisheries, even including violence, erupted. A lack of treaty education and a lack of clarity around the Supreme Court of Canada's 1999 *Marshall* decision has left the public confused about what the right is. Furthermore, while American lobster (*Homarus americanus*) productivity is at an all-time high, commercial fishers are claiming that conservation is at the heart of their opposition. Chapter 6 explores this claim through an access lens. Second, many social scientists across Canada are deeply concerned that when fishing quotas are permanently allocated to an individual fisher and transferable via the market as individual transferable quotas (ITQs), with little or no restrictions on who buys or leases them, they create serious distributional inequities by shifting benefits away from fishers and fishing communities and into the hands of investors, processors, and/or offshore entities. Documenting these inequities in Atlantic and Pacific Canada, as well as the efforts that have been and are being

made to remedy them, and what alternative approaches can teach us, was a main objective of Part 3. Third, we were motivated by a 2019 workshop with the Torngat Joint Fisheries Board, Nunavik Marine Region Wildlife Board, Makivik Corporation, Nunavut Wildlife Management Board, and Nunavut Tunngavik Incorporated. At this meeting, access to commercial fisheries was contextualized within land claims agreements and financial benefits received. Within these discussions, past injustices were recognized, along with pathways forward to improve co-management in an era of reconciliation. Document review, historical policy analysis, workshops, and current events/policy and legislative reform were all methods employed in Part 3.

Together, Chapters 6 to 8 note that despite federal legislation, regional differences in access policies and uptake across Canada exist. Additionally, access and allocation decisions are ongoing and renegotiated constantly. Finally, a generalizable conclusion across all chapters is that access needs to be understood in terms of not just a right or quota but also the enabling conditions that allow someone to actually benefit from such access.

Chapter 6 shows that the ability of Mi'kmaq to benefit from their livelihood right is limited first and foremost because the current management of the lobster fishery has not made space for Indigenous "moderate livelihood" fishing effort when accounting for allocation in the fishery. Also, uncertainties around the scale of the livelihood fishery have made developing solutions to allow more effort problematic. What might be considered enabling conditions that support access, such as being able to legally buy bait and sell catch outside the commercial season, have not been forthcoming. Additionally, despite numerous legislative documents, including the *Constitution Act, 1982* and the modernized *Fisheries Act,* Fisheries and Oceans Canada has chosen to attempt to regulate the livelihood right, further hindering the ability of Mi'kmaq to benefit from their livelihood right.

Chapter 7 describes how the owner-operator fleets emerged as the dominant socio-economic fleet in Canada's Atlantic fisheries because of policies developed independently in each Atlantic province, policies that were eventually merged because of the increasing importance of the inshore crab and lobster fisheries. Over more than four decades, the struggle to develop, enhance, and protect the owner-operator fishery from corporate ownership/control led to new government policies and a series of legal victories, culminating in the *Elson* decision of the Federal Court of Appeal (FCA) and significant amendments to the *Fisheries Act*. The 2017 Federal Court ruling in *Elson v Canada (Attorney General)*, subsequently upheld by the Federal Court of Appeal in 2019, is a significant decision in Canadian fisheries law. It unequivocally confirmed and clarified the broad scope of the Minister's authority to manage fisheries for purposes beyond conservation, and to "carry out social, cultural or economic goals or policies." The final word in the matter came later in 2019, when the Supreme Court of Canada denied Elson's request for leave to appeal the FCA decision. Weeks earlier, owner-operator fleets and fishery-dependent coastal communities had received further guarantees that their access to the fishery could be protected when the House of Commons adopted Bill C-68, which amended the *Fisheries Act* to enshrine in legislation the purposes that may be pursued by the Minister, including "social, economic and cultural factors in the management of fisheries." DFO has, for the first time, indicated a willingness to acquiesce to a long-standing request from owner-operator fleets, to commit to implementing the fleet separation and owner-operator policies through regulatory reform. Furthermore, the May 2019 report of the House of Commons Standing Committee on Fisheries and Oceans (FOPO) supported a similar direction for Pacific fisheries. This all-party consensus report recommended that DFO develop "a new policy framework for Pacific fisheries," including the establishment of "an independent commission to transition the West Coast fishery to a 'made-in-BC' owner-operator model." To assist in this proposed work, we suggested some scenarios from existing practice in fisheries management that could be considered for this kind of transitioning.

In the case of access in the Arctic, Chapter 8 shows that attempts to increase access to adjacent areas have been made in the past; some of them were successful, but progress has been slow. Eastern Arctic territories would like access to adjacent fishing areas to reflect levels that non-Inuit have access to in southern fishing areas. The result of current inequities is made clear when we consider that of the total landed value for the Inuit-adjacent

fisheries, Cdn\$566 million for 2019, only 35.3%, or Cdn\$200 million, was under the control of the three co-management boards or LCAs (Cdn\$150, \$28, and \$22 million for Nunavut, Nunavik, and Nunatsiavut, respectively). The 64.7% of cumulative commercial fisheries resources allocated through DFO processes equates to a landed value of Cdn\$366 million of Inuit-adjacent resources allocated to non-Inuit–controlled interests. In addition, the process of determining access and allocation of fisheries resources occurring between co-management boards and the Minister has been contentious in the past. Co-management board decisions (within land claims regions) or recommendations (for adjacent regions) have been varied (or modified), often without justifications or ignoring Inuit knowledge.

As Canada and Canadians grapple with the truth of past injustices toward Indigenous Peoples, it can be tempting but unrealistic to desire immediate perfection with Indigenous-led fisheries. The reality is that Canadians and the Government of Canada need to make space for Indigenous leadership and bands to find their own way. The commercial lobster fishery as we know it will have to change to support such a journey. This does not mean that both prosperous DFO-managed commercial fisheries and prosperous self-governed livelihood fisheries are not attainable concurrently. The government's commitment to nation-to-nation relations can and should be better demonstrated in its approach to Mi'kmaw fisheries, and indeed to Indigenous fisheries across the country.

Key policy recommendations emerging from this theme are linked to both policy processes as well as tools. Consultation and a genuine commitment to reconciliation in fisheries can be manifested through better access decisions by the federal government. International and domestic reconciliation strategies highlight the need for inclusive governance processes impacting Indigenous groups. This includes promoting self-reliance and self-determination in managing natural resources and the benefits derived from them. Although Canadian federal strategies align with these standards, implementing change has been slow and irregular. Moving to a more inclusive process with better understanding of each other's perspectives (co-management board versus Fisheries

Minister) provides an opportunity to improve the co-management process in the future.

At present, there appears to be a once-in-a-generation opportunity to reform Pacific fisheries management. We offer practical, well-documented scenarios for going about this. We also offer a concise integrated history of how both east and west coast management policies arrived at their present state, and how they can logically progress under new court decisions and amendments to the *Fisheries Act*. Based on the poor performance of ITQ fisheries against management and social objectives (see, e.g., Sumaila 2010), it is difficult to see how they could be considered to be in the public interest. Now that corporate ownership has been curtailed in Atlantic inshore fisheries, one would expect the government to begin seriously considering alternatives in Pacific fisheries. Now is the time for developing alternative approaches and tools that build on proven working models, such as those that we have identified, which place restrictions on ownership, do not commodify access privileges, and establish democratic ownership/control over access rights through fishers' organizations, cooperatives, community licence/quota banks, or comparable arrangements.

CHANGING GOVERNANCE

Governance refers to the process and structures through which Canadian society can make decisions about our oceans and coasts in ways that lead to just and sustainable social and ecological outcomes (Stephenson et al. 2021). A number of governance issues and themes emerge in this volume that highlight pathways for better ocean and coastal governance. For example, research on coastal communities often focuses on their vulnerability and declining conditions (Chapter 9). As the lessons from this volume show, however, we need to emphasize the strategies through which communities can proactively foster change and deliberately seek *transformations* toward sustainability. Drawing on a simple framework and cases from the Atlantic, Pacific, and Arctic in which we work, we highlight the strategies and ingredients communities might use to foster transformative change. Chapter 9 shows that a range of processes, relationships, and capacities are required to support governance transformation from the ground up. To achieve meaningful

change, moreover, it is important that the interjurisdictional engagement, leadership, and knowledge (including notably Indigenous leadership) evolve through phases of transformative change (Ban et al. 2018; Burt et al. 2020).

The need to rebuild several Canadian fisheries is also recognized as a "grand challenge" for Canadian fisheries access and governance (Teh and Sumaila 2020; Chapter 10). Opportunities to strengthen governance to rebuild fisheries are enhanced by focusing on knowledge and knowledge co-production – that is, people working together to develop and share actionable information (Alexander et al. 2019). Specifically, diverse knowledges and knowledge co-production can support governance that better fits the challenges of fisheries rebuilding (Reid et al. 2020). Theory and evidence on "governance fit" tell us how to match governance (e.g., knowledge, principles, decision making, and management) to what is known about the complexity and dynamics in a fishery, particularly in a local context (Chapter 10).

Insights from this volume also highlight how the application of integrated management (Stephenson et al. 2021) to Canada's oceans and coasts continues to be rare despite the recognized need (e.g., in the *Oceans Act*, Canada's Oceans Strategy, the *Report of the Commissioner of the Environment and Sustainable Development* to the House of Commons, the G7 Charlevoix Blueprint for Healthy Oceans, Seas and Resilient Coastal Communities, and the 2019–20 DFO departmental plan) (see Chapter 11). Yet "bright spots" in IM exist that promise to lead to positive social, economic, and ecological outcomes for ecosystem integrity and community well-being. Context is important for IM, and the diverse cases we have examined here highlight local enabling factors (including leadership, pressure applied to both federal/provincial agencies and elected officials, and adequate funding) through which actors have been able to advance more integrated initiatives. Five lessons from the cases can amplify efforts for IM in Canada:

- Nongovernment actors appear to play a critical role in IM, particularly Indigenous Peoples.
- Making progress toward IM is not necessarily a linear process.

- There is a broad spectrum of potential approaches to the implementation of IM, ranging from large-scale replacement of existing management with a different paradigm to an incremental modification of single-sector management toward a more integrated approach.
- The transition of Canada's coastal regions toward IM has inherently been a legislative process, subject to election cycles and political influence.
- Bright-spots initiatives need to be supported, while also prioritizing transformative approaches that support a broad range of perspectives.

A number of governance challenges hinder ongoing efforts to foster healthy oceans and coasts in Canada. However, *key policy recommendations* emerging from this theme are as follows:

1 A central focus on Canada's coastal communities must drive any blue economy policy shifts or initiatives, or otherwise there will be continued risks of injustice and inequities.
2 Fisheries rebuilding is important for advancing the interests of coastal communities and marine ecosystems in accordance with the diverse objectives of the *Fisheries Act*.
3 To support the transition toward IM in Canada and broader reconciliation efforts, Indigenous governments and communities are essential partners.
4 Looking forward, we suggest that governments strengthen the priority for accompanying regulations needed to guide and compel action on IM in Canada, and that research agendas emphasize the circumstances in which change needs to be transformative versus incremental, and identification of examples of governance models or mechanisms that allow for a "whole-of-government approach" (perhaps better termed a "whole-of-society approach" in the context of participatory governance and shared responsibility).

The existence of bright spots highlighted in this book, although varied in scope and history, help draw our attention to the innovative practices and governance processes that can be replicated and scaled up to other contexts (Partelow et al. 2020).

OCEAN LAWS, POLICIES, AND KNOWLEDGE MOBILIZATION

Governing our oceans and coasts demands responsive legal and policy frameworks with goals, standards, and processes that integrate climate risk in decision making, adapt to changing social-ecological systems, and allow these systems to exercise their own adaptive capacity. Across this volume, there are numerous insights on the Canadian legal and policy frameworks for selected ocean sectors, and the extent to which they integrate or are capable of integrating climate change considerations and respond to uncertainty. For example, Chapter 12 describes encouraging elements in the current legal, policy, and management framework. Key "progressions" include: (1) a clear commitment by all levels of government to promote adaptation and climate resilience; (2) investments in monitoring networks and scientific research (Sumaila et al. 2021); (3) legislated mandates to implement ecosystem and precautionary approaches (Pikitch et al. 2004); and (4) ambitious goals for marine protection through Marine Protected Areas (MPAs) (Sala et al. 2018). Moreover, there is significant adaptive capacity in existing legal frameworks. The principled discretion in the *Fisheries Act* and programmatic provisions in the *Species at Risk Act* allow flexibility in setting goals and objectives as well as means to achieve them. The provisions of these acts have not always been used effectively to respond to the risk of climate change, however. An explicit and comprehensive national vision of the role of oceans and ocean-based activities in a low-emission future is lacking. Insufficient use of strategic planning tools, including Marine Spatial Planning (MSP) or other area-based management tools, hinders horizontal and vertical coordination and prioritization of risks and responses. Fragmented, tangled, and unclear roles and responsibilities between jurisdictions, agencies, and programs hinder or delay responses to climate change or create regulatory gaps and red tape. Finally, actions to prepare for, and respond to, climate change and ocean acidification are exceedingly slow and are already out-paced by the rate of climate change.

This book also draws attention to new approaches to ocean management and conservation, including how participatory approaches can substantially help address knowledge gaps about socio-ecological dimensions and strengthen community engagement and communication (Chapter 13). In doing so, this volume offers a vision that sees both community engagement and communication as tightly linked and essential components of meaning-ful ocean research and policy (Chapter 13). For example, participatory video projects on all three of Canada's coastlines reflect ways in which we can promote engage-ment, co-create knowledge, encourage collective capacity building, and enhance communication. Such approaches are multi-dimensional, developed iteratively through respectful dialogue between researchers and community members in ways that benefit communities. Specifically, the participatory videos document community know-ledge on the socio-ecological dimensions of oceans that can inform ocean research and conservation. Fifteen communities and nearly 200 participants across Can-ada's three coastlines contributed to numerous digital media outputs that foregrounded community knowledge, perspectives, and priorities within and beyond the places where projects were developed (Chapter 13). Beyond mere knowledge production, our participatory projects and their process generated important benefits to com-munities and research:

- *Reinforced participation.* We co-developed relation-ships and processes that reinforced participation so that there were meaningful, consistent, and collabora-tive interactions between academic and community-based partners. Research was directed by, with, and for communities.
- *Equitable research.* Training and youth empower-ment, capacity building, and meaningful co-creation, synthesis, and interpretations of results were central to the work and helped ensure equitable power dynamics between researchers and community members.
- *Strategies to inform decision making.* Unlike standard approaches to research, research videos are easier to

share with policy-makers and the general public than academic texts, and video outputs can contribute to pub-lic dialogue and support political change (Chapter 13).

A number of *key policy recommendations* emerge from this volume that are relevant to law and policy:

1 Government agencies at all levels need to ratchet up their planning and management approaches to address climate change.

2 A national and comprehensive policy and action plan on "oceans and climate change" is needed for coherent, coordinated, and consistent action by all stakeholders.

3 Policies and plans need to address the many dimen-sions of oceans and climate change interdependence and steer ocean-based economies and ocean protec-tion to support climate mitigation and adaptation.

4 Strengthened governmental institutional capacity to address climate change is crucial for climate responses. Key priorities include: clarifying fragmented, tangled, and unclear roles and responsibilities between juris-dictions, agencies, and programs; and providing a clear and preferably legislated mandate for all agencies with jurisdiction or responsibility for oceans conserva-tion and management, at all levels of government, to incorporate climate change risks into decision making, with clear policy guidelines on how to do this. Wider and consistent use of climate-informed MSP and other strategic planning tools that reflect national and regional priorities for mitigation and adaptation are an essential component of the required response to climate change. Filling legislative gaps, particularly for marine aquaculture and offshore marine renew-able energies, would facilitate the development of ocean-based activities with mitigation and adaptation potential.

5 Based on our knowledge mobilization insights, par-ticipatory approaches, when tightly linked with com-munication techniques, lead to effective and ethical ocean research, and this should be mainstreamed as part of research projects and programs. It is absolutely

critical to have communities not only involved in research but actually driving the process. Doing so ensures that research benefits coastal communities and results in more relevant and durable management and policy.

FROM TURBULENT SEAS TO SAFE HARBOURS

As this volume makes abundantly clear, there are no easy pathways to achieving healthy and viable oceans and coasts in Canada (Bennet et al. 2018). The drivers of change and stressors (global and local) are many, and include significant uncertainty associated with climate change (Pörtner et al. 2019). However, the lessons that emerge from this volume are clear on some of the ingredients necessary for transitioning from turbulent seas to safer harbours and healthier oceans and coasts:

- achieving reconciliation with Indigenous partners and reimagining nation-to-nation relationships through principles of co-governance and respect for constitutionally protected access rights
- prioritizing the well-being of coastal communities – Indigenous and non-Indigenous – across the Atlantic, Arctic, and Pacific coasts
- strengthening and further supporting the diverse knowledge foundations required in order to understand and respond to ocean and coastal change
- building and implementing just and flexible policies sensitive to marine ecosystem sustainability and rapidly changing ocean and coastal conditions.

Getting these ingredients in place and navigating the pathways ahead will require collaboration. We need all hands on deck and a shared commitment to meaningful partnerships. We hope this volume reflects the spirit of such partnerships and is a source of insights to help us navigate this journey.

REFERENCES

Alexander, S.M., J.F. Provencher, D.A. Henri, J.J. Taylor, and S.J. Cooke. 2019. "Bridging Indigenous and Science-Based Knowledge in Coastal-Marine Research, Monitoring, and Management in Canada: A Systematic Map Protocol."

Environmental Evidence 8 (1): 15. https://doi.org/10.1186/s13750-019-0159-1.

Ban, N.C., A. Frid, M. Reid, B. Edgar, D. Shaw, and P. Siwallace. 2018. "Incorporate Indigenous Perspectives for Impactful Research and Effective Management." *Nature Ecology and Evolution* 2 (11): 1680–83.

Bennett, N.J., M. Kaplan-Hallam, G. Augustine, N. Ban, D. Belhabib, I. Brueckner-Irwin, A. Charles, et al. 2018. "Coastal and Indigenous Community Access to Marine Resources and the Ocean: A Policy Imperative for Canada." *Marine Policy* 87: 186–93.

Burt, J.M., K.I.B.J. Wilson, T. Malchoff, W.T.K.A. Mack, S.H.A. Davidson, and A.K. Salomon. 2020. "Enabling Coexistence: Navigating Predator-Induced Regime Shifts in Human-Ocean Systems." *People and Nature* 2 (3): 557–74.

Department of Justice Canada. 2018. *Principles Respecting the Government of Canada's Relationship with Indigenous Peoples.* Ottawa: Department of Justice Canada. https://www.justice.gc.ca/eng/csj-sjc/principles.pdf.

FOPO (House of Commons Standing Committee on Fisheries and Oceans). 2019. *West Coast Fisheries: Sharing Risks and Benefits.* Report of the Standing Committee on Fisheries and Oceans. Ottawa: House of Commons.

Partelow, S., A. Schlüter, D. Armitage, M. Bavinck, K. Carlisle, R.L. Gruby, A.-K. Hornidge, et al. 2020. "Environmental Governance Theories: A Review and Application to Coastal Systems. *Ecology and Society* 25 (4): 19. https://doi.org/10.5751/es-12067-250419.

Pikitch, E.K., C. Santora, E.A. Babcock, A. Bakun, R. Bonfil, D.O. Conover, P. Dayton, et al. 2004. "Ecosystem-Based Fishery Management." *Science* 305 (5682): 346–47.

Pörtner, H.O., D.C. Roberts, V. Masson-Delmotte, P. Zhai, M. Tignor, E. Poloczanska, K. Mintenbeck, et al., eds. 2019. *IPCC Special Report on the Ocean and Cryosphere in a Changing Climate.* Geneva: Intergovernmental Panel on Climate Change.

Reid, A.J., A.K. Carlson, D.E.L. Hanna, J.D. Olden, S.J. Ormerod, and S.J. Cooke. 2020. "Conservation Challenges to Freshwater Ecosystems." In *Encyclopedia of the World's Biomes – Reference Module in Earth and Environmental Sciences,* edited by S.A. Elias, 270–78. Amsterdam: Elsevier.

Sala, E., J. Lubchenco, K. Grorud-Colvert, C. Novelli, C. Roberts, and U.R. Sumaila. 2018. "Assessing Real Progress Towards Effective Ocean Protection." *Marine Policy* 91: 11–13.

Stephenson, L., A.J. Hobday, E.H. Allison, D. Armitage, K. Brooks, A. Bundy, C. Cvitanovic, et al. 2021. "The Quilt of

Sustainable Ocean Governance: Patterns for Practitioners." *Frontiers in Marine Science* 8. https://doi.org/10.3389/fmars.2021.630547.

Sumaila, U.R. 2010. "A Cautionary Note on Individual Transferable Quotas." *Ecology and Society* 15 (3): 36. https://doi.org/10.5751/ES-03391-150336.

Sumaila, U.R., M. Walsh, K. Hoareau, A. Cox, L. Teh, P. Abdallah, W. Akpalu, et al. 2021. "Financing a Sustainable Ocean Economy." *Nature Communications* 12 (1): 3259.

Teh, L.S., and U.R. Sumaila. 2020. "Assessing Potential Economic Benefits from Rebuilding Depleted Fish Stocks in Canada." *Ocean and Coastal Management* 195: 105289.

Index

Note: "(f)" after a page number indicates a figure; "(t)" after a page number indicates a table

Printed and bound in Canada by Friesens

Set in Myriad and Devanagari by Artegraphica Design Co. Ltd.

Copy editor and proofreader: Francis Chow

Indexer: Celia Braves

Cartographer: Eric Leinberger

Cover designer: Lara Minja

Cover image: iStock/saemilee